Urban Sustainability

The Urban Sustainability Book Series is a valuable resource for sustainability and urban-related education and research. It offers an inter-disciplinary platform covering all four areas of practice, policy, education, research, and their nexus. The publications in this series are related to critical areas of sustainability, urban studies, planning, and urban geography.

This book series aims to put together cutting-edge research findings linked to the overarching field of urban sustainability. The scope and nature of the topic are broad and interdisciplinary and bring together various associated disciplines from sustainable development, environmental sciences, urbanism, etc. With many advanced research findings in the field, there is a need to put together various discussions and contributions on specific sustainability fields, covering a good range of topics on sustainable development, sustainable urbanism, and urban sustainability. Despite the broad range of issues, we note the importance of practical and policy-oriented directions, extending the literature and directions and pathways towards achieving urban sustainability.

The series will appeal to urbanists, geographers, planners, engineers, architects, governmental authorities, policymakers, researchers of all levels, and to all of those interested in a wide-ranging overview of urban sustainability and its associated fields. The series includes monographs and edited volumes, covering a range of topics under the urban sustainability topic, which can also be used for teaching materials.

Ali Cheshmehzangi • May Tan-Mullins
• Martin de Jong • Rob Raven
Editors

Transforming Cities

Urban Growth, Resilience, and Ecological Innovation

Springer

Editors
Ali Cheshmehzangi
School of Architecture, Design and Planning, The University of Queensland, Brisbane, QLD, Australia

Martin de Jong
Rotterdam School of Management (RSM), Erasmus University Rotterdam, Rotterdam, The Netherlands

May Tan-Mullins
University of Reading Malaysia, Johor, Malaysia

Rob Raven
Monash Sustainable Development Institute, Monash University, Clayton, VIC, Australia

ISSN 2731-6483 ISSN 2731-6491 (electronic)
Urban Sustainability
ISBN 978-981-95-7377-6 ISBN 978-981-95-7378-3 (eBook)
https://doi.org/10.1007/978-981-95-7378-3

This Springer imprint is published by the registered company Springer Nature Singapore Pte Ltd.
The registered company address is: 152 Beach Road, #21-01/04 Gateway East, Singapore 189721, Singapore.

For the victims of misguided development, i.e., vulnerable communities with limited resilience, facing environmental degradation and social deprivation.

Preface

Cities are humanity's boldest experiment; they are living, breathing organisms that reflect our goals, failures, and potential rather than being static reminders of the past. The future will either be won or lost there. Urbanization, however, is all too frequently viewed as a numerical game of population expansion, building goals, and financial rewards. The city loses its social cohesion, ecological context, and global obligations in this simplistic calculus. A drastic shift from that perspective is advocated in "*Transforming Cities: Urban Growth, Resilience, and Ecological Innovation*." It advocates for cities to grow intentionally, wisely, and within the bounds that support life rather than by accident.

This book's premise is straightforward but challenging: "Transformative Growth" is about evolution rather than expansion. It is a pledge to rebalance the forces of urban growth such that each new neighborhood, street, or skyscraper enhances rather than degrades ecosystems and communities. Growth that causes soil erosion, landscape fragmentation, and a widening of disparities is not progress; rather, it is managed decline that passes for achievement. Cities must be transformed by moving away from growth driven by profit and toward stewardship motivated by purpose. It is to view development as an opportunity to restore living systems, foster trust, and ensure lasting value rather than as a zero-sum trade-off between the economy and ecological.

The pathway to such transformation is neither singular nor straightforward. It necessitates reconsidering our investment, governance, and planning practices. In an era that demands accuracy and innovation, traditional methods such as parcel-by-parcel approvals, top-down masterplans, and checklist-style sustainability are ineffective tools. Coherent ecological frameworks take the role of patchwork urbanization in landscape-first planning. Greenwashing is replaced with measured performance, which links investment to observable social and environmental results. Collaborative governance takes the place of unilateral decision-making in co-regulation, where communities, governments, and the private sector share accountability.

If cities are to prosper in a century characterized by resource restrictions and climate volatility, these changes are fundamental, not optional.

At the heart of this transformation lies resilience. The increasing frequency of climate shocks, such as heat waves, fires, floods, and sea level rise, has revealed the vulnerability of the urban systems that we previously thought were unchangeable. Resilience cannot be viewed as a technical add-on or an afterthought. It needs to be ingrained in the infrastructure, culture, and urban form. This entails planning for disruption, incorporating adaptable features into energy, water, and transportation systems, and enabling communities to take an active role in their own defence. Digital twins and citizen science projects serve as examples of how civic engagement and technology can work together to improve readiness. Resilience, as demonstrated by case studies from cities at the forefront of climate risk, is about moving ahead after each shock, becoming stronger and more equal in the process.

However, transformation is profoundly creative as well as defensive. The third pillar of this book, ecological innovation, challenges us to think of the city not as a consumptive machine but as a regenerating landscape. These interventions take us beyond sustainability as minimum compliance, from heat-mitigating green corridors and solar-optimized building forms to circular water systems and biodiversity networks interwoven throughout the urban fabric. They see cities as dynamic planetary healers that can restore habitats, absorb carbon, and bring people back in touch with nature. They also face the more sinister side of urbanization, which includes contaminated streams, deteriorated beaches, and industrial pollutants. Addressing these legacies is a moral requirement as well as an environmental one.

The following pages are neither a naïve celebration of urban creativity nor an intellectual exercise in urban theory. They serve as a model for a new urban compact that rejects decline as inevitable while acknowledging the harsh realities of our planet's situation. These lessons are derived from global design revolutions, policy innovations, and lived experiments. They are intended to challenge people, practitioners, and decision-makers to renounce antiquated beliefs and see cities as essential components of Earth's living systems rather than as separate entities.

Transformation will not be delivered by a single policy or project. It will result from a thousand interrelated acts, all driven by the belief that cities can be both vibrant and respectable, prosperous and global in their perspective. Anything less will commit us to a future of ecological collapse, growing inequality, and brittle infrastructure. However, cities may become our most powerful tools of renewal if we make a commitment to "Transformative Growth," plan for resilience, govern for inclusivity, and innovate for ecological regeneration. They have the potential to transform from representations of human overreach into catalysts for global healing.

Finally, this book is a challenge and an invitation to reconsider growth as restoration rather than accumulation, to embrace resilience as a creative strategy rather than a defensive one, and to pursue ecological innovation as a civilizational necessity rather than an aesthetic preference. Cities are changing, and this is no longer a choice. It is both the greatest opportunity of our time and a prerequisite for our collective survival.

Brisbane Ali Cheshmehzangi
Johor Bahru May Tan-Mullins
Rotterdam Martin de Jong
Melbourne Rob Raven

Acknowledgments

We gratefully acknowledge the contributions of all those involved in bringing this book to life. Our sincere thanks go to the authors, co-authors, chapter contributors, and collaborators who generously shared their expertise, passion, and creativity. We also recognize the foundation laid by our earlier collaborative research project on "smart-eco-cities" (2014–2018) and the continued dedication of those who contributed to its development. Four of us collaborated on the smart-eco-cities projects and have stayed in contact since its completion. These collaborative efforts not only shaped this publication but also helped foster new dialogues, expand research networks, and create a platform for meaningful scholarly exchange across the globe.

About This Book

Transforming Cities provides a focused exploration of how urban areas can evolve by balancing growth, resilience, and ecological innovation. Central point of argument is focused on "transformative growth," including rethinking urban growth, investment, and governance to ensure that development enhances social and ecological systems rather than undermines them. This book demonstrates how to change the paradigm from parcel-by-parcel monetization to landscape-first planning, from box-ticking sustainability to measured performance, and from top-down control to co-regulated stewardship with communities as urbanization speeds up and climate threats increase. In doing so, we explore methods for improving climate resilience, reviving urban areas, and incorporating ecological frameworks into city planning through a variety of case studies and research-driven insights.

This book provides useful strategies for creating cities that are both dynamic and ecologically conscious by tackling important topics like urban redevelopment, disaster preparedness, and sustainable land use. The result is a field manual for cities that refuse false trade-offs between prosperity and planetary limits. Therefore, transformative growth is a design, finance, and governance program that restores living systems, establishes trust, and creates long-lasting value. It is not just a catchphrase. Leaders who desire cities that are economically vibrant, culturally rooted, and climate-ready should read this book because anything less is only managed decline. In line with this perspective, the book covers three critical areas, i.e., Urban Growth, Resilience, and Ecological Innovation.

Urban Growth, when managed effectively, can drive economic vitality while fostering sustainable development. By highlighting the significance of resilient urban planning, brownfield redevelopment, and creative infrastructure design, this book explores how cities can manage growth without sacrificing ecological integrity. It draws attention to the conflict between sustainability and development and provides guidance on tactics that reduce environmental damage and advance social inclusion. Furthermore, the idea of degrowth shows up as a crucial alternative to conventional urban growth, promoting different routes for cities that are healthier and more balanced.

Resilience is fundamental in the face of climate-induced risks and growing urban vulnerabilities. This book examines how flood mitigation, digital technology integration, and earthquake-resilient design can help cities adjust to environmental shocks. To ensure that communities are actively involved in forming their adaptive capacities, the role of digital twins, citizen science, and participatory governance in enhancing urban resilience is investigated. Case studies show how cities around the world are using both community-driven tactics and high-tech solutions to increase resilience, especially in regions that are vulnerable to extreme weather events and natural disasters.

Ecological Innovation is at the forefront of reimagining urban landscapes, integrating biodiversity conservation, sustainable water management, and energy-efficient urban design. From cooling structures and solar-optimized urban forms to placemaking strategies that promote post-disaster recovery, this book explores nature-based solutions that improve climate resilience. It also promotes a more comprehensive urban ecological framework by addressing the growing problem of industrial pollution and its effects on marine and coastal environments. Cities can develop into flexible, regenerative environments that promote the welfare of people and the environment by incorporating sustainability into urban transformation.

Contents

List of Contributors

Armin Akhavan Department of Architecture, Sari Branch, Islamic Azad University, Sari, Iran

Mahdi Boroughani Research Center for Geoscience and Social Studies, Hakim Sabzevari University, Sabzevar, Iran

Yücel Caymaz Gökçen Firdevs Department of Industrial Product Design, Istanbul Aydin University, Beşyol, Küçükçekmece, İstanbul, Türkiye

Jie Chen State Key Laboratory of Subtropical Building and Urban Science, Shenzhen, China; School of Architecture & Urban Planning, Shenzhen University, Shenzhen, China; Center for Human-oriented Environment and Sustainable Design, Shenzhen University (CHESD), China

Ali Cheshmehzangi School of Architecture, Design and Planning, The University of Queensland, Brisbane, Australia; School of Architecture and Urban Planning, Hunan University, Changsha, China

Martin de Jong Erasmus Initiative Dynamics of Inclusive Prosperity, Erasmus University Rotterdam, Rotterdam, The Netherlands

Baskın Elif Architecture Master Program, Istanbul Aydin University, Beşyol, Küçükçekmece, İstanbul, Türkiye

Shabnam Galehdar Department of Urban Design, Faculty of Architecture and Urban Planning, Iran

Shadi Galehdar Faculty of Architecture and Urban Planning, Art University of Isfahan, Isfahan, Iran

Mahmoud Ghalehnoee Department of Urban Planning and Design, Faculty of Architecture and Urban Planning, Art University of Isfahan, Isfahan, Iran

Saima Gulzar Department of Architecture, University of Management and Technology, Lahore, Pakistan

Jon Kellett School of Architecture and Civil Engineering, University of Adelaide, Adelaide, Australia

Amir Reza Khavarian-Garmsir Department of Geography and Urban Planning, Faculty of Geographical Sciences and Planning, University of Isfahan, Isfahan, Iran

Yanhui Lei Design School, Xi'an Jiaotong-Liverpool University, China

Zhuoning Li Design School, Xi'an Jiaotong-Liverpool University, China

Armin Mehdipour School of Architecture and Civil Engineering, University of Adelaide, Adelaide, Australia

Vassilia Pirierros School of Architecture, Tianjin University, Tianjin, China

Sima Pourhashemi Research Center for Geoscience and Social Studies, Hakim Sabzevari University, Sabzevar, Iran

Rob Raven Monash Sustainable Development Institute, Monash University, Victoria, Australia

Amir Shakibamanesh Department of Urban Planning and Design, University of Art, Tehran, Iran

Hadi Soltanifard Department of Environment, Hakim Sabzevari University, Sabzevar, Iran

May Tan-Mullins University of Reading Malaysia, Johor Bahru, Malaysia

Tian Tang Independent Researcher, Chongqing and Sichuan, China

Amna Waheed Department of Architecture, University of Management and Technology, Lahore, Pakistan

Huan Zou School of Architecture, Tsinghua University, Beijing, China

About the Editors

Prof. Ali Cheshmehzangi is a Professor and Head of the School of Architecture, Design and Planning (ADP) at The University of Queensland. He has been the World's top 2% field leader since 2021, recognised by Stanford University. He is among the top 20 global scholars in the urban sustainability research area. Ali has a wealth of experience, expertise, and a strong commitment to innovation and excellence in the fields of architecture, design, and planning, the three core disciplines of his School at UQ. With a career spanning over two decades, he has made significant contributions to the academic and professional communities, with a focus on sustainable and environmentally conscious design. So far, Ali has published over 500 journal papers, articles, conference papers, book chapters, and reports. He also has 50 other academic books, three of which have received awards at the national, provincial, and municipal levels. He also has received international awards and recognition for his research on urban resilience studies and sustainability research, as well as the 2018 Vice-Chancellor's award for his impactful contribution to higher education.

Prof. May Tan-Mullins is a political ecologist and passionate advocate for sustainability. She is the Provost and CEO of the University of Reading Malaysia. Her previous leadership roles encompassed a wide range of academic and administrative responsibilities, including Dean International, Chief Sustainability Officer, Vice Provost of Teaching and Learning, Associate Provost for Global Engagement, Dean of Graduate School, and Director of the Institute of Asia and Pacific Studies. As a Professor of Sustainability, Professor Tan-Mullins' research interests span the political ecology of rising China, environmental and energy justice, poverty alleviation, and building resilience for marginalized communities against climate change. Leveraging her expertise in international development and human security, she has provided consultancy services to organizations such as the UNDP, the National Bureau of Asian Research, the Revenue Watch Institute, and the Chinese government.

Prof. Martin De Jong is Scientific Director of the Erasmus Initiative "Dynamics of Inclusive Prosperity," one of Erasmus University Rotterdam (EUR) flagship research initiatives. It was jointly established by Erasmus School of Law (ESL), Rotterdam School of Management (RSM) and Erasmus School of Philosophy (ESPhil) and aims to gain knowledge on the inclusion of weak or less-well represented stakeholders in governance and management. It also has an agenda setting role in the broader societal and political debate on this topic. Along with the above position he is professor at both ESL and RSM (each for 50% of his time) and has a part-time professor position at the School of International Relations and Public Affairs (SIRPA) of Fudan University in Shanghai. Given his 25-year international experience in research, education and entrepreneurship on public policy, decision making, urban studies, infrastructure development and cross-cultural management, it is his goal to make the initiative both locally relevant through knowledge dissemination in the Randstad Area and to give it strong global exposure by establishing partnerships in a number of leading universities around the world.

Prof. Rob Raven is an interdisciplinary scholar, professor of sustainability transitions and deputy director (research) at Monash Sustainable Development Institute. He is also a visiting professor at the Copernicus Institute of Sustainable Development at Utrecht University. His interest is in understanding the dynamics and governance of sustainability transitions and socio-technical innovation. Originally trained as an electrical engineer, he has turned his gaze to the social, institutional, political, and spatial factors that influence successful innovation and transitions towards sustainable development. Rob has made major contributions to multi-level theories of transformative change, socio-technical experimentation, and strategic niche management. His empirical work has covered urban, energy and mobility transition processes in Europe, Asia and Australia. Rob holds several editorial positions, including Environmental Innovation and Societal Transitions, Global Sustainability, Nature Urban Sustainability, Technological Forecasting and Social Change and Routledge Studies in Energy Transitions. He has published over 130 articles, three books and eight special issues. He has won the equivalent of more than AUD$ 18M research funds. His current research agenda is focused on the analysis of transformative change in urban contexts such as smart, sustainable, and net-zero cities and precincts.

List of Figures

List of Tables

Chapter 1
Rethinking Urban Futures: The Case for Transformative Growth

Ali Cheshmehzangi, Rob Raven, Martin de Jong, and May Tan-Mullins

Abstract This introductory chapter reframes the discussion of sustainable urbanization. It critically analyzes the shortcomings of the three main paradigms that are influencing the discourse on urban change: ecological modernization, resilience, and degrowth. Ecological modernization offers political viability through incremental eco-efficiency, but it cannot eradicate the root problems of unsustainable urbanization. Degrowth offers hope for radical ecological and social prosperity, but it is constrained by political and institutional feasibility (Schneider et al., 2010). Overall, none of these tactics provides an obvious way to balance riches with the planet's limitations.

The chapter responds by presenting "Transformative Growth" as a unifying paradigm that reinterprets growth as the extension of economic, technological, social, and ecological output in ways that reinforce one another within planetary boundaries. This rethinking incorporates adaptive innovation, inclusive governance, and regenerative natural cycles into the growth model of urban expansion. It places a strong emphasis on political viability, quantifiable advancement, and structural

A. Cheshmehzangi (✉)
School of Architecture, Design and Planning, The University of Queensland, Brisbane, QLD, Australia
School of Architecture and Urban Planning, Hunan University, Changsha, Hunan, China
e-mail: a.chesh@uq.edu.au

R. Raven
Monash Sustainable Development Institute, Monash University, Victoria, Australia

M. de Jong
Rotterdam School of Management, Erasmus University Rotterdam, Rotterdam, The Netherlands Rotterdam, The Netherlands.
Smart City Institute, HEC Liège, University of Liège, Liège, Belgium
Institute for Global Public Policy, Fudan University, Shanghai, China

M. Tan-Mullins
University of Reading Malaysia, Johor Bahru, Malaysia

A. Cheshmehzangi et al. (eds.), *Transforming Cities*, Urban Sustainability,
https://doi.org/10.1007/978-981-95-7378-3_1

integration across urban systems. As the structural pillars for a sustainable and just urban future, the four interconnected pillars of ecological productivity, social productivity, technological productivity, and economic productivity are put forth.

The chapter concludes with a description of how the three main sections of the book—Urban Growth, Resilience, and Ecological Innovation—will operationalize Transformative Growth with the help of a variety of case studies. In order to prosper within the limits of the earth while preserving social and political legitimacy, the cases demonstrate how cities can restore ecosystems, strengthen communities, use adaptive technologies, and rewire economic systems through Transformative Growth.

Keywords Transformative Growth • Urban Transformation • Ecological Productivity • Social Inclusion • Technological Innovation • Resilience • Sustainable Cities

1.1 The Limits of Current Paradigms in Urban Transformation Research and Practice

Over the last two decades or so, a relatively small number of paradigms have shaped urban development, dictating both academic discourse and practical application. These paradigms, most notably ecological modernization and degrowth (Healy et al., 2015), represent very different political and theoretical beliefs, but they all share the failure to offer a rational, scalable, and socially inclusive way for cities to flourish within the limits of the planet. Given the severity of the climate emergency, the escalating ecological problems, and the expanding socioeconomic inequities, these frameworks are insufficient as road maps for the next phase of urban development.

Ecological modernization offered a promising compromise between economic advancement and environmental responsibility (Andersen and Massa, 2000). It promised cleaner industries, more effective technologies, and progressive environmental laws that would allow the "decoupling" of ecological deterioration from economic expansion. This approach is still inextricably bound to the logic of continuous gains in material and energy throughout, despite the fact that it has resulted in notable advancements such as energy-efficient buildings, low-emission transportation systems, and improved waste recovery. This reliance limits its transformational potential. Efficiency gains have consistently been hampered by rebound effects, in which reduced unit pricing or energy savings lead to higher overall consumption. Furthermore, it usually overlooks more basic structural issues like land speculation, socio-spatial segregation, and the political economy of infrastructure investment in favor of eco-efficiency (De Marco, 2025). In summary, ecological modernization has improved the environmental performance of cities without altering the underlying patterns of unsustainable urbanization.

In contrast, the growth imperative is directly challenged by proponents of degrowth. Degrowth promotes a deliberate decrease in production and consumption in order to achieve social fairness and ecological sustainability. The degrowth literature is full with criticisms of capitalism, overconsumption, and the privatization of nature. It makes compelling arguments for sufficiency-based living, relocalized economics, and collective self-organization. Degrowth as an urban strategy, however, has many challenges. In political contexts where growth-oriented policy and investment cycles prevail, degrowth proposals are typically dismissed as unrealistic, economically unstable, or socially regressive (Xue, 2022). This is particularly true in Global South cities where urbanization is intertwined with aspirations for improved living conditions. In reality, very few towns have been successful in gaining popular support for degrowth ideas, much less turning them into coherent political systems.

Apart from these two extremes, resilience has become a crucial idea in the discussion of urban development. Resilience thinking has significant limitations, even though it has shifted the emphasis from mitigation to adaptation and systems thinking. Because the concept has been applied to many situations, from revolutionary adaptation to disaster recovery, its operational clarity is usually compromised. In addition, resilience strategies have been included into a few risk-management initiatives that strengthen existing frameworks rather than change them (Soufi et al., 2021). Resilience is too often seen as a technological capacity to endure shocks (Boh et al., 2023), and insufficient attention is given to the underlying vulnerabilities that disproportionately impact underprivileged people.

These paradigms are all prone to "partial transformation," which is defined as minor, siloed adjustments that do not successfully challenge the fundamental reasons for unsustainability (Basee and Abdulla, 2022). In addition, sufficiency principles are hard to translate into governance frameworks, technological efficiency does not disrupt the growth-consumption cycle, and resilience planning frequently fails to alter the socioeconomic and ecological baselines that generate risk.

This book takes the position that the next phase of urban transformation theory must go beyond these limitations by focused on the topic of what it means for cities to prosper. Instead of totally ignoring growth or restricting it to efficiency metrics, we argue for a shift to Transformative Growth, a concept that reframes growth in a variety of ways. In contrast to the constrained accumulation of GDP or real estate value, this concept defines growth as the extension of ecological capacity, social inclusion, and technological competence in ways that complement one another (Mikelsone et al., 2021). According to this perspective, prosperity is defined by the robustness of natural systems, the creative adaptability of urban economies, and the resilience of communities.

Transformative Growth is based on three key differences from prevailing paradigms: It first integrates ecological production into the equation for urban growth by recognizing that ecosystems are dynamic infrastructures that generate value, e.g., cooling microclimates, filtering water, storing carbon, and maintaining biodiversity. Second, it prioritizes social productivity, i.e., the capacity of collectives such as urban communities to collaboratively create solutions, equitably distribute rewards, and conserve cultural life—rather than viewing social benefits as secondary to economic ones. Third, it accepts technological productivity as a tool to attain ecological and social benefits rather than as an end in itself, through design innovation, digital tools, and adaptive governance.

This reconceptualization involves more than simply theory. In order to integrate urban growth with socially just transitions and regenerative natural cycles, it calls for new governance coalitions, investment logics, and planning metrics. In many rapidly urbanizing contexts, especially in Asia, Africa, and Latin America, it recognizes that urban growth will continue. The question is whether this expansion will spur change or make the problems we are currently facing worse.

This chapter starts with a critique of the shortcomings of current paradigms, setting the stage for a more ambitious and realistic approach. In the parts that follow, we will examine the theoretical underpinnings, historical background, and real-world applications of Transformative Growth using case studies from diverse urban environments. These examples will show that rethinking expansion is not just a theoretical endeavor but rather a necessary step to align urban futures with the reality of the Anthropocene. The difficulty lies in combining efficiency and sufficiency, as well as adaptability and transformation, into a coherent, future-ready urban layout.

1.1.1 From Ecological Modernization to Degrowth: A Stalled Debate

Over the past three decades, the discourse around urban development has been dominated by two conflicting ideologies: ecological modernization and degrowth (Escobar, 2015). Both emerged from the increasing environmental and social problems caused by conventional growth-oriented urbanization and have impacted the way that scholars, professionals, and policymakers perceive sustainable cities. Nevertheless, their trajectories have not intersected to create a workable agreement. Instead, they have slowed down their argument to the point where each framework's conceptual advantages and disadvantages are equally obvious.

Ecological modernization is predicated on the notion that economic expansion and environmental conservation could be mutually beneficial. Its fundamental idea is that market-based instruments, regulatory reform, and technological advancement may "decouple" economic growth from environmental degradation (Hermanns, 2015). Concrete results of this approach in urban environments include cleaner industrial processes, energy-efficient housing, low-carbon transit systems, and more sophisticated waste management infrastructures. It has gained political appeal due to its promise of win-win progress, i.e., prosperity without ecological disaster.

Nevertheless, this increasing eco-efficiency paradigm has structural constraints. First, while technological improvements have reduced the environmental impact of urban activities, these gains are occasionally offset by rebound effects, which negate net advantages by promoting higher overall consumption because of improved performance or lower costs. Second, ecological modernization has often remained sector-bound, optimizing buildings, transportation, or energy systems independently, without altering the interrelated metabolism of cities. Third, by prioritizing efficiency over sufficiency, it overlooks the root causes of resource extraction, urbanization, and overconsumption. The method often has the potential to prolong the sustainability of high-consumption lifestyles by perpetuating the unsustainable under the guise of environmental stewardship.

This reasoning is directly criticized by degrowth. Instead of aiming for efficiency within the existing development paradigm, degrowth promotes deliberately lowering the amount of production and consumption, especially in wealthy economies, in order to bring human activity back within the confines of the planet. This results in compact living, relocalized economies, shared resource use, and governance structures that, in terms of urban policy, put sufficiency ahead of accumulation. The conceptual power of degrowth lies in its ecological and moral clarity; it addresses the difficulty of unrestricted growth on a finite planet and asks who profits from the current routes of urban expansion.

However, degrowth faces significant obstacles when it comes to governance. Politically speaking, it is viewed as disrupting public revenue, investment flows, and employment, particularly in urban areas where growth-linked tax bases are essential for service delivery and infrastructure (Sylla and Koddenbrock, 2025). In social environments, such as the rapidly urbanizing nations of the Global South, where aspirations for opportunity and dignity are tied to material improvement, it struggles to acquire momentum. Even in affluent cities, individuals accustomed to growth-dependent social systems and established business interests oppose plans to cut back on consumption or increase infrastructure. Because of this, degrowth projects are frequently excluded from citywide policies and instead restricted to experimental niches like community gardens, cooperative housing projects, or localized energy systems.

The result is a landscape that is divided: ecological modernization offers political feasibility but falls short of significant transformation, while degrowth offers ecological sufficiency but fails to achieve political and institutional viability. This duality has held the science of urban sustainability captive for far too long, with debates often reverting to ideological arguments rather than coming up with a workable solution. This deadlock is not just an intellectual one; it also has an impact on actual policy decisions, funding priorities, and public perceptions. In the meantime, urbanization, the climate crisis, and biodiversity loss are all speeding up.

What is needed is a reframing that can maintain economic dynamism while integrating the political feasibility of ecological modernization with the ecological constraints necessary for degrowth (Asara et al., 2015). Instead of splitting the differences, it is crucial to reconsider how "growth" is understood and defined. Instead of boosting GDP (Gross Domestic Product) at all costs or abandoning expansion entirely, the objective should be to reorient urban growth toward regenerative results, which are evaluated in ecological restoration, social inclusion, and long-term adaptive capacity.

Such an approach would need adopting standards that consider ecosystem health, distributional equality, and resilience under climatic stress, in addition to energy efficiency per unit of production. Governance structures would be needed to foster local ideas without sacrificing their democratic foundations. The political economics of urban finance, infrastructure, and real estate would also need to be addressed; these are areas where structural change has been under-theorized by both degrowth and ecological modernization (Sanderson Bellamy and Ioris, 2017). Thus, the challenge is to consider an approach that is as flexible as resilience thinking when faced with ambiguity and shocks, as realistic as ecological modernization when it comes to enlisting institutional and financial backing, and as ambitious as degrowth in its ecological pledges. This is Transformative Growth's theoretical underpinning. It makes no attempt to preserve the current paradigm through marginal efficiency gains, nor does it support a drastic reduction that is politically impossible. Instead, it sees growth in terms of the expansion of ecological capacity, social well-being, and technological adaptability—dimensions that can expand without destroying planetary boundaries.

1.1.2 The Political Economy of Urban Transformation

Overcoming technical design challenges and putting new planning paradigms into practice are only two aspects of urban transformation (Inam, 2013). It is profoundly embedded in political economy, the web of institutions, market forces, budgetary constraints, and power relations that govern how resources are distributed, how infrastructure is financed, and whose interests urban development serves. Ignoring this terrain puts any sustainability strategy at risk of either being co-opted into maintaining the status quo or failing to move past isolated experimental initiatives.

The political economy of cities is influenced by three elements: capital flows, governance structures, and land and property laws. These elements support each other. Real estate markets, which usually prioritize short-term financial gain above long-term environmental stability, are the driving force behind urban growth. There is an incentive to approve projects that raise taxable value, even if doing so raises the risk of flooding, sprawl, or traffic, because a significant number of municipal budgets are funded by land sales, development levies, and property taxes. In favor of larger, conventional businesses that promise profits, financial institutions drain funds from smaller, community-driven projects that may have higher social and ecological benefits but lack typical collateral.

Governance structures can either amplify or counteract these market trends. In many cities, the fragmented jurisdictional authority among national ministries, metropolitan agencies, and municipal departments makes it more difficult to integrate land use, mobility, housing, and ecological restoration into a coherent plan. Even when sustainability goals are enshrined in laws, institutional inertia, contradictory mandates, and political cycles all affect implementation. The term "policy volatility" refers to the possibility that climate-friendly policies put in place (Ahatsi et al., 2024) during one administration could be weakened or reversed during another.

Policies pertaining to land and property further perpetuate inequality and ecological damage. By forcing lower-income individuals from well-served neighborhoods into peripheral areas with insufficient infrastructure in rapidly urbanizing environments, speculative land acquisition might push them into expensive, high-emission mobility patterns. In developed markets, zoning and planning laws may place a higher priority on property values than on density, mixed use, or affordable housing (Aurand, 2010). This hinders the creation of regenerative urban forms and perpetuate patterns of exclusion. These fundamental causes are typically unrelated to the sustainability rhetoric that local leaders may employ in public.

Both ecological modernization and degrowth stall as a result of these deeply rooted political-economic constraints. A common premise in ecological modernization is that minor policy adjustments, such as carbon pricing, green bonds, and efficiency standards, might link market incentives and technological advancement to sustainability objectives. However, if land and financial institutions are not altered, these policies are susceptible to being captured by powerful interests, which could lead to "green"-sounding but inadequate results (Carter, 2018). Although degrowth challenges the need for growth, it offers few ways to deal with the financial dependence and political connections that bind urban management to expansionary economics. Calls to cut back on production and consumption can result in political backlash, capital flight, and a drop in municipal revenue, especially in communities competing for jobs and investment.

To overcome these challenges, Transformative Growth needs to directly confront the political economy of urban development. This means developing strategies that can alter the power structures and incentives that encourage unsustainable urban expansion. For example, shifting from land-sale-based municipal funding to value capture from ecological restoration could realign fiscal priorities toward regenerative efforts. Property tax reform that favors low-carbon retrofits and penalizes speculative vacancy could alter market dynamics without a large public investment (van der Horst, 2022). Funds could be allocated to mixed-use, transit-oriented, and biodiversity-integrated projects that have both financial and environmental benefits through partnerships for public-private-community investment.

Such projects require governance coalitions that are resilient to short-term challenges. Here, political viability and ecological imperatives clash. Deep decarbonization, nature-based infrastructure, and equitable housing require enabling laws, consistent funding sources, and institutional mandates that hold up over election cycles (Cheshmehzangi and Chen, 2021). These objectives cannot be met solely by voluntary market movements or grassroots innovation. This resilience requires collaborations between civil society and progressive commercial sector organizations, as well as at the municipal, state, and federal levels.

Through the prism of political economics, one may also observe the global interdependencies of urban growth (Kantor, 2019). Capital markets, commodity flows, and climate hazards all operate on a global scale. A city's vulnerability to flooding may be impacted by international insurance markets, and the conditions of its extraction halfway over the world may have an impact on its shift to sustainable materials (Pelling, 2012). Therefore, any revolutionary program must negotiate the international trade and regulatory frameworks that define the bounds of urban economies in addition to local financing and governance.

Ultimately, the political economy of urban transformation will determine whether the "Transformative Growth" paradigm is successful or not. Without policies that take property markets, investment logics, and income systems into account, even the most compelling sustainability vision could be reduced to rhetorical flourishes. However, if cities confront these facts, they may employ economic vitality as a source of social and ecological rejuvenation rather than as a draining force.

Using examples from cities that have begun to rewire their governance structures, property regimes, and fiscal incentives toward regenerative results, the contributions to this book will explore how Transformative Growth might be operationalized within these political-economic constraints on the short-term while challenging and transforming them in the long-term. The main battles for urban futures will be fought within the contested space of urban governance and economics, despite the terrain being challenging.

1.1.3 The Risk of Resilience As a Comfort Zone

Resilience has emerged as a key idea in urban planning, urbanism, and urban policies over the past decade. Who would argue against cities becoming more resilient to stresses and shocks? Its rise has been rapid and seems to be uncontroversial. From community readiness plans to climate adaptation strategies, resilience has provided a language for action that avoids the ideological division of growth-related conversations. However, because of its universal appeal, resilience has the potential to become too much a comfort zone for conservatism. It is a notion that is so flexible that it may calm politicians and investors without necessarily raising concerns about the structural issues that cause vulnerability in the first place.

Resilience is more than merely recovering from setbacks; it is the capacity to adapt in ways that reduce future risk. This transformative resilience is directly related to the objectives of sustainability and equity for systemic change. In practice, however, many urban resilience strategies have tended to concentrate on a narrower definition: the ability to "bounce back" to pre-shock conditions (Grinberger and Felsenstein, 2014). Examples of how this framing can stabilize the very systems that generate risk include reestablishing transportation patterns that lock in high emissions, rebuilding flood-damaged homes in high-risk areas, or restoring fossil fuel-based electrical grids.

Politicians and institutions are drawn to resilience because it is manageable. Unlike radical sustainability such as degrowth, resilience does not need openly opposing the political economics of land and finance or the growth imperative (Schmid, 2019). It can be integrated into current planning frameworks, financed through traditional risk-management channels, and evaluated using incremental criteria such as infrastructure redundancy or emergency response times. For many local governments, this is a wise choice because resilience projects attract donor funding, adhere to insurance industry requirements, and offer noticeable, instant advantages.

But there is a price for this pragmatism. A resilience agenda that places too much emphasis on shock absorption runs the risk of normalizing the idea that risk and loss are inevitable and should be accepted rather than avoided. Instead of tackling structural inequalities, including the unequal vulnerability of low-income communities to environmental dangers, it can lead to a shift in focus toward a shift in focus toward stopgap measures that manage impacts without tackling the root causes of vulnerability. Worse yet, it can be used as a rhetorical defense to justify inaction on preventing climate change by asserting that we are preparing for inevitable consequences.

The commercial sector has also embraced resilience, often in ways that prioritize corporate risk-management over the welfare of the broader public. Major infrastructure and real estate actors may support resilience measures that protect their assets without taking part in more comprehensive adjustments toward ecological regeneration or fair urban development. This selective adoption reinforces the comfort zone, transforming resistance into a set of asset-centric protections within an unsustainable metropolis.

To be clear, resilience is not always enough. Important components of urban resilience are its emphasis on interdependencies, adaptive capability, and systems-thinking technique. Resilience becomes problematic when it is seen as a goal rather than a way to get there—when recovery is mistaken for progress, or in another words, when "bouncing back" is substituted for "bouncing forward." In many cases, resilience does not alter the socioeconomic metabolism of cities but rather preserves a slightly better version of the status quo.

A difficulty for Transformative Growth is integrating resilience while avoiding its comfort-zone tendencies. This means ensuring that resilience planning is directly linked to ecological regeneration, social inclusion, and long-term decarbonization. Investments in resilience should be assessed not just for their capacity to manage shocks but also for their involvement in reducing the underlying causes of risk and increasing the co-production capacity of ecosystems and communities.

Resilience must work in tandem with and be held responsible to the larger objectives of transformation if it is to be more than a politically secure stand-in for more significant change. To understand that questions of justice, power, and governance are central to determining who benefits from "resilient" systems (Lebel et al., 2006), who bears the costs, and whose vulnerabilities are prioritized, resilience must be re-politicized. The urban futures this book envisions—futures that not only endure crises but also emerge from them with greater ecological vitality, egalitarianism, and adaptability—may then be sparked by resilience.

1.2 Defining Transformative Growth

The previous sections addressed the shortcomings of the widely used paradigms in urban transformation—resilience, degrowth, and ecological modernization—as they are all too commonly implemented. None of them provides a thorough, politically viable, and environmentally sound framework for the next phase of urban growth, despite the fact that each one provides some insights. Transformative Growth, which reinterprets growth itself and offers a practical roadmap for cities to flourish within the constraints of the planet, is presented here as the book's uniting subject.

Fundamentally, Transformative Growth is not a compromise between the existing models, but a rethinking of what growth is and how it is measured. Instead of equating growth with gains in GDP, floor area, or infrastructure capacity, it defines growth as the extension of ecological, social, and technical production in ways that reinforce each other. Although this reconceptualization acknowledges that urbanization will continue, especially in rapidly expanding countries, it argues that significant changes must be made to the character, direction, and metrics of urbanization for it to be sustainable on a finite world.

1.2.1 Growth Within Planetary Boundaries

The first fundamental principle of Transformative Growth is that all urban development and revitalization must occur inside the planet's boundaries. These constraints, which relate to freshwater use, nitrogen and phosphorus cycles, biodiversity, land-system change, climate stability, and more, define the ecological "safe space" in which human societies can thrive. In conventional urban growth models, environmental thresholds have been seen as external constraints that technology can push back or transcend. Transformative Growth, on the other hand, argues that ecological restrictions are the indisputable foundations upon which all planning, investment, and governmental decisions are built.

This necessitates switching from the extract, build, discard paradigm of urban metabolism to a regenerative model in which cities restore ecosystems, cycle resources, and boost the resilience of global systems. Instead of a city's skyline height or GDP rankings, its success would now be measured by its net contributions to carbon sequestration, habitat connectivity, water security, soil health, social cohesion, and human well-being. Instead of being a struggle for material gain, this view views expansion as a strengthening of the city's capacity to live in harmony with its natural environment.

1.2.2 Expanding Productivity Beyond the Economic

The second principle is the broadening of productivity to encompass ecological, social, and technological dimensions alongside economic output. Natural systems, such as wetlands that filter water, urban woodlands that modulate microclimates, and mangroves that act as storm surge barriers, are recognized by ecological productivity as active infrastructure. Investing in these systems in urban areas enhances biodiversity and human well-being while yielding ongoing benefits that often surpass those of gray infrastructure. Social production is the process by which communities generate shared value through participation, cultural vibrancy, and equitable access to opportunities. Beyond social programs and inclusion policies, this is about incorporating co-creation into urban planning and governance so that people are change agents rather than passive recipients. Cities must be able to innovate internally, respond to crises more quickly, and maintain public trust to be resilient over the long run.

Furthermore, technological productivity recognizes that innovation may and should be directed toward regenerative effects. This means aligning digital tools, data platforms, and design innovations with ecological and social concerns. Circular construction technologies, generative AI (Artificial Intelligence) for climate adaptation, and urban digital twins are a few examples of how technology can be utilized to advance sustainability rather than consumption. Therefore, by integrating these three forms of productivity into the core idea of growth, we may argue that Transformative Growth creates a complex, well-balanced objective for investment and policy.

1.2.3 Integrating Resilience, Ecological Innovation, and Citizen Participation

The third premise is integration: Transformative Growth combines the adaptability of resilience thinking, the legitimacy-building potential of public involvement, and the forward-looking dynamism of ecological innovation. Resilience is the capacity to withstand shocks and navigate uncertainties, such as supply chain disruptions, public health catastrophes, or floods caused by climate change, without reverting to harmful pre-crisis behaviors. However, Transformative Growth particularly links resilience to transformation, ensuring that recovery processes move toward ecological regeneration and social justice.

Ecological innovation focuses on systemic breakthroughs that change the rules of urban development, like biomimetic engineering, regenerative design, and nature-based solutions. Instead of just incorporating green aspects into otherwise conventional endeavors, ecological innovation integrates environmental performance into the organizational logic of urban systems.

The participation from citizens ensures that these improvements are grounded in local knowledge, culturally appropriate, and politically sustainable. In this case, participation is more than just consultation; it is a shared governance process in which various viewpoints select objectives, consider trade-offs, and collaboratively own outcomes. Long-term change requires a growth paradigm that is innovative, flexible, and socially acceptable, all of which are produced by this combination.

1.2.4 Linking to Contemporary Political Debates

Transformative Growth also speaks directly to ongoing political debates on sustainability, green industrial strategy, and just transitions (McIlroy et al., 2022). As governments grapple with the dual imperatives of economic recovery and climate action, the manner that growth is presented is being re-examined. Green industrial strategies seek to strike a balance between decarbonization and economic competitiveness through anything from the production of renewable energy to retrofitting programs (Kolosok et al., 2024). But without a firm commitment to ecological and social production, these strategies risk becoming just "green growth" in name, replicating the extraction and inequality patterns of the fossil fuel economy.

The necessity of ensuring that workers and disadvantaged groups are not disproportionately burdened by climate action is another point of emphasis in discussions surrounding the just transition. Transformative Growth integrates justice into its notion of success by ensuring that advancements in energy, transportation, and industry create new revenue streams, expand access to urban benefits, and rectify historical injustices. This framing has political advantages. By redefining growth instead of outright rejecting it, Transformative Growth may enlist a broader coalition, ranging from environmentalists to progressive business leaders, from community organizers to pragmatic lawmakers. It offers a vision that is both practical enough to deal with political realities and bold enough to fulfill ecological demands.

In the next chapters, this concept will be dissected into useful strategies, illustrated with case studies, and assessed in various urban contexts. Beyond pointing out the flaws in earlier paradigms, the goal is to demonstrate how cities could evolve—not in ways that worsen the issues of the Anthropocene, but rather in ways that expand the ecological, social, and technological foundations that are necessary for a sustainable urban future.

1.3 The Four Pillars of Transformative Growth

One of the strengths of Transformative Growth is its ability to redefine productivity and move the forces that shape cities toward inclusive and regenerative outcomes. Its four interrelated pillars, economic, social, technical, and ecological productivity, are all essential to reconsider how urban systems create and distribute value. Instead of existing in separate silos, these are mutually reinforcing elements that work in a feedback loop where advancements in one can unleash potential in the others. Together, they produce a growth model that can meet human needs and objectives while thriving within the constraints of the earth.

The basis is ecological productivity, which recognizes that the health of urban life is ultimately determined by the vitality of the natural systems in which it is embedded. One of the most important measures of ecosystem success under Transformative Growth is their capacity to filter air and water, regulate temperature, trap carbon, and provide habitat. Ecosystems are far too often treated as management limitations or passive assets in conventional planning. Restoring wetlands, maintaining urban forests, reestablishing river systems, and bringing disparate green corridors back together are all examples of how cities are investing in the biophysical engines of long-term success. Cities must switch from extractive land use to regenerative cycles, which restore soil nutrients, recharge aquifers, and promote biodiversity, in order to sustain ecological productivity. This viewpoint maintains that a hectare of restored mangrove or rewilded parkland is equally productive as a transit hub or data center.

While ecological production provides life support, social productivity provides the human energy and unity that allow transformation. Its foundations are inclusion, trust, and participation—qualities that enable communities to innovate, adapt, and benefit from change. Social productivity recognizes that collective intelligence emerges when diverse perspectives are included as consultees and co-creators of urban futures. Participatory design processes, neighborhood governance assemblies, and community-led service models can all help restructure decision-making to leverage local knowledge and build legitimacy. Strong social networks can speed up the adoption of ecological and technical advancements because locals see themselves reflected in solutions and have a stake in their success. A more accurate measure of social production than the absence of conflict is the existence of active, egalitarian collaboration that boosts the city's resilience to shocks and its capacity for continuous improvement.

With the precision and effectiveness of modern technologies, technical productivity connects societal capacities and ecological objectives, acting as the adaptive nervous system of the Transformative Growth model. Instead of concentrating on novelty for its own purpose, it emphasizes technology that promotes adaptive and polycentric governance and regenerative results. Digital twins, AI-assisted planning platforms, and sensor-rich infrastructure allow cities to model scenarios, monitor change in real time, and make dynamic operational adjustments. Resource footprints can be significantly reduced while maintaining adaptability for future needs by using design techniques that support net-positive structures, modular construction, and material reuse. Importantly, technological productivity must be accessible and responsible; rather than creating digital divides or consolidating power in the hands of a few, technology should be designed to empower distributed actors. By concentrating technical capabilities on integrated ecological and social goals, cities can maximize the impact of each investment in data, design, and infrastructure.

Economic productivity reinterprets value creation in the context of Transformative Growth such that it is more consistent with resilience and regeneration rather than depletion and short-term gain. Market incentives, tax laws, and investment flows that support ecological restoration, low-carbon innovation, and socially inclusive outcomes are required to accomplish this. In this context, economic productivity is not solely based on growth but also on the sustainability, dispersion, and durability of the value provided. Various financing options, such as ecological value capture systems, hybrid public-private-community partnerships, and green bonds, can allocate funding to projects that meet both economic and environmental needs. Local economic ecosystems that support social entrepreneurship, green manufacturing, and circular companies reduce vulnerability to volatile global markets and promote stability and self-reliance. Economic productivity is changed from a force that pulls in the other direction to one that drives systemic change when ecological and social metrics are incorporated into investment, taxation, and procurement requirements.

These four pillars are inseparable. If ecological productivity is low, a city will still confront increasing resource risks and climate consequences, even if it is a leader in the implementation of advanced technologies. Even though strong social cohesion may be established, there won't be enough money to sustain change if there is no economic activity. Although its economy may grow, it will only exacerbate the patterns that lead to environmental degradation if it lacks ecological and technical productivity. Transformative Growth requires the development of all four, and governance frameworks need to be able to recognize and manage the trade-offs and synergies between them.

Throughout the book, these pillars will serve as both useful recommendations and critical perspectives. Case studies will show how cities are experimenting with ways to rewire their economies in a variety of contexts to encourage regeneration, restore ecological capacity, organize community agency, and use technology for the common good. The most enduring results in each case occur when the four pillars are seen as the structural foundations of a new urban model—one that emerges in ways that include rather than exclude, replenish rather than deplete, and adapt rather than fracture in the face of change.

1.3.1 Measuring Transformative Growth

For Transformative Growth to go beyond its aspirational aspect, it needs to be measurable. It risks being mired in the same vague sustainable language that has allowed incrementalism to pass for change in the absence of clear signs. According to this viewpoint, measurement is not about establishing a single composite score or placing towns in a race for prestige. It entails creating a solid framework that recognizes the boundaries of the world in which the ecological, social, technological, and economic aspects of progress operate as well as their interconnectedness.

The first step in the measurement method is to establish a baseline that appropriately represents the ecological footprint of a city. To achieve this, sector-specific efficiency indicators must be replaced with systemic metrics including net carbon flux, biodiversity richness, soil health, water retention capacity, and habitat connectivity. These indicators must be spatially explicit so that planners can track the effects of land use, governance, and infrastructure changes on ecological productivity over time. In practice, this may mean tracking the proportion of stormwater that is regulated by natural systems, the yearly expansion of rehabilitated green spaces, or the measurable cooling impacts of urban vegetation.

Social productivity can be measured using metrics related to inclusivity, cohesion, and participative agency. Metrics that show the strength of social networks, equity of access, and diversity of representation in decision-making are given precedence over merely socioeconomic information such as income levels or employment rates. Measures of how successfully the benefits of change are being spread include civic trust surveys, co-governance structure participation rates, and data on the spatial distribution of public amenities. In cities committed to Transformative Growth, these measures should increase in proportion to ecological benefits, showing that regeneration and inclusivity are moving forward together.

When assessing technological productivity, it is best to consider both purpose and capacity. Capability metrics include things like the integration of real-time environmental monitoring into municipal systems, the use of adaptive design strategies, and the penetration of data infrastructure. Purpose-driven indicators, on the other hand, ask whether the use of these technologies is promoting regenerative results. For example, figuring out the percentage of construction projects that use life-cycle carbon modeling, the percentage of infrastructure expenditures associated with the circular economy, or the degree to which urban digital twins are used for community-driven scenario planning.

Economic output requires a redefinition of value in line with resilience and regeneration (Loza Adaui, 2025). This calls for measurements that track the growth of circular and green enterprises, the integrity of regional supply networks, and the proportion of public and private investment devoted to initiatives that meet social and ecological norms (Park et al., 2010). It is necessary to measure ecological value capture—the amount of public revenue generated from restored natural assets—and assess the durability of jobs created in low-carbon industries to determine whether economic growth is advancing the goals of Transformative Growth.

Most importantly, measurement needs to be adaptable and iterative. Cities function in dynamic surroundings that are impacted by shifting climatic patterns, technological progress, and political instability (Rees, 2012). Indicators should be regularly reviewed to ensure they remain aligned with planetary boundaries and local goals. This necessitates governance systems that can coordinate across agencies, sectors, and scales to guarantee that data collection and analysis directly impact policy decisions.

In closing this discussion, the goal of evaluating transformative change is not to create a rigid checklist but to support continuous learning and accountability. By putting the idea into tangible indicators, cities may stop transformative ambitions from being diluted into incremental gestures. Measurement makes it possible to identify and act quickly when ecological restoration, social inclusion, technical innovation, and economic regeneration are not all advancing simultaneously. Integrating this evaluative discipline into municipal management is the only way to transform Transformative Growth from a theoretical objective into a lived reality.

1.3.2 Measuring What Matters: From Global Frameworks to Local Adaptation

In both national and international contexts, there has been much discussion on what should be measured. The emergence of state and federal wellbeing frameworks in Australia, which aim to capture the lived reality of communities and go beyond GDP, has heightened this discussion. In addition to conventional economic metrics, these frameworks place an emphasis on social connection, cultural identity, and environmental quality. Because they recognize that urban success cannot be only attributed to increases in property values or fiscal production, they are instructive. Such discussions serve as a reminder to cities striving for Transformative Growth that measures must be in line with ecological constraints while also speaking to people's lived realities.

The Sustainable Development Goals (SDGs) have served as a benchmark for "measuring what matters" on a global scale. However, their universal focus and scope frequently reduce their efficacy at the city level. Urban governance necessitates more detailed, geographically specific versions of these objectives, yet indicators like poverty reduction, climate action, or inclusive institutions are also vital. SDG 11 on sustainable cities, for instance, is far too broad unless it is converted into city-specific goals, like lowering average household emissions from commuting, increasing fair access to green space, or enhancing the involvement of underrepresented groups in planning procedures. When separated from local circumstances, the SDGs' abstraction presents a problem rather than their lack of relevance. The everyday geographies of neighborhoods, catchments, and communities must serve as the foundation for such global aspirations in order to achieve Transformative Growth.

The Australian context highlights how this grounding can be attempted. A move toward multidimensional measurements is indicated by programs like the Australian National Wellbeing Framework and state-based "measuring what matters" initiatives. They aim to integrate economic resilience with social cohesiveness, cultural vitality, and environmental health. This gives Transformative Growth the chance to connect the social and ecological in ways that traditional economic reporting sometimes falls short of. Measurement can incorporate social inclusiveness and environmental integrity into the notion of progress itself, eliminating the need to approach them as supplemental policies or externalities.

But there is a lesson to be learned here. In the absence of governance systems that tie outcomes to resource allocation and decision-making, wellbeing indicators, like the SDGs, run the risk of becoming merely symbolic. Without consequences, measurement creates elegant dashboards but little change. Adaptive frameworks where indicators prompt action, whether that means reinvesting funds in climate-resilient infrastructure, reworking zoning to preserve biodiversity corridors, or making sure housing policy actively mitigates vulnerability, must be the focus for cities dedicated to transformative pathways.

In this sense, "measuring what matters" refers to aligning current indicators with transformative goal rather than creating new ones. By enforcing recognition of what has hitherto been invisible, e.g., soil health, community trust, Indigenous stewardship, and cultural resilience, and integrating such recognitions into the processes of urban decision-making, such frameworks have the power to upend the status quo. Only then can measuring become a transformational engine rather than merely a representational exercise.

1.3.3 From Theory to Structure: Thematic Cohesion Across Chapters

The theoretical underpinnings presented in this chapter, i.e., systemic thinking, regenerative design, and transformative growth, are reflected in the three interconnected pillars that make up this book: urban growth, resilience, and ecological innovation. These pillars do not stand alone; rather, they form a continuum that progresses from tackling current urban issues to imagining long-term social and ecological renewal. We describe how each section and its chapters relate to these foundations below.

1.3.3.1 Part 1: Urban Growth

The first part explores how cities evolve because of redevelopment, preparedness for disasters, and liveability. The concept of adaptive reuse under systemic restrictions is demonstrated in Chapter 2, which focuses on urban transformation in Japan with a case study of brownfield redevelopment. It illustrates how converting unused property may reduce ecological footprints and spur social and economic renewal, a real-world implementation of the book's demand for integrated growth methods. This idea is extended into the field of risk-sensitive design in Chapter 3, focused on Earthquake-Resilient Urban Form, where it is demonstrated that urban growth must anticipate shocks and incorporate resilience at the morphological level. To ensure that growth does not jeopardize safety or recovery capability, this aligns with the theoretical emphasis on anticipatory governance and design for uncertainty. Micro-scale interventions are the subject of Chapter 4 with a particular focus on livable streets, which shows how human-centered design may improve urban vitality. This chapter operationalizes the idea of co-benefits, wherein physical enhancements concurrently promote inclusivity and environmental quality, by connecting street shape to social interaction and ecological performance.

1.3.3.2 Part 2: Resilience

The second part deepens the discussion by framing resilience as a multidimensional construct, i.e., thermal, spatial, and social. Configurational methods to green infrastructure are introduced in Chapter 5, focused on planning cooling structures for climate-resilient cities, which is consistent with the theoretical foundation of nature-based solutions as systemic regulators of urban microclimates. This chapter demonstrates the relationship between ecological production, a key component of transformative growth, and resilience techniques.

The idea that cultural and geographical narratives are essential to adaptive planning is reinforced in Chapter 6, which highlights the case of Dunkirk and discusses identity and resilience. This supports the book's claim that resilience is socio-spatial as well as technical, necessitating governance structures that embrace innovation while honoring regional heritage. The concepts of technical productivity with purpose are advanced in Chapter 7, focused on solar-optimized urban form, and then in Chapter 8, by highlighting adaptation and thermal resilience in subtropical communities. Echoing the theoretical emphasis on purpose-driven innovation rather than technology for its own sake, they show how design and technology may come together to provide thermal comfort and energy resilience.

1.3.3.3 Part 3: Ecological Innovation

The final part moves beyond resilience toward transformative ecological strategies, positioning cities as active agents of planetary stewardship. The theoretical connection between inclusive governance and ecological integrity is reinforced in Chapter 9 on heritage resilience in Suzhou, which illustrates participatory approaches for stormwater risk-management. A key subject in this book's critique of incrementalism is the conflict between growth and ecological balance, which is highlighted in Chapter 10's analysis of Bali's urban-ecological nexus. It promotes landscape-scale thinking, which embodies the systemic perspective previously presented and harmonizes urbanization with agro-ecological systems. As a conceptual bridge, Chapter 11 appears to integrate ideas from earlier chapters to present a vision of transformative growth via ecological innovation. By offering metrics that match ecological restoration with economic regeneration, it operationalizes the evaluative discipline covered in this introduction. Lastly, adaptive design for coastal resilience is demonstrated in Chapter 12 on floating parks, which translates theoretical concepts into practical application. This chapter's central claim, that transformative change is dynamic, context-sensitive, and collaborative, is emphasized by the iterative nature of (urban) innovation.

1.3.3.4 How About the Book's Thematic Cohesion?

This book is transformed from a collection of case studies into a cohesive vision of urban futures by three interwoven strands that construct a cohesive narrative across all chapters. First, strategies for urban expansion in Part 1 lay the foundation for resilience measures in Part 2, which in turn allow ecological innovation in Part 3. This is where "systemic integration" emerges as a fundamental principle. The theoretical contention that cities must be viewed as dynamic systems, where interventions at one scale affect outcomes at other scales, is reflected in this trend.

Second, when taken as a whole, the chapters show a "regenerative approach" that goes beyond mitigation to actively restore and rejuvenate. Each chapter shows how urbanism can revitalize natural and social systems rather than just lessen harm, whether it is through brownfield redevelopment in Japan, the design of cooling structures in Tehran, or the establishment of floating parks in coastal zones. Lastly, the evaluative framework presented in this chapter is operationalized by the book's recurring subject of "transformative measurement," particularly based on the "Transformative Growth" narrative. Chapters that monitor cooling impacts, participatory governance, and circular economy metrics provide a tangible embodiment of indicators spanning ecological, social, technological, and economic dimensions.

These strategies emphasize the value of flexibility and responsibility, preventing revolutionary goals from turning into small gestures. When combined, these threads show how the book's theoretical foundations, i.e., systemic thinking, regenerative design, and transformative growth, which can be used in a variety of settings and at various scales. The book presents a comprehensive view of urban transformation (or methods of "Transforming Cities" as the title reads), where growth, resilience, and ecological innovation come together to produce cities that are not only sustainable but also transformative. This is thoroughly done by placing separate chapters within these common principles. This thematic coherence guarantees that readers can follow a clear path from theoretical underpinnings to useful tactics, comprehending how each chapter advances the main objective of creating urban environments that flourish within global bounds while promoting social inclusivity and technological purpose.

1.4 From Concept to Practice: Reframing the Urban Growth Debate and Delving into Pathways for Urban Transformation

The preceding sections have eliminated the constraints of the prevailing urban transformation paradigms by revealing the contours of "Transformative Growth" as a unifying framework capable of integrating ecological, social, technological, and economic output within planetary boundaries. Ideas alone cannot alter cities; they must be implemented, tested in real-world environments, and refined using data. The remainder of the book tackles this job by operationalizing Transformative Growth using a paradigm that is focused on three thematic elements: urban growth, ecological innovation, and resilience. A key component of the framework is addressed by each of these elements.

The first part, *Urban Growth*, looks at how cities can grow or change without causing ecological overshoot. It accomplishes this through redevelopment or expansion that incorporates adaptive technologies, regenerates ecosystems, and fosters community agency. It questions investment patterns, planning systems, and land governance, showing how they could be rethought to produce regenerative rather than extractive outcomes. The second part, *Resilience*, examines how cities might handle uncertainty and upheaval as chances to promote change rather than as roadblocks to progress. In this context, the focus moves from defensive adaptation to the alignment of long-term ecological and social goals with social safety nets, risk governance, and adaptive infrastructure. The third part, *Ecological Innovation*, explores how innovative concepts, materials, and systems, i.e., from bio-inspired architecture to nature-based infrastructure, are reshaping the urban economy and its connection to the biosphere. These elements, when combined, demonstrate how to go from the theoretical framework laid out in this chapter to useful strategies that can guide investment and policy.

To bring this framework to life, the book draws on a set of carefully selected case studies. The case selection technique is based on three elements. First of all, each example demonstrates the direct connection between urban development and at least one Transformative Growth pillar, while also offering suggestions for advancing the other pillars. Second, the stories underscore the necessity for transformation to be adaptable to both established urban economies in the Global North and quickly urbanizing places in the Global South, since they demonstrate socioeconomic and geographic variety. Third, the stories are chosen because they may offer lessons that can be applied elsewhere, such as examples of how innovation, governance reform, or community mobilization have been effective in overcoming political, technological, or cultural barriers to change.

The organization of each case study ensures policy relevance and cross-case comparability. Each chapter will begin with the case in its local and regional context, then map out the interventions or strategies employed, and conclude with an assessment of the outcomes using the four productivity pillars as a framework. In order to give a coherent set of policy recommendations, trends and differences will be exposed in the book's last section, which is made possible by this rigorous synthesis.

Each chapter intentionally adds to the overarching framework. The case studies in the Urban Growth section will demonstrate how financial and governance structures could be changed to encourage revitalization. Those in the Resilience part will show how adaptive ability can be employed as a lever for change, as opposed to a retreat to the status quo. The section on ecological innovation will demonstrate how new paradigms for materials and design could alter the way cities operate. Through this arrangement, the book aims to bridge the gap between theoretical discourse on sustainability and realistic, politically feasible action.

In addition to a conceptual understanding of transformational growth, readers will depart with a set of operationalization techniques that have been proven effective in a range of urban contexts. This synthesis will be useful for researchers looking to further the theory of urban transformation, practitioners looking for methods that can be used elsewhere, and policymakers looking for comprehensive solutions. As the instances will show, despite the enormous task that lies ahead, cities already possess the capacity to develop in ways that repair, include, innovate, and persevere. These skills must now be developed until they define the mainstream in order to shift the discourse on urban growth from one about how much we build to one about how well and for whom we construct within the limitations of the planet we share.

References

Ahatsi E, Nie L, Olanrewaju OA (2024) Climate change mitigation and adaptation in Ghana: Strategies and challenges faced by social enterprises. Atmosphere 15(11), 1278

Andersen MS, Massa I (2000) Ecological modernization—Origins, dilemmas and future directions. J of Environ Polic and Plan 2(4):337–345

Asara V, Otero I, Demaria F, Corbera E (2015) Socially sustainable degrowth as a social–Ecological transformation: Repoliticizing sustainability. Sustain Sci 10(3):375–384

Aurand A (2010) Density, housing types and mixed land use: Smart tools for affordable housing? Urban Stud 47(5):1015–1036

Basee DH, Abdulla ZR (2022) Transformation of urban morphology, vulnerability and resilience: Haifa Street Area, as a case study. Ain Shams Eng J 13(4), 101718

Boh W, Constantinides P, Padmanabhan B, Viswanathan S (2023) Building digital resilience against major shocks. MIS Q 47(1):343–360

Carter N (2018) The politics of the environment: Ideas, activism, policy. Cambridge University Press, Cambridge

Cheshmehzangi A, Chen H (2021) China's sustainability transitions: Low carbon and climate-resilient plan for carbon neutral 2060. Springer Nature, Singapore
De Marco M (2025) Neoliberal cities. Misrecognition and urban crisis. In: De Marco J (ed) Counteracting the neoliberal city: Towards the integral and recognitive regeneration of public spaces. Springer Nature Switzerland, Cham, pp 105–147
Escobar A. (2015) Degrowth, postdevelopment, and transitions: A preliminary conversation. Sustain Sci 10(3):451–462
Grinberger AY, Felsenstein D (2014) Bouncing back or bouncing forward? Simulating urban resilience. Proc of the Inst of Civ Eng-Urban Des and Plan 167(3):115–124
Healy H, Martinez-Alier J, Kallis G (2015) From ecological modernization to socially sustainable economic degrowth: Lessons from ecological economics. In: Bryant RL (ed) The international handbook of political ecology. Edward Elgar Publishing, Cheltenham, pp 577–590
Hermanns H (2015) Ecological modernization and the development of emissions trading schemes in Australia and South Korea. Pac Focus 30(1):103–125
Inam A (2013) Designing urban transformation. Routledge, New York and London
Kantor P (2019) The dependent city revisited: The political economy of urban development and social policy. Routledge, New York and London
Kolosok S, Myroshnychenko I, Gries T (2024) Energy and sustainability: A balanced energy transition strategy. In: Spataru C, Lv X, Carvalho P, Nowbuth MD, Ameli N (eds) The Elgar companion to energy and sustainability. Edward Elgar Publishing, Cheltenham, pp 239–257
Lebel L, Anderies JM, Campbell B, Folke C, Hatfield-Dodds S, Hughes TP, Wilson J (2006) Governance and the capacity to manage resilience in regional social-ecological systems. Ecol and Soc 11(1)
Loza Adaui CR (2025) The principle of regeneration in circular economy: Revitalising for resilience. In: Fröhlich E, Steinbiß K, Schmiedeknecht MH, Adaui CRL (eds) Circular economy in sustainable supply chains: A global perspective on challenges, concepts and cases. Springer Nature Switzerland, Cham, pp 49–60
McIlroy D, Brennan S, Barry J (2022) Just transition: A conflict transformation approach. In: Pellizzoni L, Leonard E, Asara V (eds) Handbook of critical environmental politics. Edward Elgar Publishing, Cheltenham, pp 416–430
Mikelsone E, Atstaja D, Koval V, Uvarova I, Mavlutova I, Kuzmina J (2021) Exploring sustainable urban transformation concepts for economic development. Stud of Appl Econ 39(5)
Park J, Sarkis J, Wu Z (2010) Creating integrated business and environmental value within the context of China's circular economy and ecological modernization. J of Clean Prod 18(15):1494–1501
Pelling M (2012) The vulnerability of cities: Natural disasters and social resilience. Routledge, New York and London
Rees WE (2012) Cities as dissipative structures: Global change and the vulnerability of urban civilization. In: Weinstein MP, Turner RE (eds) Sustainability science: The emerging paradigm and the urban environment. Springer New York, New York, pp 247–273
Sanderson Bellamy A, Ioris AA (2017) Addressing the knowledge gaps in agroecology and identifying guiding principles for transforming conventional agri-food systems. Sustain 9(3), 330
Schmid B (2019) Degrowth and postcapitalism: Transformative geographies beyond accumulation and growth. Geogr Compass 13(11), e12470
Schneider F, Kallis G, Martinez-Alier J (2010) Crisis or opportunity? Economic degrowth for social equity and ecological sustainability. Introduction to this special issue. J of Clean Prod 18(6):511–518
Soufi HR, Esfahanipour A, Shirazi MA (2021) Risk reduction through enhancing risk management by resilience. Int J of Disaster Risk Reduct 64, 102497
Sylla NS, Koddenbrock K (2025) Delinking and global reparations: Alternative perspectives on the socio-ecological crisis. Transcript Verlag, Bielefeld, Germany, p 228
van der Horst D (2022) Low carbon commons: Changing property rights for urban retrofits. In: van der Horst D (ed) ECEEE 2022 summer study on energy efficiency: Agents of change. European Council for an Energy Efficient Economy, pp 733–742
Xue J (2022) Urban planning and degrowth: A missing dialogue. Local Environ 27(4):404–422

Part 1
Urban Growth

Urban growth, when steered by *transformative growth* principles, becomes a catalyst for ecological renewal rather than depletion.

It redefines expansion to prioritise blue-green networks, cultural continuity, and climate resilience. This approach transforms cities into living systems where economic vitality and environmental stewardship advance together.

Chapter 2
Urban Transformation in Japan: The Challenge of Brownfield Redevelopment

Armin Mehdipour and Jon Kellett

Abstract As urbanization accelerates, cities worldwide face the dual challenge of sustaining economic growth while mitigating environmental degradation. Brownfield redevelopment has emerged as a key strategy within the sustainable urban development agenda, particularly in industrialized nations seeking to revitalize underutilized or contaminated land. While extensively studied in Western contexts, brownfield redevelopment in Japan remains under-explored in English-language scholarship. This paper examines the Musashi Kosugi area in Kawasaki City as a case study to analyze Japan's evolving approach to brownfield regeneration. Once a major industrial hub, Musashi Kosugi experienced significant economic restructuring in the 1990s, leaving behind large tracts of brownfield land. In response, local and national governments implemented a series of legislative and planning interventions to reclaim and repurpose these sites for residential and commercial development. Through a mixed-methods approach that integrates secondary policy analysis with primary data from stakeholder interviews and field observations, this study investigates the key policy mechanisms, governance structures, and socio-economic outcomes of brownfield redevelopment in Japan. The findings highlight the effectiveness of hierarchical planning frameworks, land reclamation policies, and investment incentives in facilitating urban transformation while also revealing persistent challenges related to environmental remediation and stakeholder coordination. Weak coordination between land-use and environmental policies, limitations in the Soil Contamination Countermeasures Act (SCCA), and inadequate financial support mechanisms hinder effective remediation efforts. By

A. Mehdipour (✉) • J. Kellett
School of Architecture and Built Environment, Adelaide University, Adelaide, Australia
e-mail: armin.mehdipour@adelaide.edu.au

A. Cheshmehzangi et al. (eds.), *Transforming Cities*, Urban Sustainability,
https://doi.org/10.1007/978-981-95-7378-3_2

contextualizing Japan's brownfield regeneration efforts within broader global urban sustainability discourses, this research provides valuable insights for policymakers and planners navigating similar land recycling challenges in industrialized and rapidly urbanizing regions.

Keywords Urbanization • Brownfield redevelopment • Policy and practice • Musashi Kosugi • Japan

2.1 Introduction

Urbanization has been one of the most significant global trends over the past two centuries, and as the world moves further into the twenty-first century, it remains a focal point of economic, social, and environmental discourse. Cities are not only centers of population growth but also engines of economic development, generating over half of global GDP (UN-Habitat, 2013). However, while urban areas drive economic prosperity, primarily through industry and services, they also pose major environmental challenges, contributing to pollution, carbon emissions, climate change, and the depletion of natural resources (Huang et al., 2010; Newman and Jennings, 2012; Yang et al., 2023). In response, policymakers worldwide are increasingly prioritizing sustainable urban development strategies to balance economic growth with environmental stewardship. Urban regeneration has emerged as a key strategy within this broader sustainability agenda, emphasizing the efficient use of land and natural resources (Newton and Glackin, 2014; Mehdipour and Rashidi Nia, 2013a). Since the 1990s, brownfield redevelopment, the process of repurposing previously industrialized or contaminated urban sites, has become a critical component of urban planning policies in many developed nations (Mehdipour and Rashidi Nia, 2013b). However, the extent and effectiveness of such policies vary across different national contexts. While brownfield redevelopment has been well established in Western countries, it remains an evolving approach in others, such as China (Mehdipour and Kellett, 2024).

Japan, as one of the most industrialized nations in Asia, provides a compelling case study of brownfield redevelopment within a unique socio-economic and historical framework. The country experienced rapid industrialization during the post-war period, particularly in the wake of the so-called Japanese Economic Miracle (Liao and Katada, 2022). This period witnessed substantial decentralization, with populations shifting from central urban areas to suburban and provincial towns, facilitated by extensive railway expansion (Ohashi et al., 2022). Many industrial facilities and infrastructure projects were developed along these railway corridors, contributing to Japan's economic boom. However, since the early 1990s, prolonged economic stagnation and structural recession have led to the decline of heavy manufacturing industries, leaving behind large swaths of underutilized brownfield sites in major cities (Mehdipour, 2020; Otsuka et al., 2021). These sites often present significant environmental concerns, particularly in terms of soil and groundwater contamination due to decades of intensive industrial activity.

The Musashi Kosugi area in Kawasaki City exemplifies this phenomenon. Once a thriving industrial hub, the area experienced extensive brownfield emergence following the economic downturn and industrial restructuring of the 1990s. Over the past two decades, a series of legislative and planning interventions have aimed to revitalize these underutilized spaces (Martinez-Fernandez et al., 2016; Kadomatsu, 2019). In response, the Kawasaki local government launched a comprehensive redevelopment plan in 1993, seeking to transform brownfields into functional urban spaces through targeted policy measures and land-use strategies. This paper examines the Musashi Kosugi redevelopment project as a case study to explore key issues in Japan's brownfield policy and practice, including the historical drivers of brownfield emergence, the legislative framework governing redevelopment, and the socio-economic and environmental impacts of these transformations.

Given the relative scarcity of English-language literature on brownfield redevelopment in Japan, this study employs a qualitative, site-based approach to bridge this knowledge gap. A case study analysis of the Musashi Kosugi redevelopment project is conducted, drawing on field observations and interviews with key stakeholders, including local government officials, urban planners, and community representatives. These qualitative insights complement a review of relevant policy documents, regulatory frameworks, and redevelopment plans to provide a comprehensive understanding of Japan's approach to brownfield regeneration. By examining the trajectory of redevelopment in Musashi Kosugi, this study sheds light on the key policy mechanisms, planning tools, and governance structures that have shaped the transformation of brownfield sites in Japan. Specifically, it explores the role of national and local government policies in facilitating land reclamation, investment incentives, and environmental remediation efforts. Furthermore, the study assesses the socio-economic and environmental impacts of redevelopment, considering how former industrial sites have been repurposed for residential, commercial, and mixed-use developments to meet contemporary urban needs.

Beyond its local significance, the findings from Musashi Kosugi contribute to broader discussions on sustainable urban development in Japan and other industrialized nations facing similar challenges of urban land recycling. By assessing the successes and limitations of Musashi Kosugi's transformation, this research offers insights into the effectiveness of regulatory frameworks, stakeholder engagement strategies, and innovative land-use planning approaches, particularly in rapidly urbanizing regions where land scarcity and ecological concerns demand integrated solutions. Situating Musashi Kosugi's redevelopment experience within a broader international context, this research contributes to a deeper understanding of how brownfield sites can be successfully repurposed into vibrant, resilient, and inclusive urban spaces. Ultimately, it aims to enrich the discourse on sustainable land-use planning, providing valuable guidance for policymakers, urban planners, and scholars striving to create more adaptable and livable cities worldwide.

2.2 Research Methods

This study employs a mixed-methods approach, integrating both secondary and primary data to examine brownfield policy and regeneration in Japan. Secondary data sources include government reports, policy directives, and academic literature, which were systematically selected based on their comprehensiveness, credibility, and relevance to the research objectives. However, given the relative immaturity of brownfield policy in Japan, secondary sources alone were insufficient, necessitating the collection of primary data to fill critical knowledge gaps. Primary data were collected through semi-structured interviews and in-depth field surveys conducted at a specific case study site. The interviewees included city officials, professionals, representatives from public development agencies, academics, and urban researchers, all of whom offered valuable perspectives on the traditional policy drivers and hierarchical planning approaches to brownfield regeneration within the context of Japan's political and socio-economic framework. A total of 25 face-to-face interviews were conducted in Kawasaki, Tokyo, and Tsukuba between 2017 and 2020. Participants and case study sites were identified using the Snowball Sampling Method (also known as chain-referral sampling), with guidance from a research team at Tsukuba University, who were well-connected with industry and academia. To maintain confidentiality, all interview participants were anonymized in the data reporting. Due to the semi-structured format of the interviews, the analysis concentrated solely on data pertinent to the study's research objectives.

2.3 Case Study Overview: Musashi-Kosugi Area, Kawasaki

Musashi-Kosugi is a district located in the central part of Kawasaki City, Kanagawa Prefecture (Fig. 2.1). Historically, this area was known as *Industrial City Station* due to its high concentration of business sites and large-scale industrial enterprises. However, with the rapid urban expansion and factory relocations of the 1990s, Musashi-Kosugi underwent large-scale, high-speed redevelopment, transforming it from an industrial hub into a dynamic, multi-functional urban center. Kawasaki City, situated between the Tokyo Metropolitan Area and Yokohama, is administratively divided into seven wards: Kawasaki, Asao, Miyamae, Nakahara, Saiwai, Takatsu, and Tama. The Musashi-Kosugi district is located in Nakahara Ward, the most populous ward in Kawasaki, with a population of 251,248 (Kawasaki City, 2016b). Strategically positioned along the Tamagawa (Tama) River, Musashi-Kosugi serves as a key transportation hub, where multiple railway lines intersect (Fig. 2.2). These include three national railway lines, the JR Nambu Line, JR Shonan-Shinjuku Line, and JR Yokosuka Line, as well as two privateAudioVolumeMute railway lines, the Tōkyū Tōyoko Line and Tōkyū Meguro Line. In addition, the Tsunashima Highway, which connects Shinagawa Ward in Tokyo to Kanagawa Ward in Yokohama, further enhances the district's accessibility. These transport links provide seamless

Fig. 2.1 Musashi-Kosugi Redevelopment Area in the heart of Kawasaki city (Source: Authors' elaboration on Kawasaki City, 2017).

connectivity between Musashi-Kosugi and the central districts of both Tokyo and Yokohama, reinforcing its role as a vital urban node.

2.3.1 Kawasaki and Musashi-Kosugi: From Industrial Hub to Urban Center

The closure of manufacturing industries such as NEC (electronic), TKS (machinery), and Fujisash (aluminum sashes) in the Musashi-Kosugi area, along with the district's broader redevelopment, has been driven by two significant trends: (1) the city's transport-oriented growth, which began in the 1920s and accelerated after the 1950s, and (2) prolonged economic recessions since the 1990s, which led to the relocation of industries from the central city.

2.3.1.1 Transport-Driven Urban Expansion in Kawasaki

The city of Kawasaki exemplifies the rapid urban and industrial expansion characteristic of many Japanese cities following the post-war recovery and economic boom. However, Kawasaki's urbanization began well before World War II, in the 1920s, when several municipalities and villages were merged to form the city. Nakahara Ward was also established during this period, in 1925, through the consolidation of Sumiyoshi and Nakahara villages (Kawasaki City, 2016a). This early development was closely tied to improvements in Japan's national railway system,

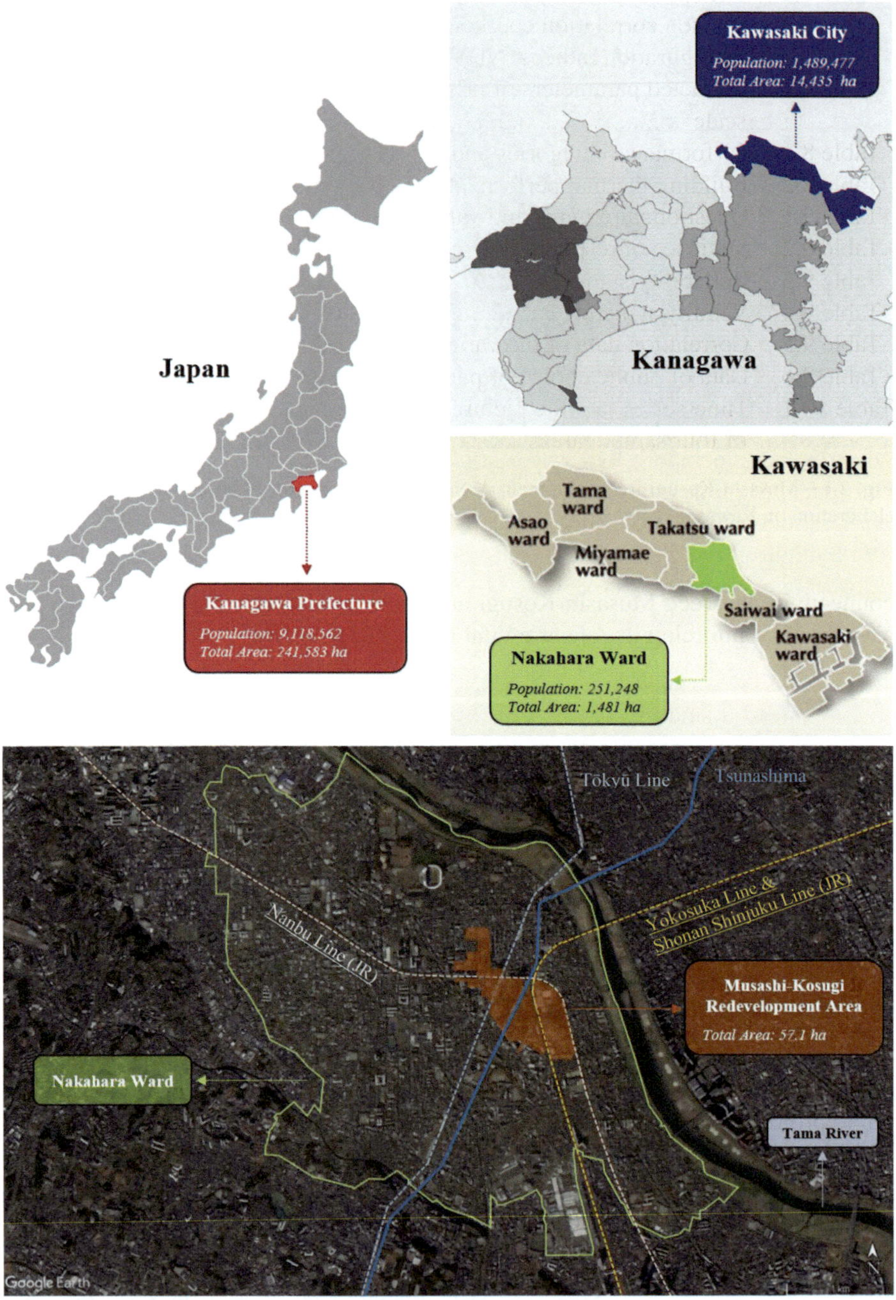

Fig. 2.2 Regional overview of the Musashi-Kosugi Area (Source: Authors' construct with data collected form Kawasaki City, 2016b, Kanagawa Prefecture, 2025).

particularly along the industrial corridor connecting the Keihin Industrial Region (Tokyo, Kawasaki, and Yokohama) with the Hanshin Industrial Region (Osaka and Kobe). A significant milestone was the construction (1926) and opening (1927) of the Tokyo-Yokohama Electric Railway, now known as the Tōkyū Tōyoko Line, which linked major cities in the Greater Tokyo Area, including Tokyo Metropolis and the neighboring prefectures of Saitama, Kanagawa, and Chiba. Around the same time, in 1927, the JR Nambu Line also commenced operations, serving as a crucial transport link between western Tokyo (Tachikawa Station) and Kawasaki Station, passing through Musashi-Kosugi Station (Kawasaki City, 2016a).

Over the past century, the area around Musashi-Kosugi Station has undergone a remarkable transition, shifting from a freight transportation hub to a residential and commercial center. The expansion of transport infrastructure and railway networks led to a dispersed residential development pattern around the station. Meanwhile, in the pre-war years, large-scale industries increasingly moved into the area. Among them were highly specialized assembly, electric machinery, steel, and chemical industries, including the NEC Tamagawa Plant (1936), Fuji Communication Equipment Manufacturer (1938), Manufacturer of Tokyo Radio Equipment, Ebara Corporation (1939), and Daido Steel (1941). The establishment of Musashi-Kosugi Station on the Tōkyū Tōyoko Line in 1943 was closely linked to the strong industrial activity in the surrounding area. In 1953, the station was integrated with the nearby Industrial City Station (opened in 1939), forming the current station complex (Kawasaki City, 2015). Despite the rapid conversion of rural land to residential and industrial zones driven by transportation growth, the Musashi-Kosugi area remained a small suburban town before World War II, with agriculture still dominating the landscape. However, during the war years, portions of agricultural land and several industrial sites, such as the NEC Tamagawa Plant, were repurposed for military production, training, and related facilities.

According to a report by Kawasaki City (2015), much of the land development around Musashi-Kosugi Station took place after World War II, particularly during the Japanese Economic Miracle from the 1950s to the early 1970s (see Fig. 2.3). This period was marked by significant economic restructuring, as numerous heavy manufacturing industries and enterprises relocated to the Kawasaki region, driving rapid private investment and urban development. During the 1960s and 1970s, Kawasaki emerged as a key driver of Japan's high-speed economic growth (WorldBank, 2001). Its bay area became the industrial core of the Keihin region, evolving into one of Japan's largest and most important manufacturing centers (Kanada et al., 2013). The concentration of industry not only fueled economic expansion but also attracted a significant influx of new residents, accelerating the pace of residential land development. Between 1955 and 1975, Kawasaki's population grew from 445,520 to 1,014,951, an increase of over 56% (Kawasaki City, 2016b). Since 1975, population growth has slowed but remained steady, with an increase of nearly 15% from 1975 to 1995 and a further 19% from 1995 to 2015 (Kawasaki City, 2016b).

However, in the 1990s, economic recession, corporate bankruptcies, and physical constraints on land supply led to the gradual relocation of manufacturing facilities away from Kawasaki. This shift left behind large tracts of idle and underutilized land

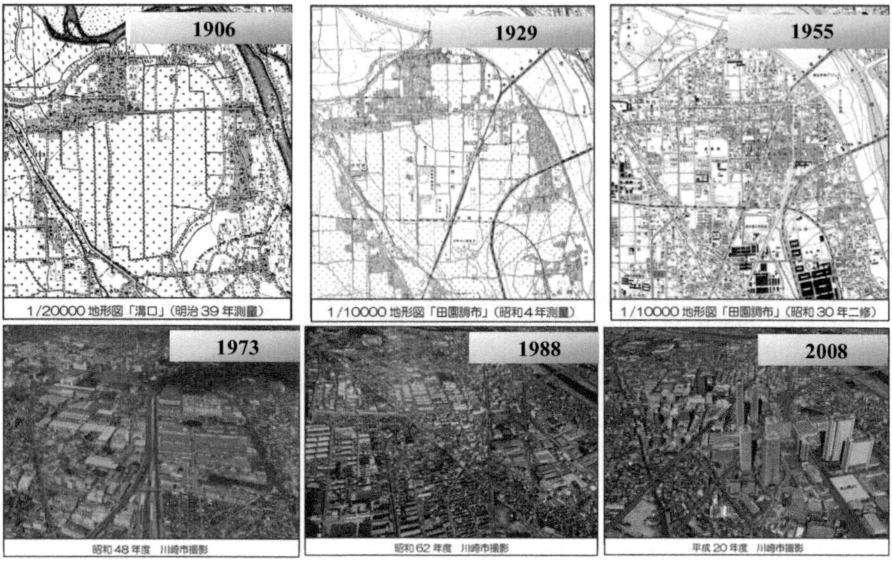

Fig. 2.3 Evolution of the Musashi-Kosugi Area from 1906 to 2008 (Source: Authors' elaboration on Kawasaki City, 2015).

around Musashi-Kosugi Station. In response, the Ministry of Land, Infrastructure, and Transport (MLIT) introduced revitalization policies to repurpose these areas. In 1993, Kawasaki City developed a comprehensive plan for the Kosugi Station area, aiming to transform it into a compact, vibrant, and multi-functional urban district. This redevelopment strategy will be explored further in the following sections.

2.3.1.2 Industrial Relocation

Following the rapid industrial-led economic growth and urbanization after World War II, Kanagawa Prefecture, particularly Kawasaki, emerged as a key manufacturing hub in Japan. However, since the 1990s, the Japanese economy has experienced two major downturns: the bursting of the economic bubble in the early 1990s and the global financial crisis (GFC) of 2008–2009. Both economic shocks had significant consequences for industrial relocation and manufacturing activities in major Japanese cities. As a heavily industrialized region, Kanagawa Prefecture exemplifies the hollowing-out process of industry in Japan, where a substantial number of manufacturing enterprises, particularly in steel, electric machinery, and petrochemicals, relocated from the region following the economic downturn. Between 1987 and 1993, for example, the prefecture lost 138 manufacturing establishments and 16,010 jobs in the electric machinery and appliance sector across Yokohama, Kawasaki, and Yokosuka (Guelle, 2001). This industrial decline continued beyond 1993, with Kanagawa, along with Tokyo, Aichi, Hyogo, and Saitama, recording the highest levels of job losses in manufacturing between 1993 and 2001 (WorldBank, 2001).

The transformation of the Musashi-Kosugi area offers a representative case study of different models of industrial relocation in Japan following the economic bubble collapse. The NEC Tamagawa Plant exemplifies large-scale manufacturers that reduced domestic production capacity and shifted their primary operations overseas, particularly to China and ASEAN (Association of Southeast Asian Nations) countries, in response to increasing global competition. Despite downsizing, NEC has retained some manufacturing activities in the Musashi-Kosugi area. Conversely, the closure of the Fujisash plant illustrates the fate of many medium-sized manufacturers that either scaled down or ceased operations entirely. After reducing its production capacity in the early 1990s, Fujisash ultimately sold its land in Musashi-Kosugi to the construction firm Haseko Co. in 2007 for 12.2 billion yen (over 100 million USD), reinvesting the proceeds into its business (Iwami, 2013).

Another common relocation pattern is exemplified by Tokyo Kikai Seisakusho (TKS), a small-scale manufacturer that maintained its core production but moved operations to a non-urban area in a neighboring prefecture, Chiba, in this case. This trend was largely driven by the financial benefits of selling valuable urban land, coupled with the lower costs and greater availability of space in suburban and rural areas. Notably, while Kawasaki and other urbanized industrial regions in Kanagawa experienced manufacturing decline between 1987 and 1993, the western inland area of Kanagawa (Nishi-Kanagawa) simultaneously saw industrial expansion, gaining 308 companies and 18,225 jobs (Horaguchi, 2004).

2.3.1.3 Soil Pollution Challenges

Since the local government of Kawasaki formulated the redevelopment master plan in the mid-1990s, a significant number of buildings and structures from the former industrial sites were required to be removed to facilitate new construction. These included industrial facilities, residential buildings, derelict commercial properties, and substandard infrastructure near the station, many of which had not been well-maintained or effectively utilized. Before redevelopment, much of the land and built environment belonged to former industrial areas occupied by NEC, TKS, and Fujisash Plants. In total, 34.2 hectares, nearly 60% of the 57.1 hectare redevelopment zone, were previously used for industrial activities. Given this high concentration of manufacturing, all three factory sites faced soil contamination issues following their closure or relocation (Iwami, 2013, Kawasaki City, 2025). During the demolition and construction phases, harmful substances and toxic materials were detected at various depths and across different time frames. Soil contamination incidents at Musashi-Kosugi Station occurred throughout the 20-year redevelopment period, including:

- **NEC Tamagawa Renaissance City Towers (1997–2005):** High-rise office buildings constructed on the former NEC factory site (4 ha).
- **District 4 (2005–2011):** High-rise residential and commercial buildings, as well as new parks and green spaces, developed on the former Fujisash factory site (9 ha).

- **Tsunashima Highway Widening (2009–2010):** Expansion project along the former TKS factory site.
- **Fuchu Highway Widening (2011–2012):** Expansion project affecting both the TKS and Fujisash factory sites.
- **City Tower Musashi Kosugi (2012–2016) & Grand Tree Shopping Complex (2012–2014):** Residential and commercial developments on the former TKS factory site (4.1 ha).

Unlike Western practices, which typically require comprehensive site cleanup before development begins, this situation seems to have involved addressing contamination piecemeal as different buildings were developed. Despite the extensive redevelopment, information on soil contamination and the measures taken for remediation remains scarce. In Japan, the development of contaminated land is often stigmatized, influenced by deeply rooted cultural attitudes toward environmental pollution (Otsuka and Abe, 2008; Dixon et al., 2010). The fear of reputational damage discourages landowners and developers from publicly disclosing contamination, even after remediation and redevelopment. Under the Soil Contamination Countermeasures Act (SCCA) of 2002, local governments must remove sites from contamination records once they are certified as fully remediated. As a result, access to brownfield-related data remains a challenge. In the Musashi-Kosugi redevelopment, several key issues related to soil contamination and remediation remain unclear, such as the nature and extent of pollution, responsible parties for treatment, and the specific remediation techniques employed. However, certain aspects can be reasonably affirmed:

- **High Contamination Risk:** Given the intensity of industrial activities, all three former factory sites were highly likely to be contaminated. According to an Australian assessment framework (DSE, 2005) and a Japanese industrial contamination index (Yasutaka et al., 2007), the likelihood of soil contamination across these sites was approximately 67%.
- **Variability in Contamination and Remediation Methods:** The type and level of soil contamination differed across the NEC, TKS, and Fujisash sites due to variations in manufacturing processes. Different remediation techniques were applied based on the specific pollutants identified. At the TKS site, for example, tetrachloroethylene and lead were detected, likely resulting from the chemical-intensive nature of printing machinery operations. Heavy metals like these can become airborne through contaminated dust, which may explain why vinyl sheets were used to cover soil surfaces during the Tsunashima Highway expansion, preventing hazardous material dispersion.
- **Responsibility for Remediation Costs:** The responsibility for remediation varied depending on ownership, institutional structure, and financial capacity. NEC retained and redeveloped its land, commissioning voluntary soil and groundwater assessments. When contamination was identified, designated construction firms handled cleanup under NEC's direction. In contrast, TKS sold its land to a real estate developer, transferring full responsibility for site investigation, remediation, and construction to the new owner. Nonetheless, in

all cases, the original industrial companies bore the financial burden of remediation. According to TKS' 2013 business report, a substantial sum was allocated for soil decontamination as part of the land transfer process.

- **Regulatory and Policy Frameworks:** The regulatory landscape evolved during the redevelopment process, affecting different sites in different ways. NEC's remediation took place in the late 1990s, before the enactment of national legislation on soil contamination. Until 2002, Japan lacked a comprehensive legal framework for contaminated land remediation. However, Kawasaki City and Kanagawa Prefecture, with their long histories of industrial activity, had already introduced local environmental policies addressing soil contamination. As a result, NEC's remediation was conducted under local government oversight. In contrast, remediation at the TKS and Fujisash sites occurred after the introduction of the SCCA in 2002 meaning these projects had to comply with a national regulatory framework, which was operational from 2003. Section 6 further explores the implications of the SCCA and its operational system.

2.4 Brownfield Definition in Japan

Over the past two decades, numerous research surveys (Murayama et al., 2006; Otsuka and Abe, 2008; Dixon et al., 2010; Dixon et al., 2011; Sato, 2007; Otsuka et al., 2021; Mehdipour et al., 2024) have contributed to Japanese policy discussions by identifying and comparing key issues associated with brownfield development. This research activity was particularly pronounced in the early 2000s when brownfields remained undefined in official terms, and scholars increasingly recognized the need for a comprehensive legislative framework to address brownfield-related challenges.

A crucial observation in the Japanese literature and governmental surveys is the narrow scope through which brownfields have been conceptualized, with a predominant focus on environmental aspects. Even before an official definition was established in 2007, the prevailing academic and policy discourse in Japan, strongly influenced by the U.S. approach, primarily regarded brownfields as contaminated or potentially contaminated sites. For instance, (Murayama et al., 2006) defined brownfields as "contaminated land potentially to be developed," while (Yasutaka et al., 2007) described them as "sites caused by the actual presence of a contaminant." Sato (2007) offered a broader perspective, defining brownfields as "land parcels that have become significantly underutilized or abandoned due to soil contamination or concerns over potential contamination."

In March 2007, the Ministry of the Environment (MoE) released an interim report titled, The Current Status of the Brownfields Issue in Japan. This report was based on a comprehensive survey conducted in February 2007 among member companies of the Geo-Environmental Protection Center, including research and investigation firms, construction consultants, and construction companies (MoE, 2007). The primary objective of the report was to assess the scale of contaminated or potentially contaminated land and explore viable approaches for redevelopment.

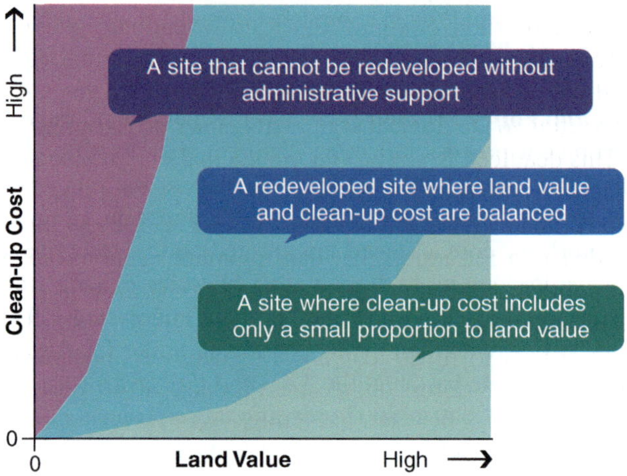

Fig. 2.4 Classification of brownfield sites in Japan, based on the cleanup cost and intrinsic land value (Source: Authors' elaboration on Kokusai Kokusai Co, 2016; Kanagawa Prefecture, 2025).

More significantly, this report provided, for the first time, an official nationwide definition of brownfields for legislative purposes. Based on the MoE, brownfields were defined as: "Lands which are unused or with extremely limited use compared to their intrinsic value because of the existence or potential existence of soil contamination."

This definition remains the most widely used in Japanese urban studies to date. Three key aspects can be derived from this definition. First, it emphasizes the physical condition of the site, defining brownfields as either unused or significantly underutilized, thus highlighting issues of land vacancy and dereliction. Second, it underscores the environmental dimension of brownfields by identifying actual or potential soil contamination as a fundamental characteristic. Finally, it acknowledges the economic implications of brownfield redevelopment by implicitly linking the feasibility of land reuse to the relationship between contamination cleanup costs and the site's intrinsic economic value. This economic consideration plays a crucial role in determining the viability of brownfield regeneration efforts (see Fig. 2.4).

2.5 Policy Framework for Brownfields in Japan

In Japan, brownfield-related activities are governed by two distinct groups of legislative acts and policies (see Fig. 2.5):

1. **Urban Land-Use Planning and Development Acts:** Overseen by the Ministry of Land, Infrastructure, Transport and Tourism (MLIT), these include the *City Planning Act* and its supplementary laws, such as the *Land Readjustment Act*

and the *Urban Renewal Act*. These regulations primarily focus on the redevelopment of underutilized land, providing the legal framework for urban renewal projects.

2. **Environmental Acts:** Administered by the Ministry of the Environment (MoE), the primary legislation addressing land contamination is the SCCA. This remains the only national law in Japan specifically designed to manage soil contamination issues, outlining requirements for site investigation, remediation, and risk management.

While these two legislative frameworks operate in parallel, their interaction in practice remains complex, often requiring coordination between urban development objectives and environmental protection measures.

Japan's land-use planning system is notably complex, as it involves multiple regulations and laws that collectively govern urban development. Broadly, the system operates on three hierarchical layers of policy and regulation.

1. **First Layer: National Land-Use Policies**
   At the highest level, the national government establishes overarching land-use strategies through foundational laws such as the Comprehensive National Land Development Act and the National Land-Use Planning Act. These laws provide long-term, large-scale development frameworks covering diverse land categories, including farmland, forests, natural parks, and urban areas.
2. **Second Layer: City Planning Act (1968)**
   The City Planning Act of 1968 serves as the core legislative framework for urban land-use planning in Japan. This policy-oriented law establishes key principles and regulatory measures for managing land-use in urban areas, guiding both redevelopment and new development initiatives.

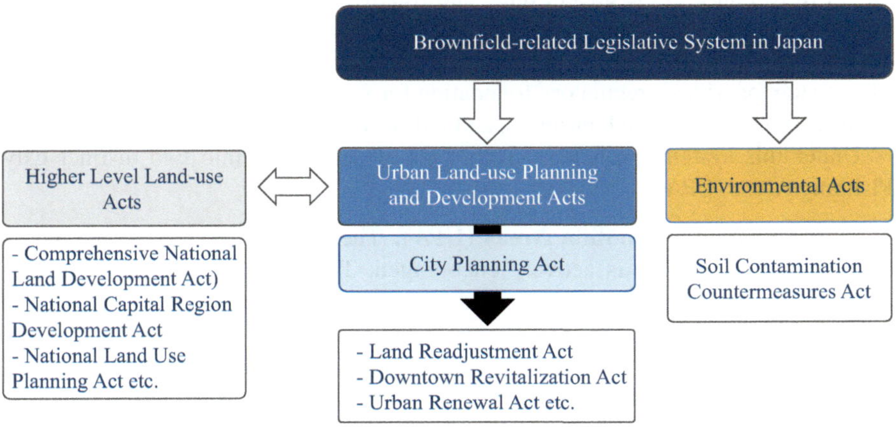

Fig. 2.5 Overview of Japanese regulatory framework for brownfield redevelopment (Source: Authors).

3. **Third Layer: Implementation Through Individual Development Laws**
   The third layer consists of specific development laws that provide implementation mechanisms for the City Planning Act of 1968. These laws translate the broad policy directives into practical regulations, ensuring that land-use measures are effectively enforced at the local level.

Given the focus on redevelopment of underutilized land, this discussion will primarily examine the second and third layers of Japan's legislative framework, as they directly influence land-use transformation and urban renewal.

2.5.1 Urban Development and Land-Use Policies

2.5.1.1 City Planning Act of 1968

In 1968, the Japanese central government introduced the *City Planning Act*, a comprehensive and fundamental planning law that came into effect in 1969, replacing the *City Planning Act of 1919*. The primary motivation behind this legislative overhaul was the rapid economic growth and population expansion of the post-war period, which had led to significant structural issues and inadequate urban infrastructure across Japanese cities. While the Act has undergone partial amendments almost annually since its enactment, its fundamental structure has remained largely intact. It establishes a national framework for urban planning, which is administered and implemented at the local level through Japan's 47 prefectures and approximately 1,800 municipalities (Otsuka and Abe, 2008; Bureau of Urban Development Tokyo Metropolitan Government, 2025a).

The *City Planning Act of 1968* is widely recognized as a pivotal legal framework aimed at curbing urban sprawl and guiding sustainable urban development (Kidokoro, 2008). In response to the uncontrolled and fragmented land-use expansion of Japanese cities during the 1960s and 1970s, the Act introduced a strategic urban planning system, known as the *Urban Land-Use Planning System*, which remains in use today. This system provides a regulatory foundation for various land-use categories, including residential, commercial, business, and industrial zones (MLIT, 2025).

Under this system, prefectural territories are classified into two distinct City Planning Areas (CPA):

1. **Urbanization Promotion Areas (UPA):** These are designated zones where urban development is actively encouraged. The UPA includes both already built-up areas and regions expected to undergo urbanization within a 10-year period. Development in these areas follows land-use regulations and urban planning guidelines to ensure coordinated growth.
2. **Urbanization Control Areas (UCA):** In these zones, urban development is strictly regulated to prevent unplanned expansion. However, exceptions may be granted through the *Development Permission System*, allowing certain projects to proceed under specific conditions.

The Act also includes a key provision known as the Policy for Improvement, Development, and Preservation of City Planning Areas, which encompasses measures for long-term urban redevelopment. A notable component of this policy is the *Urban Redevelopment Policy* (URP), codified in Article 7-2, Section 1, Chapter 2 of the *City Planning Act No. 100 of June 15, 1968*. The URP establishes a framework for guiding and promoting urban redevelopment projects through a structured master planning approach (Bureau of Urban Development Tokyo Metropolitan Government, 2025b). Since its inception, this policy has been revised multiple times, integrating provisions from four key legislative acts:

- Urban Renewal Act (Act No. 38 of 1969): Governs general urban redevelopment within UPA, addressing land-use improvements and revitalization.
- Act on Special Measures Concerning the Promotion of Housing and Residential Land Supply in Major Urban Areas (Act No. 67 of 1975): Focuses on the development of residential areas within UPA.
- Act Concerning the Promotion of the Development of Regional Core Urban Areas and the Relocation of Facilities for Industrial Business (Act No. 76 of 1992): Regulates the establishment and enhancement of business districts in UPA.
- Act Concerning the Promotion of Disaster Prevention Block Improvement in Concentrated Urban Areas (Act No. 49 of 1997): Provides guidelines for redeveloping disaster-prone urban areas within UPA.

2.5.1.2 Supplementary Laws for Urban Redevelopment

Following the enactment of the *City Planning Act of 1968*, the Japanese government introduced several supplementary legislative instruments to facilitate its implementation. These additional laws were formulated under the broader framework of the Act's *Policy for Improvement, Development, and Preservation of City Planning Areas*, specifically reinforcing the *Urban Redevelopment Policy*. Recognizing the need for more detailed planning mechanisms, the government developed additional legislative layers to provide technical and operational support for the planning, improvement, and redevelopment of UPA. Some of the key supplementary acts that directly or indirectly address the redevelopment of brownfield or underutilized land include:

- **Land Readjustment Act (1954):** Introduced to streamline land acquisition, facilitate urban development, and enhance the quality of existing urban land and properties (Yanase, 2013).
- **Urban Renewal Act (1969, revised in 1995):** Focuses on the improvement and reuse of built-up areas in major Japanese cities. It establishes regulations for redeveloping urban land, buildings, infrastructure, and public facilities within UPA.
- **Downtown Revitalization Act (1998, revised in 2006):** While the *Urban Renewal Act* primarily addresses large cities, this act targets the redevelopment

of central urban areas in smaller cities and local municipalities. Unlike the broader redevelopment projects permitted under the *Urban Renewal Act*, the *Downtown Revitalization Act* mainly focuses on preserving and strengthening residential neighborhoods.

- **Urban Renaissance Special Measures Act (2002, revised in 2004):** Designed to enhance economic activity by encouraging private-sector investment in urban redevelopment projects. The Act established the *Urban Renaissance Headquarters*, a government body tasked with identifying *Priority Urban Development Areas (PUDA)* where private investment should be concentrated, based on population density and industrial potential (PMJHC, 2025). Till date, 63 PUDA areas, covering approximately 6,424 hectares of urban land, have been developed or are under development, with an estimated private investment of $120 billion (PMJHC, 2025).
- **Compact City Act (2014):** The most recent urban planning legislation, introduced as a revised version of the *Urban Renaissance Special Measures Act*. This act seeks to promote higher-density development and stimulate both public and private investment in specific urban areas, leveraging public transportation networks to support compact urban growth (Terada et al., 2016). The need for this act arose partly due to pressure from landowners on local authorities following the implementation of the *City Planning Act of 1968* and its zoning regulations.

Under the *Compact City Act*, the **UPA** was reorganized into a new planning framework called the *Plan for Deployment Adjustment of Urban Functions*, which categorizes urban areas into three sub-zones:

- **Urban Function Promotion District (UFPD):** Areas designated for essential urban functions, such as commercial, institutional, and transport hubs.
- **Residence Promotion District (RPD):** Zones where residential development is actively encouraged.
- **Possible Area for Residence Control District (RCD):** Areas where residential development may be restricted by local governments to prevent urban sprawl.

By integrating these supplementary acts into the broader framework of the *City Planning Act of 1968*, Japan has developed a multi-layered urban planning system aimed at balancing redevelopment, environmental sustainability, and economic revitalization. These legal instruments continue to shape the urban landscape, ensuring that redevelopment efforts align with national policies and regional urban planning objectives.

2.5.2 Legal Framework for Contaminated Sites

During the 1990s and early twenty-first century, Japan's economy underwent significant shifts in manufacturing activity, driven by increasing international investment from multinational corporations (Dixon et al., 2011). The relocation and closure of heavy manufacturing plants during this period raised concerns regarding the environmental conditions of former industrial sites (Mitsunari, 2009). In particular, the growing awareness of soil contamination and its potential risks to public health and safety prompted Japanese policymakers to take legislative action. In response to these concerns, the Japanese National Diet enacted the Soil Contamination Countermeasures Act (SCCA) in 2002, which came into force in 2003. This comprehensive environmental law was designed to investigate and remediate soil contamination, thereby mitigating health and environmental risks. To enhance its effectiveness, the Act was subsequently amended in 2009 to strengthen regulatory measures and improve implementation.

The SCCA of 2002 establishes a structured framework for investigating, remediating, and managing contaminated sites. The first step in this process is site investigation and assessment, typically carried out by designated investigation institutes. This assessment determines whether real or suspected contamination is present due to past site activities. To facilitate this process, the SCCA has compiled a list of 25 contaminants, each with defined regulatory limits (Dixon et al., 2010). If a site exceeds any of these limits, it is officially designated as contaminated and registered in the records of prefectural and municipal governments (Mitsunari, 2009). The final stage involves removal and remediation as the responsibility of the original polluter which could be the landowner, manager, occupant, or any entity responsible for the contamination. The prefectural governor identifies the polluter, who is then required to bear the cost of remediation. However, if the polluter is unidentifiable or absent, the landowner assumes responsibility for contamination remediation. In such cases, landowners may apply for financial assistance through designated environmental remediation funds. Once the site is fully remediated, its contaminated designation is lifted, and it is removed from governmental records.

To support landowners lacking sufficient financial resources for remediation, the SCCA also introduced a financial support scheme (Aoki, 2003). Under this scheme, if a landowner or polluter can demonstrate an inability to cover remediation costs, they are only required to pay one-quarter of the total expenses, with the remaining amount covered by the prefectural government. To fund these efforts, prefectural governments can apply for subsidies from the Designated Support Corporation, which operates under the Ministry of Environment (MoE). The Designated Support Corporation itself receives financial backing from both governmental and non-governmental sources. As outlined in Article 47, Chapter VI of the SCCA (No. 53, May 29, 2002), *"the Government may subsidize the funds to be appropriated for the Fund to the Designated Support Corporation within the scope of the budget."*

2.6 Discussion

Understanding the relationship between urban-industrial transitions and brownfield emergence is central to this study. In the Japanese context, two key periods following World War II have played a crucial role in shaping brownfield development: the steady urban expansion since the 1990s and the two significant industrial-led economic recessions that began in the early 1990s. The area surrounding Musashi-Kosugi Station exemplifies the substantial impact of these trends on the proliferation of underutilized and contaminated land in Japan. Additionally, as widely discussed in the literature, the mid-1990s marked the onset of a new phase of re-urbanization, driven by anti-sprawl planning policies that have stimulated inner-city redevelopment, particularly in industrial-based cities such as Tokyo and Osaka.

The redevelopment of Musashi-Kosugi underscores the importance of strong collaboration between public and private sectors in brownfield regeneration. The involvement of local government, private landowners, developers, and real estate companies has facilitated the transformation of former industrial sites into a well-integrated urban environment with modern infrastructure, services, and facilities. This project stands as a noteworthy example of brownfield regeneration, demonstrating improvements in environmental conditions, urban structure, economic viability, and social cohesion. However, due to data limitations, there remains considerable uncertainty regarding the effectiveness of environmental remediation efforts, making it unclear whether a total cleanup or a suitable-for-use approach is being used. Key concerns include the extent and nature of soil contamination, the actual condition of remediated land, and the adequacy of the applied remediation techniques.

Institutional barriers between environmental and land-use policies have created significant challenges for brownfield management in Japan. Despite Japan's long-standing history of land-use planning, environmental countermeasures for land contamination, particularly under the SCCA, are relatively new within the country's legislative framework. In practice, land-use and environmental policies remain poorly coordinated, leading to inefficiencies in brownfield remediation and redevelopment. A clear example of these policy inconsistencies is the Tsukiji Fish Market relocation project in Tokyo. In 2008, severe soil and groundwater contamination was discovered at the proposed relocation site, delaying the project for over a decade. The primary cause of this delay was the lack of a communicative and integrated decision-making process among high-ranking officials responsible for both the construction and decontamination efforts.

Under the SCCA, local governments bear primary responsibility for managing brownfield sites, particularly large-scale redevelopment projects such as Musashi-Kosugi. Municipal authorities administer local regulations, not to override the SCCA, but to facilitate site investigation and remediation. However, the SCCA lacks a comprehensive funding mechanism to incentivize land cleanup. Although the act provides financial support for landowners unable to afford remediation costs, this system has proven largely ineffective. In most cases, contaminated sites are or were previously owned by large-scale manufacturing companies, which are ineligible for financial aid under the existing scheme. In Musashi-Kosugi, for instance, none of the private landowners or developers received financial support for remediation, highlighting the limitations of the current funding model.

The complexity of liability allocation further exacerbates brownfield remediation challenges in Japan. The application of the Polluter Pays Principle (PPP) has been particularly difficult due to two key factors. First, Japan's legal and regulatory framework assigns environmental liability predominantly to local authorities, such as prefectural governments, rather than establishing clear accountability at the national level. Second, the rapid pace of urban redevelopment complicates liability tracing. Given the compressed timeline between a site's active use and its redevelopment, local authorities often struggle to identify the original polluter, resulting in current landowners or developers bearing the full cost of remediation.

A final yet significant issue in Japan is the social and cultural perception of land contamination. Many scholars and stakeholders (Otsuka and Abe, 2008; Dixon et al., 2010) have observed that Japanese society exhibits heightened sensitivity to public health risks associated with human-made environmental hazards, in contrast to a more resilient attitude toward natural disasters like earthquakes. Consequently, land contamination is highly stigmatized, posing considerable challenges to brownfield redevelopment. This stigma has led to substantial restrictions on public access to data regarding previously contaminated sites. Under pressure from landowners, sites deemed fully decontaminated must be removed from government records, making it difficult to track the history and extent of contamination. As seen in the Musashi-Kosugi case, public anxiety surrounding contamination issues has significantly influenced government policy approaches to brownfield regeneration.

The issue of stigma is not unique to Japan; similar challenges were observed in the United Kingdom during the 1990s when contaminated land designation adversely impacted property values. However, many developed nations, including the United States, Germany, and the Netherlands, have successfully addressed this issue through decades of public awareness campaigns and progressive policy development regarding soil and groundwater contamination. The Japanese experience highlights the critical need for enhanced transparency, regulatory coordination, and public engagement to overcome the social and institutional barriers impeding effective brownfield remediation and redevelopment.

2.6.1 Transformative Growth and Brownfield Redevelopment

It is worthwhile to think about the concept of Transformative Growth as something that can conceptually bring the diverse dimensions of this chapter together into a single narrative, while also integrating notions of productivity to align with contemporary political debates. This book introduces the concept of Transformative Growth as a way to move urban transformation scholarship beyond the binary of ecological modernization and degrowth. Whereas ecological modernization reduces change to incremental eco-efficiency, and degrowth often risks political and practical dead-ends, Transformative Growth redefines what it means for cities to prosper within planetary boundaries.

The concept of Transformative Growth provides a particularly valuable lens for understanding brownfield redevelopment. Redevelopment efforts, such as those in Musashi-Kosugi, demonstrate that the regeneration of underutilized and contaminated land is not simply about economic revitalization or physical transformation. On one level, redevelopment stimulates new economic activity by attracting investment, creating employment, and enhancing land values. Yet it also generates ecological benefits through soil remediation, the integration of green infrastructure, and the restoration of degraded landscapes. Put bluntly, brownfield regeneration shows how urban redevelopment can simultaneously deliver environmental repair and long-term resilience, broadening the criteria by which cities define and measure prosperity.

Furthermore, brownfield redevelopment generates social productivity by providing new housing opportunities, improving urban amenities, and fostering more inclusive and resilient communities. In Japan, however, the social dimension is complicated by the persistent stigma surrounding land contamination. Contaminated sites are strongly associated with public health risks and reputational damage, making residents and developers cautious about reuse. This stigma has shaped the country's registry system under the SCCA, which requires sites to be removed from official contamination records once certified as remediated. While this approach protects property values and reduces public anxiety, it also undermines transparency, limits long-term monitoring, and constrains public debate.

Finally, redevelopment increasingly relies on technological productivity, as innovative remediation methods, digital planning tools, and design-led approaches enable more adaptive and sustainable outcomes. Taken together, these dimensions underscore that brownfield redevelopment is not merely a technical or regulatory exercise. It can serve as a platform for Transformative Growth, positioning urban regeneration as a catalyst for resilient, equitable, and ecologically grounded futures. By reframing productivity in this way, brownfield regeneration in Japan contributes not only to local redevelopment but also to global debates on how cities can prosper within planetary limits.

2.7 Conclusions

Brownfield redevelopment in Japan, as exemplified by the Musashi Kosugi case, presents a complex interplay of historical, economic, regulatory, and environmental factors. The transformation of former industrial sites into vibrant urban spaces highlights the effectiveness of hierarchical planning frameworks, public-private partnerships, and transport-oriented development in facilitating urban regeneration. However, the findings also reveal persistent challenges, particularly in the coordination of land-use and environmental policies, financial support mechanisms for remediation, and the social stigma associated with land contamination. The limitations of the SCCA coupled with weak institutional integration between environmental and urban planning policies, have hindered more efficient and transparent brownfield regeneration efforts.

Despite these challenges, the case of Musashi Kosugi demonstrates that Japan's approach to brownfield redevelopment has evolved through adaptive policy frameworks and strategic interventions. Lessons from this case suggest that integrating land-use planning with robust environmental governance, increasing financial incentives for remediation, and improving public engagement in redevelopment processes can enhance future urban regeneration projects. Moreover, the study underscores the broader relevance of Japan's experience for other industrialized and rapidly urbanizing nations grappling with similar land recycling challenges. As urban sustainability becomes an increasingly critical policy goal, fostering a more coordinated, transparent, and financially viable approach to brownfield redevelopment will be essential in ensuring resilient and inclusive urban transformations.

Future research should further investigate policy tools that facilitate the integration of land-use and environmental planning in brownfield redevelopment, with a particular focus on regulatory instruments, financial incentives, and institutional frameworks that balance economic viability with environmental sustainability. Comparative analyses with international cases, especially in East Asia and Europe, could offer deeper insights into best practices and policy innovations that enhance risk-informed decision-making for urban infrastructure management. By addressing contamination risks, improving land-use coordination, and fostering collaboration among government agencies, private developers, and local communities, Japan continues to refine its strategies for revitalizing underutilized urban land. Further examination of these advancements and their long-term socio-economic and environmental impacts can deepen our understanding of effective brownfield management approaches and inform future policy development.

Acknowledgments

This research was supported by funding from the Mawson Lakes Fellowship Program (Grant No. 2016/4495), the Australian Housing and Urban Research Institute (Grant No. 2015/4819), and the University of Adelaide. The authors extend their sincere appreciation to the research team at the Land-Use Lab of the University of Tsukuba for their valuable contributions to the field study in Japan. Special thanks also go to the officials of Kawasaki City, opinion leaders, and various experts who generously shared their insights.

Ethics Declarations

This chapter is largely based on a chapter from the primary author's PhD dissertation, which was completed under the supervision of the co-author.

References

Aoki S (2003) Soil contamination countermeasures law the present status of legislation in Japan and contamination risk management. Japan Tappi J 57:1475–1493, 021

Bureau of Urban Development Tokyo Metropolitan Government (2025a) Outline of the city planning, chapter 1 transition of Tokyo's urban planning. [Online]. http://www.toshiseibi.metro.tokyo.jp/. Accessed 17 Mar 2025

Bureau of Urban Development Tokyo Metropolitan Government (2025b) Outline of the city planning, chapter 3 city planning, section 3 development and redevelopment of urban areas. [Online]. http://www.toshiseibi.metro.tokyo.jp/. Accessed 17 Mar 2025

Dixon T, Otsuka N, Abe H (2010) Cities in recession: Urban regeneration in Manchester (England) and Osaka (Japan) and the case of 'Hardcore' Brownfield sites. Project Report. Oxford Brookes University, Oxford

Dixon T, Otsuka N, Abe, H (2011) Critical success factors in urban brownfield regeneration: an analysis of 'hardcore'sites in Manchester and Osaka during the economic recession (2009–2010). Environ and Plan A 43:961–980

DSE (2005, June) Potentially contaminated land; general practice note. Department of Sustainability and Enivornment. Victoria

Guelle F (2001) The links between Japanese investment in Asia and de-industrialization in Japan. In: Andreosso-O'Callaghan B, Bassino JP, Jaussaud J (eds) Changing economic environment in Asia: Firms' strategies in the region. Palgrave Macmillian UK, London, pp 52–61

Horaguchi H (2004, March) Hollowing-out of Japanese industries and creation of knowledge-intensive clusters. Globalization and Revitalization of Industrial and Regional Employment: Comparison of Germany and Japan. The Japan Institute for Labour Policy and Training, Tokyo, Japan

Huang SL, Yeh CT, Chang, LF (2010) The transition to an urbanizing world and the demand for natural resources. Curr Opin in Environ Sustainability 2:136–143

Iwami R (2013) Kawasaki City's Urban Planning and Development Strategy (in Japanese). Kawasaki City. https://www.jcp-kawasaki.gr.jp/wpcontent/uploads/2015/01/iwami-houkoku.pdf. Accessed 11 Apr 2026

Kadomatsu N (2019) Legal management of urban space in japan and the role of the judiciary. In: Rose-Ackerman S et al. (ed) Comparative administrative law (2nd edition, Edward Elgar, 2017), pp 497–512

Kawasaki City (2015) Kawasaki City Urban Planning Promotion around Kosugi Station (in Japanese) [Online]. http://www.city.kawasaki.jp. Accessed 20 Mar 2015

Kawasaki City (2016a) History of Nakahara-ku (in Japanese) [Online]. Kawasaki City. http://www.city.kawasaki.jp. Accessed 17 Mar 2025

Kawasaki City (2016b) The Principal Statistics of Kawasaki City; Population & Number of Households (in Japanese) [Online]. http://www.city.kawasaki.jp. Accessed 17 Mar 2025

Kawasaki City (2017) Redevelopment of Musashi Kosugi (in Japanese). [Online]. http://www.city.kawasaki.jp. Accessed 17 Mar 2025

Kawasaki City (2025) Town Development in the Area around Kosugi Station (in Japanese). [Online]. http://www.city.kawasaki.jp. Accessed 17 Mar 2025

Kidokoro, T (2008) Urban regeneration and the shift of planning approaches: the case of Japanese regional cities. In: Kidokoro T, Harata N, Subanu LP, Jessen J, Motte A, Seltzer EP (eds) Sustainable city regions: Space, place and governance. Springer, Tokyo, pp 145–162

Kokusai Kokusai Co (2016) Redevelopment of contaminated sites in Japan (in Japanese) [Online]. http://biz.kkc.co.jp/kankyo_wp/brownfield/redevelop/. Accessed 18 Mar 2025

Kanagawa Prefecture. (2025). Profile of Kanagawa Prefecture (in Japanese) [Online]. Kanagawa Prefecture. http://www.pref.kanagawa.jp/. Accessed 19 Mar 2025

Liao JC, Katada SN (2022) Institutions, ideation, and diffusion of Japan's and China's overseas infrastructure promotion policies. New Political Econ 27:944–957.

Martinez-Fernandez C, Weyman T, Fol S, Audirac I, Cunningham-Sabot E, Wiechmann T, Yahagi H (2016) Shrinking cities in Australia, Japan, Europe and the USA: From a global process to local policy responses. Prog in Plan 105:1–48

Mehdipour A (2020) Understanding brownfield regeneration in Iran through the lens of international experience. University of Adelaide Adelaide, SA, Australia

Mehdipour, A., & Kellett, J. (2024). Brownfields in China: The case of Liaocheng chemical plant. In: Siew G, Allam Z, Cheshmehzangi A (eds) Diversity as catalyst: Economic growth and urban resilience in global cityscapes. Springer, Singapore, pp 145–161

Mehdipour A, Kellett J, Palazzo E, Larbi M (2024) Policy integration for Brownfield regeneration: An analytical tool. Sustain Horiz 10, 100085
Mehdipour A, Rashidi Nia H (2013a) Industrialization and city change; the concept and historical evolution of urban regeneration. Int J of Sci: Basic and Appl Res 12:176–181
Mehdipour A, Rashidi Nia H (2013b) The role of brownfield development in sustainable urban regeneration. J of Sustain Dev Stud 4(2):78–87
Mitsunari M (2009) Japan's Rising Phase I Market. Mizuho Information and Research Inc. [Online]. https://mizuho-ir.com/. Accessed 11 Mar 2025
Ministry of Land, Infrastructure, Transport and Tourism (2025) Introduction of urban land use planning system in Japan [Online]. http://www.mlit.go.jp/. Accessed 18 Mar 2025
Ministry of Environment (2007) Current status of the Brownfields issue in Japan [Online]. http://www.mizuho-ir.co.jp/. Accessed 18 Mar 2025
Murayama A, Banno K, Ishii S, Kurose T, Sato T (2006) Toward the integration of brownfield rehabilitation and planning methodologies: Case study of Keihin Industrial Area, Tokyo, Japan. WIT Transactions on Ecol and the Environ 94
Newman P, Jennings I (2012) Cities as sustainable ecosystems: principles and practices. Island Press, Washington, DC
Newton P, Glackin S (2014) Understanding infill: Towards new policy and practice for urban regeneration in the established suburbs of Australia's cities. Urban Polic and Res 32:121–143
Ohashi H, Phelps NA, Tomaney J (2022) Between decentralization and recentralization: conflicts in intramunicipal and intermunicipal governance in Tokyo's shrinking suburbs. Urban Plan 7:98–114
Otsuka N, Abe H (2008) Challenges for brownfield regeneration: A comparison of English and Japanese approaches. WIT Transactions on Ecol and the Environ 107:33–42
Otsuka N, Abe H, Isehara Y, Miyagawa T (2021) The potential use of green infrastructure in the regeneration of Brownfield sites: Three case studies from Japan's Osaka Bay Area. Local Environ, 26:1346–1363
Prime Minister of Japan and His Cabinet (2025) Urban renaissance headquarters [Online]. http://japan.kantei.go.jp/policy/tosi/index_e.html. Accessed 18 Mar 2025
Sato H (2007) General Description of Japanese Measures for Soil Contamination, and Plan for Future Measures. International Committee on Contaminated Land: Stockholm Meeting.
Terada T, Tanaka T, Yokohari M (2016) Directions for vacant lot management in the outer suburbs of the Tokyo Metropolitan region. Urban and Reg Plan Rev 3:66–84
UN-Habitat (2013) State of the world's cities 2012/2013: Prosperity of cities, Routledge
Worldbank (2001) Study of Japanese experiences on sustainable urban development including Pollution Control and Management, Resource/Energy Efficiency and GHG Reduction. World Bank. http://hdl.handle.net/10986/12815. Accessed 11 Apr 2025
Yanase N (2013, June 24) Understanding kukaku-seiri (land readjustment). Conference Paper on JICA's Urban Development Course. Japan International Cooperation Agency (JICA), Japan
Yang M, Chen L, Wang J, Msigwa G, Osman AI, Fawzy S, Rooney DW, Yap PS (2023) Circular economy strategies for combating climate change and other environmental issues. Environ Chem Lett 21:55–80
Yasutaka T, Matsuda H, Makino M (2007) An estimating method for potentially contaminated brownfield sites: Lessons from Japan. The 2nd International Conference on Managing Urban Land 25–27

Chapter 3
Earthquake-Resilient Urban Form: Design Strategies for Emergency Access, Rescue Operations, and Post-Disaster Recovery

Amir Shakibamanesh and Shabnam Galehdar

Abstract With the escalating frequency and intensity of environmental hazards—driven by climate change and anthropogenic pressures—urban resilience has become a central imperative in contemporary urban planning and design. Cities worldwide face diverse disasters, including earthquakes, floods, and wildfires, each posing distinct challenges to urban systems. However, the growing complexity and unpredictability of these events demand proactive, multidimensional strategies to strengthen cities' capacities for prevention, adaptation, response, and recovery. Crucially, urban resilience is not merely a function of emergency management but is deeply rooted in the spatial and morphological characteristics of urban form. Although resilient urban form has been extensively studied, critical gaps persist in understanding how urban morphology specifically influences seismic resilience, particularly at the block scale. The absence of integrated urban design guidelines that systematically incorporate seismic resilience exacerbates these risks. This chapter systematically investigates the spatial and morphological attributes underpinning earthquake-resilient urban form, focusing on three key dimensions: (1) block configuration, (2) open space networks, and (3) street systems. Drawing on comparative analysis of major earthquake events, this study develops a set of actionable urban design guidelines that prioritize life safety, enable efficient evacuation and rescue operations, and support rapid post-disaster recovery. By advancing principles such as flexible and modular block layouts, distributed and accessible open spaces, and highly connected street networks, the findings offer a practical framework for enhancing both the resilience of existing neighborhoods and the planning of new urban areas. This integrated approach aims to systematically embed seismic

A. Shakibamanesh (✉) • S. Galehdar
Department of Urban Planning and Design, School of Architecture and Urban Studies, Iran University of Art, Tehran, Iran
e-mail: a.shakibamanesh@art.ac.ir

A. Cheshmehzangi et al. (eds.), *Transforming Cities*, Urban Sustainability,
https://doi.org/10.1007/978-981-95-7378-3_3

resilience into urban form, thereby reducing vulnerability, safeguarding human life, and supporting long-term urban adaptability.

Keywords Seismic resilience • Earthquake-resilient urban form • Disaster risk reduction (DRR) • Urban morphology • Design strategies

3.1 Introduction

In recent years, natural hazards have impacted many regions worldwide, causing devastating consequences for human life. Certain areas, due to their environmental conditions, are more prone and vulnerable to natural hazards, facing more severe impacts at both urban and rural levels. The interactions among these hazards can lead to intensified and more widespread effects (Arvin et al., 2023). Examples of such natural disasters include tsunamis, wildfires, floods, earthquakes, and pandemics. The occurrence of any of these unpredictable disasters can disrupt the normal operational functions of urban systems (Afrin and Yodo, 2019).

The ability to withstand (before the event), respond (during the event), and recover (after the event) is essential for resilient cities. Achieving this requires comprehensive policies (Cheshmehzangi, 2025). Neglecting these aspects may result in uncontrolled expansion of the physical urban fabric and additional constructions, potentially exacerbating future disaster-related damages (Parizi et al., 2022). Disaster management combined with strengthening urban resilience is recognized as an effective approach for disaster prevention and response (Jamali et al., 2023). Enhancing urban resilience offers a long-term solution that prepares urban systems for any future disruptions (Alawneh and Rashid, 2022). Moreover, resilience provides a dynamic and developmental approach to urban disaster management (Dong et al., 2025). Therefore, integrating resilience thinking into urban form theories and design techniques is necessary to avoid adverse outcomes (Sharifi, 2019a).

Urban form has always exhibited some form of order-whether chaotic or planned, weak or strong, bourgeois or plebeian, unfeasible or efficient, stable or unstable, flexible or even inelastic. Furthermore, there exists an intertwined relationship between urban form and its inhabitants, which shapes both the public realm and the city itself, influencing how cities appear and function. When this dialectic ceases, cities decline; when it thrives, cities become resilient. Consequently, studying urban form ultimately leads to examining urban resilience and resilient cities (Aguilar, 2014).

This study focuses specifically on earthquakes as an unpredictable and powerful natural crisis. Earthquake impacts extend beyond building destruction and casualties; it can trigger cascading effects such as physical and economic incidents (subsidence, soil liquefaction, damage to urban infrastructure), human and psychological consequences (death, disappearance, homelessness, migration), disease outbreaks, and secondary crises like chemical accidents and tsunamis (Tang et al., 2019).

Initially, this research examines urban form prior to earthquake crises, followed by a review of experiences, studies, and lessons learned from earthquake events in urban forms globally. Subsequently, a framework for resilient urban form indicators against earthquake crises is proposed. The physical characteristics of cities are categorized into three groups: street networks as the structural framework, blocks representing urban buildings, and open spaces within neighborhoods. The choice of blocks over individual buildings was motivated by the need for faster and more accurate data processing, although detailed building-level analysis could provide more precise insights into physical resilience potential. Finally, measurement methods for each indicator in cities are described. This framework can guide urban designers and planners in assessing urban physical form and serve as a reference for designing earthquake-resilient cities in the future. The final indicators derived from this study not only evaluate potential vulnerabilities of any urban form to earthquakes, but also assist in implementing effective measures aimed at saving lives, facilitating emergency rescue operations, and promoting rapid recovery in the aftermath of seismic disasters.

3.2 Background Knowledge

3.2.1 Theoretical Foundations of Urban Resilience

The concept of socio-ecological resilience emerged in the late 1990s, reflecting a new paradigm aimed at overcoming the separation between society and natural sciences and establishing a novel intellectual foundation to address modern environmental challenges. Although resilience has roots in various disciplines such as physics, psychology, and ecology, its relatively recent application in urban planning, design, and development marks an important evolution (Fouda and Elkhazendar, 2023). Given the rapid growth of urban populations worldwide, the physical infrastructure of cities is expected to expand significantly over the next three decades. Since the physical structure of cities has a long lifespan and represents a primary channel for planners and designers to intervene in urban systems, this can create inertia in planning efforts aimed at fostering resilient cities (Sharifi, 2018).

Brian Walker and David Salt (2006), in their book *Resilience Thinking*, use metaphors like "repetition," "delay," and "room to maneuver" to explain resilience. They detail several "resilience characteristics," including diversity, modularity, tight feedbacks, innovation, overlapping governance layers, ecosystem services, social capital, and allowance for variation. These metaphors are powerful because they help explain how resilience functions. These characteristics are crucial for research as they bridge the gap between resilience theory and urbanism and serve as a link between general and specific resilience features. They relate to urbanism by clarifying the spatial dimensions of urban resilience (Allan et al., 2013).

Accordingly, addressing resilience strategies in planning is vital to overcoming related challenges. According to Ahern (2013), resilience design strategies fall into five main categories: biodiversity, networks, multifunctionality, modularity, and adaptive design (Fouda and Elkhazendar, 2023). Sharifi and Yamagata (2018) identify the objectives of urban physical resilience against natural and human-made disasters as robustness, sustainability, redundancy, prudence, modularity, complexity, flexibility, multifunctionality, self-organization, adaptability, and efficiency. Resilience, by its nature-absorbing disturbances and digesting crises before major destruction-can be viewed in urban planning either as a goal or as a tool to achieve more resilient, desirable, and higher-quality cities (Alawneh and Rashid, 2022). Enhancing urban resilience requires continuous efforts across all phases of the disaster management cycle: mitigation, preparedness, absorption, recovery, response, and adaptation (Sharifi and Yamagata, 2018).

Achieving an urban society capable of absorbing internal and external disturbances requires attention to social and human, economic, institutional, infrastructural, and environmental dimensions. Examining these dimensions demands interdisciplinary expertise to gather diverse knowledge and skills for adequately understanding stresses and pressures on a city (Kapucu et al., 2024). However, this study focuses specifically on the physical dimension of urban resilience.

Over the past decade, resilience has rapidly gained prominence, becoming a key term in international policy and academic discourse related to disaster and crisis management. Governments and institutions facing threats such as environmental hazards, technological accidents, climate change, and terrorism increasingly recognize resilience as a critical policy response. Emphasizing resilience is among the most important factors in urban management. By developing resilience, a system can reduce the impact of shocks and more quickly resume normal functions after a disaster (Coaffee et al., 2018).

3.2.2 *Empirical Analysis of Earthquake-Induced Urban Damage*

Since the 1960s, the annual number of reported natural disasters worldwide has quadrupled, reaching over 400 events by 2010. This rise is partly due to global warming and poor resource management, but the main factor is the increasing exposure of people to hazards, especially in rapidly urbanizing areas of developing countries. As urbanization accelerates, cities have become disaster hotspots. Disaster management strategies have also evolved from reactive relief to proactive risk reduction (León and March, 2016).

Natural hazards are divided into two main categories: natural disasters (such as earthquakes, floods, and storms) and human-made disasters (such as epidemics, infrastructure failures, and wars). Natural events can also trigger human-made disasters, as seen in the Fukushima nuclear accident. Human activities can further influence the frequency and intensity of disasters (Sharifi and Yamagata, 2018; Fouda and Elkhazendar, 2023).

As illustrated in Fig. 3.1, the number of deaths by disaster type in 2023 dramatically highlights the critical importance of disaster risk reduction, especially in urban environments where vulnerability is heightened. According to recent EM-DAT[1] data

[1] The Emergency Events Database.

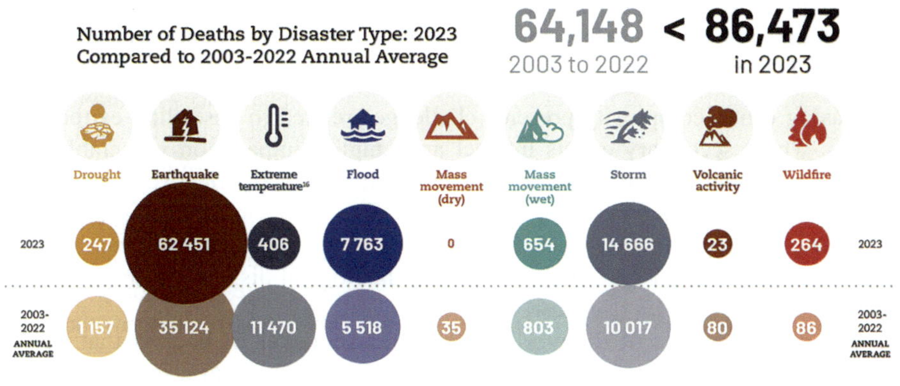

Fig. 3.1 A Comparative overview of disaster-related deaths (Source: Adapted from CRED, 2024).

(CRED, 2024), earthquakes overwhelmingly accounted for the highest mortality, with over 62,000 deaths recorded in 2023 alone—nearly double the annual average of previous decades. This figure, visually dominant in the comparative chart, emphasizes how seismic disasters not only result in immediate loss of life, but also trigger widespread social and infrastructural disruptions across cities. Other hazards such as floods and storms also caused significant fatalities, often exacerbated by urban density and inadequate risk reduction measures. The clear escalation in both the frequency and impact of natural disasters, as depicted in the figure, demonstrates that piecemeal approaches—such as isolated building codes—are insufficient. Instead, comprehensive urban planning that integrates resilience across systems is urgently needed to safeguard lives and livelihoods in rapidly growing urban areas.

- **Case Studies of Urban Damage from Past Earthquakes in the World**
  Earthquakes, as one of the unpredictable natural disasters, are always surrounded by uncertainty and speculation (Habibi and Alesheikh, 2024). They are among the most frequent and devastating natural hazards, with over 1.4 million occurring annually, causing significant human and economic losses. The countries with the highest earthquake fatalities include China, Indonesia, Iran, Turkey, and Japan (Parizi et al., 2022). China, having undergone rapid urbanization in the past three decades, now experiences about one-third of the world's earthquakes, making it one of the most seismically active regions globally. Over 50% of Chinese cities and 70% of large and medium-sized cities lie in areas with seismic intensity greater than VII. Notable events include the 2008 Wenchuan earthquake (Ms 8.0), which damaged over 23 million homes and caused nearly 69,000 deaths; the 2010 Yushu earthquake (Ms 7.1), which damaged approximately 90% of homes and resulted in 2,698 fatalities; and the 2014 Ludian earthquake (Ms 6.5), which destroyed over 80,000 homes and caused 617 deaths. The cascading effects of earthquakes significantly exacerbate damage to the natural environment, infrastructure, society, and economy in urban areas. Rapid urban expansion and economic growth in China have elevated

seismic risk to unprecedented levels. Although precise earthquake prediction remains impossible, recent governmental measures have effectively minimized damage following crises (Liu et al., 2023; Xu et al., 2023; Tang et al., 2019).

Increasing urban complexity presents challenges related to cascading earthquake effects, where secondary events interact and amplify impacts on vulnerable elements. Addressing these requires systemic thinking and effective cooperation among responsible sectors (Tang et al., 2019). The 2011 Great East Japan Earthquake and ensuing tsunami, which triggered the Fukushima Daiichi nuclear disaster, exemplify how natural hazards can induce technological disasters with prolonged impacts on communities (Suppasri et al., 2021).

Based on the table of the world's largest earthquakes, it is clear that earthquake occurrence and impact are strongly influenced by natural factors such as topography, fault depth, seismic intensity, and epicenter location-all of which are beyond human control. However, the shape and physical structure of each city can greatly intensify post-earthquake disasters or trigger secondary crises such as tsunamis, landslides, or fires. Ultimately, a city's physical resilience following a major earthquake is largely determined by its urban form. Key components including infrastructure, street networks, building blocks, and open spaces can significantly influence whether lives are saved or lost. While natural factors dictate the occurrence and intensity of earthquakes, it is the design and configuration of the urban environment that critically shape the extent of damage, the effectiveness of emergency response, and the city's capacity to recover and safeguard its population.

The profound impacts of earthquakes on urban environments are often best understood through direct observation of the aftermath. Fig. 3.2 captures an aerial view of a city devastated by a major earthquake, displaying collapsed buildings, severely disrupted infrastructure, and the resulting spatial disorder. This image starkly illustrates how vulnerabilities in the built environment—such as dense or poorly connected urban fabrics, inadequate building standards, and limited open space—can amplify destruction and complicate emergency response efforts.

While this visual provides a striking example of earthquake damage on the urban landscape, it is crucial to examine a wider scope of international experiences to identify common patterns and lessons for resilience. In the following part of this chapter, major earthquakes from recent years are summarized.

3.2.3 Assessing Resilient Urban Form

Urban morphology, which involves the design and composition of urban form and its shaping processes, is essential for urban planners to understand local development patterns (Shakibamanesh and Ebrahimi, 2020). Urban form elements are classified into three hierarchical scales—macro, meso, and micro-facilitating better

Fig. 3.2 Sichuan Earthquake, China, 2008. In this image, the boundaries between the street network and collapsed buildings, reduced to rubble, are barely distinguishable (Source: Adapted from South China Morning Post, n.d).

understanding of spatial distribution and interrelations within cities (Sharifi and Yamagata, 2018; Sharifi, 2019a; Alawneh and Rashid, 2022).

Resilience, closely linked to urban form, varies depending on intervention scale, risk type, and objectives. For example, adding transportation networks to aid evacuation is critical but may reduce overall system efficiency (Sharifi and Yamagat, 2018). The COVID-19 pandemic highlighted the importance of resilience at meso (neighborhood) and micro (public space) scales, emphasizing the need to assess urban form resilience at the "resilient place scale" amid rapid urban change (Shafiei and Lak, 2023).

3.3 Framework for Resilient Urban Form Indicators

An optimal urban form plays a significant role in strengthening the city's economy and enhancing the health and well-being of its residents. Therefore, it can be argued that interventions by urban designers in shaping the physical structure of cities should be regarded as strategic approaches through which advancements in urban resilience can be achieved (Sharifi and Yamagata, 2018, p. 167). Urban form directly influences a city's capacity to withstand and recover from adverse events, making a comprehensive understanding of "resilient urban form" and its typologies essential for progress in resilience-oriented urban planning.

3.3.1 Analysis of Urban Form Attributes at the Block Scale

Disruption or crisis in one part of the urban fabric can lead to inefficiencies across the entire city. To analyze this, the characteristics of urban form are examined at three scales: the urban street network, blocks, and public spaces.

- **Street Networks**
  Urban street networks serve as fundamental elements, forming the backbone and overall structure of urban form (Aguilar, 2014, p. 341), as they drive the growth and evolution of cities. Streets are among the longest-lasting components of urban morphology, often persisting for decades or even centuries in their original locations. Consequently, their design and configuration can have significant positive or negative impacts on urban systems. Poorly designed streets undermine a city's capacity to adapt to changing conditions arising from adverse events or social and technological shifts (Sharifi, 2019b).
  Effective street design plays a key role in enhancing accessibility, permeability, vitality, walkability, and urban safety. However, conventional planning predominantly prioritizes vehicular movement, often neglecting other essential street functions. Disasters and adverse events substantially affect street and street network performance; therefore, street structures must be designed to facilitate resilience to variable conditions with minimal need for major modifications. Optimal street widths for improving urban resilience vary depending on context, including street canyon geometry, land use intensity, disaster type, and disaster risk management phase. Following rapid-onset disasters like earthquakes, which likely block parts of the street network, ensuring access to affected areas becomes a primary priority for emergency responders (Sharifi, 2019b).
  Centrality and connectivity are additional critical attributes in street networks that maximize resilience potential by ensuring improved access to open spaces and vital land uses during crises. Specifically, closeness centrality (measuring accessibility to a location) and betweenness centrality (indicating a location's relative importance) can guide the strategic placement of critical facilities based on access needs under varying conditions (Sharifi, 2019b). Blockage of nodes with high centrality values can cause significant disruptions in accessing damaged zones. Moreover, street connections enhance urban fabric permeability (Sharifi, 2018). Because high centrality locations lie between many other sites, their role in maintaining street network functionality is vital. Disruptions at such nodes and links have widespread consequences across the system. Urban nodes with the shortest access paths exhibit higher visual centrality; thus, damage to nodes with high betweenness centrality notably diminishes network resilience (Sharifi, 2018). Betweenness centrality quantifies how often a node lies on the shortest path between other nodes (Boeing, 2017) and effectively predicts the importance of specific intersections in urban transport networks (Sevtsuk & Mekonnen, 2012).
  In summary, resilient urban street networks require thoughtful design and analysis of topological and spatial properties to sustain urban functionality during and after disruptions, supporting efficient emergency response and long-term adaptability.

- **Block Configuration**
  An urban block (French: îlot, etymologically meaning "small island") refers to a section of an urban area delineated and separated from adjacent zones by streets. Thus, a block is not an architectural form per se but rather a collection of interconnected building parcels whose meaning is only fully realized through their dialectical relationship with the street network (Panerai et al., 2004). The functionality of urban blocks is generally linked to their geometric parameters and metric characteristics (Alawneh and Rashid, 2022).
  It is important to note that the physical resilience of urban form depends on multiple factors; for instance, smaller plot size does not necessarily imply lower building resistance. Therefore, specific contextual conditions of the urban fabric must be examined in studies aiming to explore the relationship between plot size and seismic resilience. Plot shape also plays a critical role, with simpler geometries being preferable as they enhance urban form flexibility, improve connectivity and accessibility, and facilitate easier aggregation (Sharifi et al., 2021). Block size and structure significantly influence the evolutionary pattern of urban form. From a size perspective, small to medium-sized blocks with regular shapes are more suitable for creating resilient urban forms capable of adapting to changing conditions. A block may be subdivided into one or more plots; hence, discussions on block and urban land resilience are closely interconnected. As previously noted, appropriate block design is essential to prevent entrapment in rigid, undesirable patterns, which has implications for urban physical resilience and overall well-being during crises. Subdividing blocks into smaller plots with multiple access points enhances permeability, thereby connecting blocks to streets and open spaces usable for social interaction and, importantly, for improving disaster recovery and response (Sharifi, 2019). The height of buildings adjacent to streets is another critical factor influencing street blockage and obstruction during crises and should be carefully considered. Consequently, when examining the relationship between street width and evacuation capacity, the aspect ratio (height-to-width, H/W) must also be accounted for. In the event of collapse of tall buildings following a major earthquake, surrounding passages are likely to become blocked. Therefore, wider streets facilitate more effective evacuation processes (Sharifi, 2018).
- **Open Spaces**
  Urban open spaces are essential elements of urban form that play a critical role in disaster resilience by providing accessible areas for shelter, evacuation, and coordination of emergency operations (Sharifi, 2019; Del Pinto, 2021). Although definitions vary, publicly accessible open spaces (UPOS[2]) contribute significantly to urban functionality and social activities, especially during and after disasters. Their design features-such as connectivity, diversity, flexibility, multifunctionality, and redundancy-enhance a city's capacity to absorb shocks and facilitate recovery (Sharifi, 2019c).

[2] Urban Public Open Space.

Following earthquakes, many displaced individuals use these spaces for extended periods, preferring to remain close to their homes and neighborhoods (Shrestha et al., 2018; Villagra et al., 2014). Open spaces also support social cohesion and aid recovery efforts during the post-crisis phase (Brand and Nicholson, 2016). They serve multiple functions including timely evacuation, temporary shelter, building restoration, and staging areas for rescue teams (Parizi et al., 2022). The urban structure influences evacuation effectiveness; suburban areas with wide roads and large open spaces allow faster recovery, whereas dense historic urban centers with limited open space require evacuation to safer zones (Allan et al., 2013). Additionally, the ratio of building height to open space width is crucial to minimize risks from collapsing structures during disasters, underscoring the need for adequate street and open space dimensions to ensure safety and efficient open space (Shrestha et al., 2018). In summary, the strategic distribution, design, and security of open spaces are vital for enhancing urban resilience, facilitating emergency response, and supporting long-term recovery in disaster-prone cities. The strategic distribution, design, and security of open spaces are vital for enhancing urban resilience, facilitating emergency response, and supporting long-term recovery in disaster-prone cities.

To further clarify how these spatial qualities—and other urban form indicators—relate to key resilience characteristics, the following Sankey diagram (Fig. 3.3) illustrates the multifaceted connections between resilience traits and urban elements such as street networks, blocks, and open spaces.

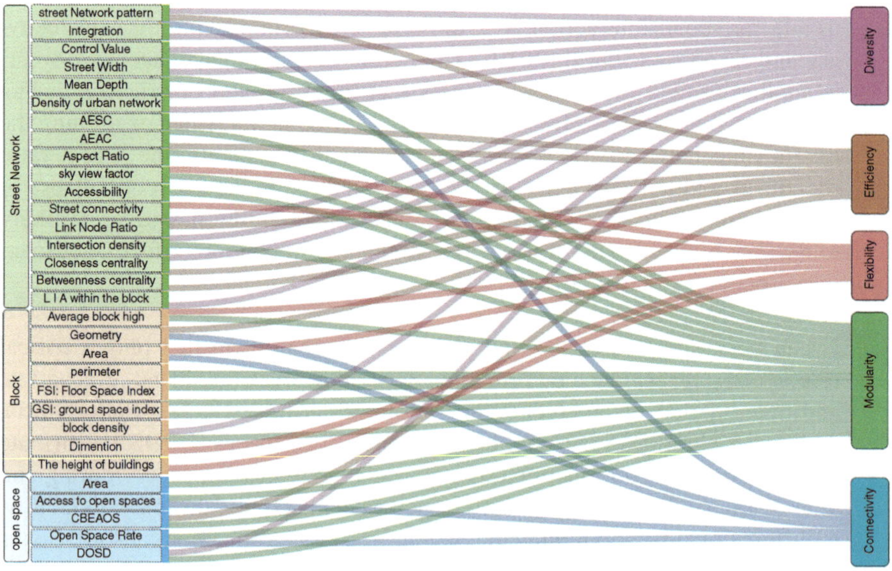

Fig. 3.3 Sankey diagram mapping urban form indicators (street networks, blocks, open spaces) to resilience characteristics.

This diagram shows the multidimensional connections between key resilience traits and relevant urban planning indicators, highlighting the integrative nature of resilient city design. The colors in this Sankey diagram represent five core resilience characteristics—connectivity, modularity, flexibility, efficiency, and diversity—and trace their connections to various urban form indicators. Each colored flow shows how these specific resilience traits influence elements of street networks, blocks, and open spaces, underscoring the importance of integrating multiple qualities in urban resilience planning.

3.3.2 Methodology for Measuring Resilience Indicators

Assessing a city's physical resilience requires sensitivity to local context, culture, and conditions, making broad generalizations impractical. This study adopts a two-step approach: first, reviewing general concepts of urban resilience; second, analyzing urban form features most relevant to earthquake impacts, as identified in major global case studies (Table 3.1). The resilience criteria developed are grouped into three principal categories: street networks, urban blocks, and open spaces. For each criterion, its relationship to resilience and its methods of measurement are clarified, providing a robust foundation for targeted risk reduction strategies tailored to local needs.

To operationalize this framework, Table 3.1 presents each indicator in a structured manner: the **"Indicator"** column lists the key attributes, the **"Explanation"** column details definitions and significance, the **"Connection to Resilience"** column explains its role in disaster risk reduction, and the **"Measurement Method"** column describes practical ways to assess each attribute. This systematic organization offers researchers and practitioners a comprehensive, actionable guide for evaluating and strengthening earthquake resilience in urban form.

Table 3.1 distills the transition from broad resilience goals to specific, actionable design strategies for street networks, blocks, and open spaces. By outlining clear objectives and practical guidelines, it provides a systematic reference for enhancing earthquake resilience in urban areas. This framework helps cities address essential qualities like connectivity, accessibility, and multifunctionality—enabling effective preparedness, rapid recovery, and robust urban safety.

3.4 Discussion and Key Insights

Growing attention to urban resilience reflects cities' roles as economic hubs facing increasing natural and human-made hazards, intensified by climate change (Sharifi et al., 2021). Since the ultimate goal of resilience-oriented urban form is to facilitate life-saving emergency operations, minimize casualties, and ensure rapid recovery of affected communities, thereby contributing to longer-term sustainability, urban

Table 3.1 Research indicators framework for evaluating Earthquake-resilient urban form.

Indicators			Explanation	Connection to resilience	Measurement method	References
Street Network	Urban network	Grid Organic Radial Spiral Curve	The geometry of the city is determined by the form of its street network, which in most cases corresponds to one of the patterns.	Simpler geometries are much more desirable because they improve the flexibility of the urban form, enable better connectivity and accessibility, and facilitate easier aggregation.	Depends on Geometry	(Parizi et al., 2022). (Sharifi et al., 2021, p. 7). (León and March, 2016).
	Integration		This index indicates the degree of integration of a space with its surrounding spaces.	High connectivity indicates high accessibility, while low connectivity signifies low accessibility. Therefore, high connectivity reflects the high resilience of a network.	$\frac{1}{\sum_k d_{ik}}$ Where d_{ik} refers to the shortest path between line ***i*** and line ***k***.	(Sharifi et al., 2021).
	Control Value		The control rate of each street indicates the level of importance and strategic significance of that passage compared to other passages.	The axis with the highest control value has the greatest evacuation potential during a crisis because it connects to many other passages (A high control rate means high physical resilience).	$=\text{CTRL}_i \sum_{j=1}^{ki} = 1\frac{1}{cj}$ Sum of the inverse values of Ci for the ki axes connected to the i-th axis.	(Wang and Huang, 2019).
	Street width		Space allocated to pedestrians and the movement of cars.	The wider the streets and sidewalks, the higher the physical resilience.	–	(Sharifi et al., 2021). (Parizi et al., 2022). (Poursharifi et al., 2020).
	Mean depth			Higher street depth values indicate lower accessibility and are undesirable. (Mean Depth (MD) and Total Depth (TD)).	$\text{MD} = \frac{TD}{k-1}$	

	Density of urban network			The greater the area of the communicative network relative to the total area of the neighborhood, the less dense the neighborhood space is. This indicates higher resilience.	The ratio of the communicative network area of the neighborhood to the total area of the neighborhood.	(Network connectivity)/ (Total neighborhood area)	(Nikpour et al., 2021).
	Access Emergency	Service center	Health centers	The number of each station or user, the radius covered by each per capita standard, the position and capacity of each must be considered separately.	The more services center and the same dispersion, the higher the resilience of the region.	Network Analysis	(Arvin et al., 2023). (Poursharifi et al., 2020).
			Fire Department				
			Radio, television and telecommunications centers				
		Accommodation center	Indoor sports space	The number of each station or user, the radius covered by each per capita standard, the position and capacity of each must be considered separately.	The more accommodation center and the same dispersion, the higher the resilience of the region.		(Parizi et al., 2022).
			Mosques				
			The parking lots				
			Parks and open spaces				
			Schools				

(continued)

Table 3.1 (continued)

Indicators			Explanation	Connection to resilience	Measurement method	References
	Visibility of space	Aspect Ratio	The sum of the heights of the buildings in a block divided by the width of the adjacent street.	The greater the street width relative to the building height, the more effective the evacuation process will be during emergencies.	$\frac{high}{width}$	(Sharifi, 2018).
		Sky View Factor	It is calculated using the aspect ratio of all streets at the urban block scale, assuming the presence of people in the space.	This index can indicate the width of the passages based on the street canyon geometry for evacuation.	skyline, skyline graph	(Sharifi et al., 2021). (Del pointo, 2021).
	Accessibility		The maximum distance among all the shortest distances between two intersections in the network.	The shorter the distance between intersections, the greater the resilience of the space.	–	(Parizi et al., 2022).
	Street connectivity		Number of Connections (NC) is a local indicator that shows the number of adjacent elements to a specific element.	A high CN value (a large number of immediately connected elements) indicates high accessibility and high centrality of that element.	$C_i = K_i$ K is the number of the lines without intermediaries. Number of axes connected to the i the axis.	(Wang and Huang, 2019). (Sharifi et al., 2021).
	Link Node Ratio		It refers to the ability of a line or space to control the surrounding lines or spaces.	A high connectivity rate indicates high resilience.	$\frac{links}{nodes}$	(Alawneh and Rashid, 2022). (Alawneh and Rashid, 2022). (Parizi et al., 2022).

	Intersection Density	Intersection density refers to the number of intersections within a given area. It is a measure that indicates how many intersections exist in a specific neighborhood or urban area.	Higher intersection density indicates a more interconnected street network with smaller blocks. The greater the intersection density in an area, the higher the resilience of that neighborhood.	$\frac{\text{number of intersections}}{\text{area}}$	
	Closeness Centrality	It indicates how close an intersection (node) is to all other accessible intersections in the city along the shortest network paths.	To enhance the flexibility of urban form in terms of accessibility, considering proximity centrality when deciding on the location of services and facilities is essential.	$C_i^c = \frac{N-1}{\sum_{j=1;j\neq i}^{N} dij}$	(Boeing, 2017). (Sharifi, 2018).
	Betweenness Centrality	Nodes that are traversed by a larger portion of the shortest paths between all pairs of nodes in the network exhibit higher betweenness centrality.	This index indicates the relative importance of a node within the entire system. It highlights strategic points during crisis events and identifies important land uses or intersections.	$\theta = (x)\,B\theta \frac{\sum_{i=1}^{n}\sum_{j=1}^{n}\sigma\,(i, xmj)}{(n-1)(n-2)/2}$	(Sharifi et al., 2021). (Sharifi, 2018).
	The level of internal accessibility within the block	Access within each block indicates its high permeability.	The more accessible and permeable the block, the higher the flexibility in responding in times of crisis.	–	

(continued)

Table 3.1 (continued)

Indicators			Explanation	Connection to resilience	Measurement method	References
Block	Average Block High		Average building High of a block.	The higher the block height, the less resilience it has.	–	(Parizi et al., 2022). (Lu et al., 2020).
	Geometry (Regular or Irregular block shape)		Block form shape.	Simpler geometries are much more desirable because they improve urban form flexibility, enable better connectivity and accessibility, and facilitate easier aggregation.		
	Area		–	Large blocks have lower resilience.	–	(Lu et al., 2020) (Sharifi et al., 2021)
	Perimeter		–	Large blocks have lower resilience.	–	(Shafiei and Lak, 2023).
	Density	Ground Space Index (GSI)	The intensity of block density refers to the ratio of the total floor area within the block to the block's land area, which indicates the compactness of the fabric.	Block density implies low resilience of an urban fabric.	(Total floor area)/(Total block area).	(Baldea and Dumitrescu, 2013).

		Floor Space Index (FSI)	In this index, the amount of built-up space, or in other words, the mass of space, can be identified.	The less space built in the block, the more resilience it is.	(Built-up area of the block)/(Total area of the block).	(Baldea and Dumitrescu, 2013).
		Block Density	It means the number of blocks counted per hectare.	The more blocks, the higher the connectivity. Which means higher resilience.	(Number of blocks in the area)/(Total area of the neighborhood).	
	Dimention		Ratio between length and width and height of block (maximum and minimum).	Blocks with large block proportions are called mega blocks and have low resilience.	–	(Parizi et al., 2022).
	The height of buildings adjacent to streets		The pattern of building height arrangement on the facade affects many other buildings and components.	The clustering of tall buildings at block corners significantly multiplies damage during crises; moreover, irregular height arrangements of buildings may cause mutual damage during earthquakes.	Drawing the shape of the form.	
	Area		Measuring the area of open space.	The High area of any public open space indicates the high resilience of that space.	–	(Sharifi et al., 2021).

(continued)

Table 3.1 (continued)

Indicators		Explanation	Connection to resilience	Measurement method	References
Open space	Access to open space	Connection to open space.	High connectivity in the open space indicates the high level of accessibility to any open city space, which also speaks of the high resilience of that space.	Kernel Density	(Shrestha et al., 2018). (Vilagra, 2014). (Sharifi et al., 2021). (Parizi et al., 2022).
	Density of the built environment around open spaces	This indicator shows the number of floors around the open space.	The high confinement around open spaces reduces the resilience of open spaces.	Balance Index	(Parizi et al., 2022). (Alawneh and Rashid, 2022). (Vilagra, 2014).
	Open space Ratio (OSR)	Open space size relative to the number of floors. Ratio of unbuilt surfaces to total floor area.	Open spaces indicate high resilience in any environment.	Open space area in the block)/ (Total area of the number of floors.	(Baldea and Dumitrescu, 2013).
	Distribution of open space	Spatial distribution of open spaces in blocks and neighborhoods.	High dispersion of open spaces increases spatial resilience (Spatial version of Simpson's Diversity Index) ***Q***: *number of districts.* ***A***: *area of open space.* ***S***: *area of district.*	$= D_{\text{spatial distribution}} \frac{Q}{Q-1}$ $= [1- \sum_{district=1}^{Q} (\frac{Ai}{Si})\ 2$	(Alawneh and Rashid, 2022).

environments subjected to unsustainable practices may face greater risks (Amirzadeh et al., 2022). Therefore, enhancing capability of the physical form of cities is crucial to transform vulnerabilities into opportunities during disasters (Jamali et al., 2023). Urban form significantly affects a community's ability to cope with crises, making flexibility, adaptability, and sustainability key planning priorities for creating high-quality, safe living environments (Allan et al., 2013; Cheshmehzangi, 2025). Effective resilience requires integrated risk management, infrastructure improvement, and public awareness, tailored to local contexts. Addressing climate change impacts remains essential for sustainable and safe urban development amid growing hazards.

3.4.1 Strategic Policies for Earthquake-Resilient Urban Form

The absence of integrated planning and coordinated decision-making has long posed significant challenges for effective disaster management and the development of resilient urban environments. To address these obstacles, adaptive interventions supported by intelligent decision-making systems are essential for fostering urban resilience (Sharifi and Yamagata, 2018). Thoughtful urban planning and design are fundamental to long-term risk reduction, as they govern the spatial and physical arrangements that determine how a city grows, adapts, or declines in response to hazards (León and March, 2016).

As the frequency and impact of both natural and human-made disasters rise, embedding resilience into future urban planning becomes increasingly critical. A modular approach to resilience is particularly important, ensuring that cities remain operational even if certain urban subsystems or components fail (Fouda and Elkhazendar, 2023). Urban design enhances systemic flexibility by shaping resilience elements and by providing evidence-based guidelines that inform effective interventions (Feliciotti et al., 2016). Proactive planners are thus tasked with envisioning possible disaster scenarios, especially in seismic regions, in order to devise strategies that can reduce damage and facilitate rapid recovery (Lu et al., 2020; Allan et al., 2013).

Design **guideline** for **street networks**, **blocks**, and **open spaces** focus on improving physical resilience. Simpler street layouts with wider roads enhance permeability but excessive intersections may cause congestion and risks. Smaller, lower blocks reduce vulnerability, while compatible land uses during crises can save lives. Open spaces are critical for evacuation and shelter, requiring accessibility, flexibility, multifunctionality, and security. These principles should guide new neighborhood designs, with emergency scenarios tested for evacuation and relief. Resilience assessment must consider local physical, cultural, and economic contexts to effectively strengthen urban fabric before disasters occur. A holistic, interdisciplinary approach is necessary, as urban elements are interdependent; disruptions in one area can affect others. Yet, each component must maintain connectivity, efficiency, modularity, diversity, and coordination to ensure overall urban resilience, exactly like Fig. 3.3 (*Sankey diagram*).

Building on the analytical framework and empirical lessons discussed above, it is crucial to translate resilience principles into practical, context-sensitive design actions. Table 3.1 synthesizes these insights by presenting a clear set of design strategies and operational guidelines tailored to enhance the physical earthquake resilience of urban environments. By mapping broad resilience objectives to actionable goals and site-specific interventions for street networks, urban blocks, and open spaces, this table provides planners, designers, and policymakers with a practical reference for implementation. The strategies prioritize not only structural performance and connectivity but also flexibility, accessibility, and integration with the local context—enabling cities to proactively strengthen their ability to withstand, respond to, and recover from seismic disturbances. In doing so, Table 3.3 serves as a critical link between theory and practice, ensuring that recommendations drawn from international experience and rigorous analysis are translated into effective measures at the neighborhood and city scale. The following presents a set of strategies and guidelines for earthquake resilience, outlined in the *Lessons and Policy Implications* section, derived from global earthquake experiences and the characteristics of urban form.

3.5 Lessons and Policy Implications

All the strategies, policies, and design guidelines presented in this section are directly derived from the research methodology adopted in this study, which systematically analyzed both international earthquake case studies and empirical evidence. By examining documented lessons from major seismic events across the world, the research distilled a set of practical, evidence-based recommendations tailored to enhance urban resilience. Thus, the resulting conclusions and proposed interventions reflect not only theoretical insights but also proven responses to real-world disaster scenarios. Table 3.2 examines major global earthquakes that have occurred since 2010. It presents the extent of overall damage to life and infrastructure, secondary disasters such as fires and tsunamis, and, finally, the key factors influencing the severity of structural destruction.

The international case studies summarized in Table 3.2 highlight not only the scale and diversity of earthquake impacts but also reveal recurring vulnerabilities and risk factors in urban environments worldwide. Analyzing these past disasters provides invaluable lessons for both policy makers and urban designers. Key patterns that emerge include the critical importance of enforcing building codes, the vulnerability of densely built or poorly connected neighborhoods, the dangers posed by inadequate emergency access, and the cascading failures resulting from uncoordinated infrastructure. These global experiences emphasize that enhancing seismic resilience requires not just technical improvements in construction, but integrated urban planning that prioritizes connectivity, redundancy, accessibility, and flexible open space. The following section distills essential lessons from these real-world events, offering actionable guidance for shaping urban forms that are better prepared to withstand and recover from future earthquakes.

Key Lessons for Earthquake-Resilient Urban Form:

- **Enforce modern building codes and standards:** Failure to comply with up-to-date seismic regulations was a recurrent factor in catastrophic building collapses.
- **Reduce urban density where possible:** Extremely dense or compact areas with little open space suffered higher casualties and greater structural damage.
- **Increase and connect open and accessible spaces:** Parks, plazas, and open spaces played vital roles in evacuation, emergency shelter, and disaster recovery.
- **Ensure redundancy and connectivity in road networks:** Well-connected grid or networked street patterns facilitated effective emergency access and post-disaster logistics, whereas dead-ends and inadequate routes hindered rescue operations.
- **Integrate risk zoning and land use planning:** Strategic zoning (such as buffer zones, and avoiding construction in high-hazard areas) mitigated risk in some cities.
- **Promote community preparedness and awareness:** Cities with stronger public disaster education and preparation had more coordinated, effective responses.

Building on the analytical framework (Table 3.1) and empirical lessons discussed in Table 3.2, it is crucial to translate resilience principles into practical, context-sensitive design actions. Table 3.3 synthesizes these insights by presenting a clear set of design strategies and operational guidelines tailored to enhance the physical earthquake resilience of urban environments. By mapping broad resilience objectives to actionable goals and site-specific interventions for street networks, urban blocks, and open spaces, this table provides planners, designers, and policymakers with a practical reference for implementation. The strategies prioritize not only structural performance and connectivity but also flexibility, accessibility, and integration with the local context—enabling cities to proactively strengthen their ability to withstand, respond to, and recover from seismic disturbances. In doing so, Table 3.3 serves as a critical link between theory and practice, ensuring that recommendations drawn from international experience and rigorous analysis are translated into effective measures at the neighborhood and city scale.

Finally, strategies and frameworks for developing a resilient urban form in the face of destructive earthquake events were presented. This table serves as a guide for urban designers and planners in strengthening the physical resilience of cities. "Integrating these lessons and strategies within local planning systems can transform reactive disaster management into proactive resilience-building, embedding seismic safety into everyday urban growth."

Table 3.2 Significant earthquakes after 2010: Case studies of physical damage and contributing factors.

Location and year	MS (M_w)	Disasters Resulting from Earthquake Occurrence	Extent of Physical Damage	Factors Influencing the Intensification of Physical Damage
Myanmar, 2025	7.7	Fire	• Over 4,900 fatalities were reported. • Extensive damage occurred to infrastructure, including residential, historical, governmental, and religious structures.	• Liquefaction caused the collapse of many buildings. • Bridge failures complicated damage control and rescue operations. • The ongoing internal conflict in Myanmar hindered relief efforts. • The earthquake had three hazardous characteristics: high magnitude, shallow depth (10 km), and occurrence in a densely populated area with vulnerable buildings and structures.
Turkey, Gaziantep city, 2023	7.7	Fire	• More than 214,000 buildings and structures completely collapsed in the earthquake-affected areas, with many historical sites destroyed. • Over 2,000 people lost their lives due to this disaster. • Hospitals, public buildings, residential and commercial blocks-approximately 25,000 buildings in total-were either destroyed or damaged.	• This earthquake shook the surrounding areas for 45 seconds. • The destruction and collapse of debris caused many casualties. • Due to this earthquake, tsunamis occurred without casualties along the coasts of Famagusta in Cyprus, Iskenderun, and Erdemli.
Philippines, 2019	6.9	Fire	• Buildings and infrastructure suffered extensive damage. Over 26,000 homes and nearly 400 schools were affected. Thirteen people lost their lives.	• The physical damage caused by poor engineering practices was the improper design of the building.

Indonesia, Sulawesi, 2018	7.5	Tsunami	• The earthquake in Donggala caused the displacement of oil tanks. • More than 68,000 buildings were damaged in this event. • Due to the intensity of the tsunami, many people went missing and over 4,000 were killed.	• The tsunami waves generated by this earthquake reached heights of up to 6 m.
Mexico, 2017 (Twice)	8.1, 7.1		• Severe damage was inflicted on infrastructure, buildings, and bridges. • Approximately 184,000 residential units were damaged in these two earthquakes. • Around 230 people lost their lives.	• The epicenter of the first earthquake (magnitude 8.1) was located off the Pacific coast, while the second earthquake (magnitude 7.1) occurred in central Mexico.
Iran, Kerman-shah, 2017	7.3	Landslide	• Many buildings, including historical structures, were damaged. • One thousand houses were completely destroyed, and 3,000 houses require reconstruction. • Six hundred and twenty people lost their lives.	• The highest number of fatalities occurred in recently constructed Maskan-e Mehr buildings, which were expected to have high resistance; however, only about 250 deaths were reported in these units due to the earthquake
Nepal, 2015	7.9	Avalanche	• Over 8,510 people were killed. • Two powerful earthquakes measuring 7.8 and 7.3 magnitude occurred within a two-week period, with aftershocks felt for more than a month.	• The earthquake triggered an avalanche in the Everest region. • The compaction of the urban fabric due to the organic layout of the city led to significant physical damage in residential areas. • As a result of the earthquake, many roads were completely blocked, which complicated rescue operations.

(continued)

Table 3.2 (continued)

Location and year	MS (M_W)	Disasters Resulting from Earthquake Occurrence	Extent of Physical Damage	Factors Influencing the Intensification of Physical Damage
Turkey, Van, 2011	7.2		• This earthquake caused the destruction of thousands of buildings in Erciş, Van, and other surrounding cities. • More than 644 people lost their lives.	• The factors affecting the physical damage caused by that earthquake include the earthquake's magnitude, focal depth, distance of the epicenter from densely populated areas, and the quality of buildings and infrastructure in the affected regions.
New Zealand, Christchurch, 2011	6.3	Migration of residents—rockfalls—landslides	• Fifty percent of the buildings in the city center were destroyed. • 180 people were killed. • Approximately 12,000 houses and 1,500 commercial buildings were demolished.	• Many buildings were demolished for economic reasons, as the cost of repair exceeded the cost of replacement. • Due to the destruction of commercial centers in the city, many jobs were lost, which in turn led to the migration of people from the city.
Japan, Tohoku, 2011	9	Tsunami, fire Nuclear disaster	• The earthquake caused the collapse of over 123,000 houses and damage to nearly one million more. • With damages estimated at 300 billion dollars, it became the costliest disaster in human history. The only catastrophe with higher costs than this earthquake was World War II. • Approximately 20,000 people lost their lives. At that time, due to the failure of tsunami warning systems, the height of the tsunami waves was underestimated, which led to a major humanitarian disaster.	• The 2011 earthquake and tsunami caused explosions in three Fukushima Daiichi nuclear reactors due to hydrogen gas buildup, leading to radioactive contamination and mass evacuations. Despite Japan's advanced earthquake preparedness, significant damage occurred, though buildings with seismic isolation technology, like MT Sendai and Sendai Twin Towers, remained unharmed. The event highlights the ongoing challenges of managing nuclear safety amid natural disasters and the importance of resilient infrastructure.

Chili, Bio Bio, Santiago,	8.8	Tsunami	• The earthquake caused severe damage to homes, schools, 70% of hospitals, and urban infrastructure. • More than 500 people lost their lives.	• The damage sustained was predominantly structural in nature. Considering Chile's high seismicity, most buildings possess only standard levels of resistance; however, the country remains highly vulnerable to large-scale earthquakes.
New Zealand, Christchurch, 2010	7.1		• The Christchurch earthquakes caused extensive damage to infrastructure and buildings, resulting in the collapse of numerous structures. The city experienced significant disruptions due to ruptured water supply networks, flooding, and power outages. The port sustained damage, and many roads developed cracks. Additionally, over 4,500 vehicles remained abandoned in the city center for an extended period.	• This earthquake triggered numerous aftershocks that gradually approached the city, culminating in the 2011 Christchurch earthquake, which, with a shallower depth and a magnitude of 6.3, caused more severe damage. • The fault in this region had not been previously identified.
Haiti, 2010	7		• The earthquake resulted in over 300,000 fatalities, displaced more than one million people, damaged over 50,000 homes, and injured approximately 77,000 individuals.	• Failure to comply with construction standards resulted in the collapse of many buildings, and the lack of disaster preparedness planning hindered timely emergency response efforts.

Source: Adopted from Shahzada et al. (2025); United Nations Office for the Coordination of Humanitarian Affairs [OCHA] (2023); Hu et al. (2025); Reinoso et al. (2021); Kahrizi and Tahamouli Roudsari (2025); Goda et al. (2015); Zare and Nazmazar (2013); Reyners (2011); Suppasri et al. (2021); Allan et al. (2013); Brand and Nicholson (2016); Turkey–Syria earthquakes (2024) and Cooner et al. (2016).

Table 3.3 Design strategies and guidelines for earthquake-resilient urban form.

Overall Goal	Specific Objective	Strategy	Design Guideline
Increased resilience of urban form to earthquakes.	Increasing the resilience of the **street network.**	Increased connectivity, integration and permeability of network.	• Connecting local dead-end streets to adjacent main roads to ensure at least two access points: Enhances the connectivity and cohesion of the street network. This improvement is critical during emergencies, as it facilitates faster evacuation and more efficient emergency response. • Increasing the width of streets narrower than 5 m to facilitate emergency and rescue operations: Increasing the width of streets can improve traffic flow, debris clearance, and accelerate emergency response efforts. Streets should not be designed with widths less than 8 m. In cases where narrow streets exist, their widths should be expanded as much as possible, except in historic districts where modifications may be restricted. • Enhancing clarity for rapid evacuation by connecting secondary pedestrian pathways to main streets: The street network should be continuous, well-connected, and free of obstacles. Directly linking pedestrian pathways to main streets can improve route clarity, facilitating quicker and safer evacuation during emergencies. • Organizing intersections to ease traffic flow and prevent congestion, especially during crises: Intersections should be designed and managed to not only avoid causing problems during crises but also to improve traffic conditions in emergency situations. Both overly complex and excessively simple intersection designs can either help or hinder crisis management. One effective approach to managing traffic flow during emergencies is the implementation of intelligent traffic control systems. These systems use real-time data and advanced technologies to optimize signal timing, prioritize emergency vehicle movements, and reduce congestion, thereby facilitating quicker and safer responses during crises. • Establishing connections between open spaces and neighborhood centers: Open spaces should be connected to neighborhood centers to ensure that individuals can quickly access these areas for emergency services during crises and for temporary shelter after disasters. These spaces must be situated at minimal distances from neighborhood centers to facilitate rapid and efficient use in both emergency response and recovery phases.

		Increased modularity and efficiency.	• Improving the hierarchy of access: Open spaces should be connected to neighborhood centers to ensure that individuals can quickly access these areas for emergency services during crises and for temporary shelter after disasters. These spaces must be situated at minimal distances from neighborhood centers to facilitate rapid and efficient use in both emergency response and recovery phases. • Regulating height restrictions considering neighborhood: To increase the efficiency and modularity of access networks, the hierarchy of access routes must be clearly defined. This includes prioritizing access routes based on their function and capacity to handle emergency traffic during crises. One of the most critical factors for accelerating recovery operations after a disaster is the efficiency and modular structure of the access network. • seismicity, especially in old textures with narrow streets: In deteriorated or historic neighborhoods with narrow streets, access routes should be made as permeable as possible. Enhancing the permeability of these pathways is crucial to improve evacuation efficiency and emergency response during seismic events, given the inherent vulnerabilities of such urban fabrics.
	Accessibility	Enhance pedestrian mobility and access during crises.	• Increasing the width of sidewalks on main streets to facilitate pedestrian movement, especially during crises: Sidewalks on main streets should be designed with minimum width standards (e.g., at least 3 m) to accommodate a large volume of pedestrian traffic during emergency evacuations. Additionally, clear signage and wayfinding systems should be installed to guide pedestrians safely, while obstacles or barriers must be minimized to ensure unobstructed movement. • Creating public land uses around neighborhoods to facilitate emergency sheltering nearby: Public spaces should be designed with a dual purpose: serving as recreational areas during normal times and as emergency shelters during crises. These spaces must be located close to residential areas to enable rapid access for residents seeking emergency services and temporary shelter after disasters. Proximity to essential services, such as healthcare and food distribution points, is critical. Additionally, these spaces should be equipped with basic infrastructure to support evacuees and aid workers during disasters.

(continued)

Table 3.3 (continued)

Overall Goal	Specific Objective	Strategy	Design Guideline
			• Planning short routes to vital land uses: In urban resilience planning, short and direct routes must be established to ensure rapid access to critical infrastructures such as hospitals and fire stations during emergencies. These routes should be designed to be safe, clear, and free of obstacles to facilitate swift evacuation and unimpeded access to emergency services. • Proper distribution of vital land uses across the city to ensure comprehensive access for rapid emergency response. Spatial equity should be ensured in the distribution of vital land uses, making certain that these facilities are situated within 5 to 10-minute walking distance—from residential areas. This ensures that all individuals can reach essential services promptly during emergencies, facilitating faster response and recovery. • Designing emergency and rescue routes within neighborhoods to accelerate aid delivery during incidents: Emergency and rescue routes within neighborhoods should be clear, wide, and unobstructed to allow for the swift passage of emergency vehicles and aid delivery. These routes should connect key areas like residential zones, hospitals, and shelters. In densely populated areas, one-way systems or priority lanes for emergency vehicles can help reduce delays during crises.
	Increasing the resilience of the **urban block.**	Flexibility.	• Increasing permeability of large-scale blocks: Large urban blocks can impede movement and reduce access during emergencies. To increase flexibility and resilience, large-scale blocks should be subdivided into smaller, more permeable sections. This can be achieved by creating pedestrian-friendly pathways, internal streets, and open spaces that allow for easy movement and access for emergency services. • Renovation and reconstruction of damaged and deteriorated buildings: Building retrofitting is critical for increasing resilience. Old and damaged buildings must be renovated using earthquake-resistant materials such as reinforced concrete, steel bracing, and shear walls. In areas with high seismic risk, this technique is essential. • Defining localized thematic projects to stimulate areas of neighborhoods with low resilience: Thematic projects such as community centers, green spaces, or micro-zones can help enhance the resilience of areas that are vulnerable to natural disasters. These projects should focus on low resilience neighborhoods and be designed with community participation to ensure that the specific needs of the population are met. These projects also enhance local engagement and increase social cohesion, both of which are vital for during and post-disaster recovery.

	Increasing the resilience of the **open space.**	Accessibility.	• Creating public land uses or open spaces around neighborhoods, especially in areas with the highest physical and population density. "Open spaces and public land uses are essential for urban resilience. To enhance accessibility, parks, plazas, and green spaces should be strategically located in high-density areas. These spaces must be situated close to residential neighborhoods to ensure accessibility during crises. Additionally, they should feature multifunctional designs to address various community needs in disaster situations." • Establishing pedestrian access to open spaces for emergency sheltering and evacuation during crises: In disaster-prone areas, pedestrian access to open spaces should be designed with clear, unobstructed pathways to ensure swift evacuation. These paths should be well-lit and marked with directional signs to guide people quickly and safely. Moreover, spatial planning should ensure that these spaces are connected to major roads to facilitate rapid transport of emergency supplies and medical teams.
		Multifunctionality.	• Considering the access radius of public land uses and open spaces in cities and neighborhoods for emergency shelter use: Public land uses and open spaces should be strategically located within a 400–800-m radius of residential neighborhoods. This ensures quick and easy access during emergencies, allowing open spaces to function as emergency shelters or safe zones. These spaces should be clearly accessible through well-marked pedestrian routes and equipped with basic facilities. • Creating multifunctional green spaces within neighborhoods: Green spaces should be designed not only as recreational areas but also as spaces for emergency sheltering, food distribution, and community gathering during times of crisis. Research in urban planning for resilience emphasizes that green spaces play a dual role in cities—providing environmental benefits, such as carbon sequestration and temperature regulation, while also serving as critical evacuation points during disasters.
		Increased security.	• Creating regulations to prohibit high-rise buildings around open spaces aims to enhance the resilience of these spaces for public use during crises: High-rise buildings situated close to open spaces can obstruct natural light, ventilation, and emergency access, thereby increasing vulnerabilities during crises. To address this, regulations should be implemented to limit building heights around public spaces, ensuring unobstructed access and preserving the functionality of these areas as emergency shelters. By restricting building heights, open spaces remain accessible for evacuation, sheltering, and relief efforts, which is particularly crucial in densely populated urban environments.

3.6 Resilience and Transformative Growth

Aligned with the central theme of the book on "Transformative Growth," this section elaborates how the findings of this chapter contribute to reimagining urban productivity within planetary boundaries. The concept of Transformative Growth provides a broader interpretive lens through which earthquake-resilient urban form can be understood as more than a technical or structural pursuit. While resilience seeks to maintain or restore urban function in the face of shocks, transformative growth reframes resilience as a generative force—one that enhances the ecological, social, and technological productivity of cities within planetary limits.

The design strategies and indicators developed in this study—focused on street networks, urban blocks, and open spaces—demonstrate how urban morphology can evolve toward such transformation. Resilient street networks that ensure connectivity and accessibility also foster new forms of urban productivity, enabling efficient mobility, inclusive access to services, and adaptive public space use. Likewise, modular and flexible block structures support incremental renewal and participatory redevelopment, strengthening both the physical and social fabric of cities. Open spaces, when designed as multifunctional assets for recreation, ecology, and emergency response, exemplify the synthesis of resilience and productivity—serving simultaneously as life-saving infrastructure and as everyday contributors to health and environmental quality.

In this sense, earthquake resilience becomes a catalyst for *transformative growth*: it redefines urban prosperity not solely as expansion or economic gain, but as the capacity of the built environment to sustain life, adapt creatively, and regenerate ecological and social systems. Integrating resilience-oriented design into planning practice thus represents a pragmatic yet ambitious pathway for achieving transformative growth—one that aligns urban development with safety, inclusivity, and sustainability in an era of increasing environmental uncertainty.

3.7 Conclusion and Future Research Directions/Concluding Remarks

Despite the valuable contributions of this study, several limitations should be recognized. The research concentrated primarily on the physical aspects of urban resilience—specifically street networks, urban blocks, and open spaces—without fully addressing equally important social, economic, and cultural factors that can shape resilience outcomes. For instance, variables such as community awareness, poverty levels, and local governance may substantially affect a city's capacity to withstand and recover from earthquakes, but were outside the scope of this work. Additionally, the analysis was confined to the urban block scale; thus, finer building-level characteristics (such as material quality or structural system) were treated only broadly. As a result, the nuanced impacts of individual vulnerable buildings within otherwise resilient blocks may not be fully captured. Expanding the analysis to encompass both finer (building) and broader (city-wide) scales, as well as integrating long-term urban development trends, is suggested for future research.

In summary, this chapter contributes not only by categorizing urban form components (streets, blocks, open spaces) but also by translating them into a set of practical lessons for planners and designers. While context variability limits universal prescriptions, the guidelines distilled from past disasters provide a structured reference for integrating resilience in urban form. Future research should focus on testing these lessons in real-world pilot projects, quantifying their effectiveness in saving lives and accelerating urban recovery after earthquakes.

Earthquake crises continue to threaten cities globally, especially densely populated urban centers near active faults. Systematic identification of vulnerabilities and targeted assessment of seismic risk remain vital for preventing human and economic losses. Leveraging advanced technologies and interdisciplinary expertise—involving not only urban planners and designers but also structural and earthquake engineers—is crucial for effective disaster preparedness and the improvement of urban physical structures before, during, and after seismic events.

Finally, adapting resilience criteria to accommodate diverse city contexts, monitoring innovative construction, and fostering public disaster awareness will be essential for reducing future risks. Tailoring strategies to local cultural conditions and building types, and emphasizing evidence-based interventions, can further strengthen both community safety and urban recovery capacity. Adopting this holistic perspective ensures that resilience efforts remain responsive, inclusive, and effective in safeguarding urban populations from earthquake hazards.

References

Afrin T, Yodo N (2019) Resilience-based recovery assessments of networked infrastructure systems under localized attacks. Infrastruct 4(1), 11

Aguilar JR (2014) Resilient cities: an analysis of resilient urban form. Doctoral dissertation, Georgia Institute of Technology

Ahern J (2013) Urban landscape sustainability and resilience: The promise and challenges of integrating ecology with urban planning and design. Landsc Ecol 28:1203–1212

Alawneh SM, Rashid M (2022) Revisiting urban resilience: A review on resilience of spatial structure in urban refugee neighborhoods facing demographic changes. Front in Sustain Cities 4, 806531

Allan P, Bryant M, Wirsching C, Garcia D, Teresa Rodriguez M (2013) The influence of urban morphology on the resilience of cities following an earthquake. J of Urban Des 18(2):242–262

Amirzadeh M, Sobhaninia S, Sharifi A (2022) Urban resilience: A vague or an evolutionary concept?. Sustain Cities and Soc 81, 103853

Arvin M, Beiki P, Hejazi SJ, Sharifi A, Atashafrooz N (2023) Assessment of infrastructure resilience in multi-hazard regions: A case study of Khuzestan Province. Int J of Disaster Risk Reduct 88, 103601

Boeing G (2017) OSMnx: New methods for acquiring, constructing, analyzing, and visualizing complex street networks. Comput, Env and Urban Syst 65:126–139. https://doi.org/10.1016/j.compenvurbsys.2017.05.004

Brand D, Nicholson H (2016) Public space and recovery: Learning from post-earthquake Christchurch. J of Urban Des 21(2):159–176

Cheshmehzangi A (2025) Urban planning for the contemporary age: Navigating complexities and shaping urban futures. Encycl 5(1), 19. https://doi.org/10.3390/ encyclopedia5010019
Coaffee J, Therrien MC, Chelleri L, Henstra D, Aldrich DP, Mitchell CL, ... & Participants (2018) Urban resilience implementation: A policy challenge and research agenda for the 21st century. J of Conting and Crisis Manag 26(3):403–410
Cooner AJ, Shao Y, Campbell JB (2016) Detection of urban damage using remote sensing and machine learning algorithms: Revisiting the 2010 Haiti earthquake. Remote Sens 8(10):1–17, 868. https://doi.org/10.3390/rs8100868
CRED. (2024). 2023: Disasters in Numbers. Brussels: Centre for Research on the Epidemiology of Disasters (CRED), UCLouvain. https://files.emdat.be/reports/2023_EMDAT_report.pdf
Del-Pinto M (2021) Urban form and disaster risk: The role of urban public open spaces in vulnerability of earthquake-prone settlements. Doctoral dissertation, Loughborough University
Dong D, Yu Z, Xu J (2025) A study on the characteristics and system construction of urban disaster resilience in Shanghai: A metropolis perspective. Sustain 17(1), 248
Feliciotti A, Romice O, Porta S (2016) Design for change: five proxies for resilience in the urban form. Open House Int 41(4):23–30
Fouda YE, ElKhazendar DM (2023) Achievement of resilience in urbanism: A prototype for a simulative methodology. Alexandria Eng J, 70:145–168
Goda K, Kiyota T, Pokhrel RM, Chiaro G, Katagiri T, Sharma K, Wilkinson S (2015). The 2015 Gorkha Nepal earthquake: Insights from earthquake damage survey. Front in Built Environ 1(8).
Habibi R, Alesheikh AA (2024) Uncovering implicit Seismogenic associated regions towards promoting urban resilience. Resil Cities and Struct 3(4):83–94
Hu Y, Barberopoulou A, Koch M (2025) Tracing the 2018 Sulawesi Earthquake and Tsunami's Impact on Palu, Indonesia: A Remote Sensing Analysis. J of Marine Sci and Eng 13(1):178
Jamali A, Robati M, Nikoomaram H, Farsad F, Aghamohammadi H (2023) Urban resilience assessment using hybrid MCDM Model based on DEMATEL-ANP Method (DANP). J of the Indian Soc of Remote Sens 51:1–23
Kahrizi M, Tahamouli Roudsari M (2020) Seismic performance of school buildings in 2017 Ezgeleh Earthquake, Iran. Bull of the New Zealand Soc for Earthq Eng 53(2):70–82
Kapucu N, Ge Y, Rott E, Isgandar H (2024) Urban resilience: Multidimensional perspectives, challenges and prospects for future research. Urban Gov 4(3):162–179
León J, March A (2016) An urban form response to disaster vulnerability: Improving tsunami evacuation in Iquique, Chile. Environ and Plan B: Plan and Des 43(5):826–847
Liu Y, Zhang X, Liu W, Lin Y, Su F, Cui J, ... Gross L (2023) Seismic vulnerability and risk assessment at the urban scale using support vector machine and GIScience technology: a case study of the Lixia District in Jinan City, China. Geomat, Nat Hazards and Risk 14(1), 2173663
Lu Y, Zhai G, Zhou S, Shi Y (2020) Risk reduction through urban spatial resilience: A theoretical framework. Hum and Ecol Risk Assess: An Int J 27:921–937
Mehdi ZARÉ, Nazmazar B (2013) Van, Turkey earthquake of 23 October 2011, mw 7.2; an overview on disaster management. Iranian J of Public Health 42(2):134
Nikpour Lotfi S, Mansour Lakuraj Z (2021) Studying the impact of form on the sustainability of neighborhoodsevaluating the effect of form on the sustainability of the urban neighborhoods (Case study: Amol) (Case study: Amol). Geogr Stud of Coast Areas J 2(2):5–30 (In Persian Language)
Panerai P, Castex J, Depaule JC, Samuels I (2004) Urban forms: The death and life of the urban block. Routledge, Oxford
Parizi SM, Taleai M, Sharifi A (2022) A GIS-based multi-criteria analysis framework to evaluate urban physical resilience against earthquakes. Sustain 14(9), 5034
Reinoso E, Quinde P, Buendía L, Ramos S (2021) Intensity and damage statistics of the September 19, 2017 Mexico earthquake: Influence of soft story and corner asymmetry on the damage reported during the earthquake. Earthq Spectra 37(3):1875–1899
Reyners M (2011) Lessons from the destructive Mw 6.3 Christchurch, New Zealand, earthquake. Seismol Res Lett 82(3):371–372

Sevtsuk A, Mekonnen M (2012) Urban network analysis: A new toolbox for ArcGIS. Revue internationale de géomatique 22(2):287–305. https://doi.org/10.3166/rig.22.287-305

Shafiei-Dastjerdi M, Lak A (2023) Towards resilient place emphasizing urban form: An assessment framework in urban design. Sustain Cities and Soc 96, 104646

Shahzada K, Noor UA, Xu ZD (2025) In the Wake of the March 28, 2025 Myanmar Earthquake: A Detailed Examination. J of Dyn Disasters 1(2), 100017

Shakibamanesh A, Ebrahimi B (2020) Toward practical criteria for analyzing and designing urban blocks. In: Almusaed A, Almssad A, & Truong-Hong L (eds) Sustainability in urban planning and design (pp. 185–207). IntechOpen. https://doi.org/10.5772/intechopen.90504

Sharifi A (2019a) Resilient urban forms: A macro-scale analysis. Cities 85:1–14

Sharifi A (2019b) Resilient urban forms: A review of literature on streets and street networks. Build and Environ 147:171–187

Sharifi A (2019c) Urban form resilience: A meso-scale analysis. Cities 93:238–252

Sharifi A, Roosta M, Javadpoor M (2021) Urban form resilience: A comparative analysis of traditional, semi-planned, and planned neighborhoods in Shiraz, Iran. Urban Sci 5(1):18

Sharifi A, Yamagata Y (2018) Resilience urban forms: A concept framework. In: Yamagata Y, H. Maruyama (eds) Resilience-oriented urban planning: Theoretical and empirical insights. Springer, pp 167–179

Shrestha SR, Sliuzas R, Kuffer M (2018) Open spaces and risk perception in post-earthquake Kathmandu city. Appl Geogr 93:81–91

South China Morning Post (n.d.) Retrieved August 14, 2023. From https://multimedia.scmp.com/news/china/article/sichuan-earthquake/

Suppasri A, Maly E, Kitamura M, Pescaroli G, Alexander D, Imamura F (2021) Cascading disasters triggered by tsunami hazards: A perspective for critical infrastructure resilience and disaster risk reduction. Int J of Disaster Risk Reduct 66, 102597

Tang P, Xia Q, Wang Y (2019) Addressing cascading effects of earthquakes in urban areas from network perspective to improve disaster mitigation. Int J of Disaster Risk Reduct 35, 101065

Turkey–Syria earthquakes (2024, April 20) In: Wikipedia. https://en.wikipedia.org/w/index.php?title=2023_Turkey%E2%80%93Syria_earthquakes&oldid=1176543210. Accessed 7 May 2025

United Nations Office for the Coordination of Humanitarian Affairs (2023) Philippines: 6.9-magnitude earthquake, Davao del Sur Flash Update No. 1 (As of 16 December 2019). https://www.unocha.org/publications/report/philippines/philippines-69-magnitude-earthquake-davao-del-sur-flash-update-no-1-16-december

Villagra P, Rojas C, Ohno R, Xue M, Gómez K (2014) A GIS-base exploration of the relationships between open space systems and urban form for the adaptive capacity of cities after an earthquake: The cases of two Chilean cities. Appl Geogr 48:64–78. https://doi.org/10.1016/j.apgeog.2014.01.010

Xu Z, Zhang F, Wu Y (2023) A dual⊠scale system dynamics prediction method for post⊠earthquake repair of urban building groups. Earthq Eng & Struct Dyn 52(5):1423–1444

Chapter 4
Towards Livable Streets: Physical Factors for Enhanced Urban Growth and Street Design

Armin Akhavan and Amir Reza Khavarian-Garmsir

Abstract This study aims to identify and evaluate the physical factors that impact the livability and quality of urban streets, with a primary focus on how these factors drive, enable, and constrain urban growth. This study utilized an exploratory sequential mixed-methods approach to examine in detail the relationship between key physical street design parameters and patterns of urban expansion and transformation. Initially, qualitative data were gathered through a comprehensive literature review that identified key themes related to physical street layout and its direct impact on urban development. Field surveys were conducted to collect quantitative data from street users using structured questionnaires. To analyze the physical structure and visual aspects of the streets, data collection methods included both observational interviews and survey-based techniques. The target population for this study comprised residents, pedestrians, and workers on the selected streets in 2024.

The findings reveal that several physical challenges, such as inadequate lighting, weak integration between streets and buildings, insufficient pathway width, narrow sidewalks, parking shortages, traffic congestion, inefficient public services, unimproved building façades, limited planting spaces, poorly managed signage and boards, and a lack of seating areas, not only reduce the perceived livability of urban streets but also directly hinder the capacity for urban growth, redevelopment, and densification.

The study shows that fixing these physical issues is crucial not only for enhancing user experience and quality of life but also for eliminating barriers to urban growth and helping urban areas adapt better to increasing populations and economic activity. It advocates a holistic approach to urban street design that prioritizes the

A. Akhavan (✉)
Department of Architecture, Sari Branch, Islamic Azad University, Sari, Iran
e-mail: akhavan.ar@gmail.com

A. R. Khavarian-Garmsir
Department of Geography and Urban Planning, Faculty of Geographical Sciences and Planning, University of Isfahan, Isfahan, Iran

A. Cheshmehzangi et al. (eds.), *Transforming Cities*, Urban Sustainability,
https://doi.org/10.1007/978-981-95-7378-3_4

functional, spatial, and aesthetic qualities of street environments as foundational strategies for stimulating and supporting sustainable urban growth. By fostering effective collaboration among urban planners, local governments, and communities, cities can implement targeted physical design interventions to create more livable urban spaces and actively facilitate urban development. This chapter serves as a vital resource for understanding how improving street design through key physical factors is directly linked to enhanced urban livability and robust growth.

Keywords Urban growth • Urban livability • Street Design • Physical Factors • Livable streets

4.1 Introduction

In contemporary urban planning, there has been a growing interest in the "livable streets" movement, which envisions streets as more than simple machine movement corridors, but rather as lively public spaces that promote the well-being of all people using the transportation system (Ewing and Cervero, 2010; Forsyth, 2015; Appleyard et al., 1981). This approach prioritizes the comfort and convenience of pedestrians, cyclists, and public transportation users while encouraging sociability and safety, thereby contributing to a strong sense of community (Gehl, 2010; Mehta, 2014). Urban streets are the essential arteries that define the daily experience of city dwellers, acting as the lifeblood of urban life, creating movement, and fostering social connections. With an increasing number of cities facing rapid expansion and transformation, the concept of streets being truly "livable" has become increasingly central in the field of urban planning and design (Southworth, 2014; Stevens, 2017).

Street livability significantly contributes to the quality of urban life. Beyond their function as transportation corridors, livable streets are vital public spaces that enhance the social lives of urban communities (Carmona, 2019; Montgomery, 2013). They are also instrumental in driving community vibrancy, public health, and overall well-being. Crucially, with rapid urbanization and densification, the demand for well-planned, multi-use streets has never been greater, as these spaces are fundamental to accommodating and shaping urban growth.

Traditional street design practices have often prioritized the unimpeded flow of vehicular traffic, frequently resulting in environments that are less conducive to walking, cycling, and social interaction. However, there is a growing recognition of the urgent need to rebalance our streets to better serve all users, including pedestrians, cyclists, and public transportation users (UN-Habitat, 2020). This crucial shift in focus aligns with the broader aims of creating more sustainable, resilient, and human-focused urban environments that can effectively manage and support ongoing urban growth.

This study aims to identify and assess the physical factors that influence the livability and quality of urban streets, focusing on how these factors directly promote, enable, or limit urban growth and redevelopment. Specifically, it aims to pinpoint the critical physical design parameters of urban streets that negatively affect perceived livability, such as inadequate lighting, weak integration between streets and buildings, insufficient pathway width, narrow sidewalks, parking shortages, traffic congestion, inefficient public services, unimproved building façades, limited planting spaces, poorly managed signage and boards, and a lack of seating areas. The study further demonstrates that effectively addressing these physical shortcomings is essential not only for improving the user experience and quality of life but also for removing barriers to urban expansion and enabling urban environments to better adapt to increasing population and economic activity.

This study utilized an exploratory sequential mixed-methods approach. This methodology combines qualitative data gathered through a comprehensive literature review with quantitative data collected from street users using structured questionnaires and observational techniques. By examining both the physical structure and visual aspects of streets, as well as user perceptions and experiences, this study aims to provide a holistic understanding of the physical factors that contribute to street livability and their profound impact on urban development. The significance of this comprehensive approach lies in its potential to address multiple urban challenges simultaneously, particularly those related to accommodating and enhancing urban growth. By focusing on thoughtful physical design, cities can create spaces that encourage social interaction, support local businesses, promote active transportation modes, and integrate essential infrastructure, all of which are vital for a thriving and expanding urban environment.

4.2 Literature Review

Urban street design is a central element in shaping the livability, functionality, and sustainability of a city. Streets are not only conduits for movement; they are also vital public spaces that influence social behavior, economic activity, and urban growth patterns. A well-designed street enhances the quality of life by creating environments that are accessible, attractive, and socially vibrant, and by fostering compact, connected, and resilient urban development. Achieving such outcomes requires an integrated approach that considers the physical, environmental, and social dimensions of urban spaces.

Foundational work by Gehl (2010) emphasizes the importance of human-centered design, which prioritizes pedestrian comfort, accessibility, and opportunities for social interaction. This approach challenges traditional car-dominated planning paradigms by advocating streets that support vibrant public life. Jacobs (1961) similarly highlighted the role of pedestrian-friendly mixed-use streets in creating lively, safe, and economically vibrant neighborhoods. Carmona (2019) expands on these ideas, emphasizing how street layout, public space quality, and design equity shape neighborhood livability and influence broader urban development trajectories.

Walkability is among the most widely cited physical design factors contributing to street livability. Ewing and Handy (2009) identified five essential urban design qualities—imageability, enclosure, human scale, transparency, and complexity—as key to promoting pedestrian activity. These qualities enhance the experiential dimensions of urban streets and are strongly linked to increased pedestrian traffic and local economic vitality. Speck (2012) further argues that walkable streets are economically and environmentally advantageous, boosting local business performance and property values while reducing automobile dependency. Kasraian et al. (2021) provide empirical support for these claims through a stated-preference study in Toronto, showing that people consistently favor street designs that include transit lanes, cycle paths, generous sidewalks, and outdoor dining—even when trade-offs such as reduced sidewalk width are involved. The concept of visual perception and spatial enclosure is closely related to walkability. Ashihara (1983), Lynch (1960), and Gehl and Koch (2011) highlighted the psychological impact of visual cues in urban environments, including the ratio of street width to building height, façade transparency, and edge definition. Ewing and Clemente (2013) found that streets with strong enclosures—characterized by coherent skylines, active frontages, and vegetative canopies—were perceived as safer and more inviting than those lacking such features, even after controlling for socioeconomic variables.

Street connectivity plays a critical role in shaping urban livability and growth. Connected street networks reduce reliance on automobiles, improve accessibility, and support efficient land use. Millard-Ball and Barrington-Leigh (2015) found that grid-based street layouts outperform cul-de-sacs in promoting walkability, transit use, and overall urban adaptability. Complementing this, Boeing (2017) demonstrated that in most cities, pedestrian networks are more direct and accessible than vehicle-oriented networks, facilitating compact urban growth and efficient mobility.

Another essential set of physical interventions is traffic-calming strategies, which include narrower vehicle lanes, raised crosswalks, curb extensions, and daylighted intersections. These measures improve pedestrian safety and comfort by slowing down traffic and prioritizing non-motorized users. Appleyard's (1981) seminal study showed that reducing traffic volumes leads to higher levels of social interaction and neighborhood cohesion. Recently, urban street design manuals (e.g., NACTO, 2013) have advocated for intersection daylighting and sidewalk extensions to reduce vehicular conflicts and enhance pedestrian visibility.

Streetscape elements, such as benches, lighting, shopfronts, and active façades, are also key to cultivating street life. Ewing et al. (2006) found that the presence of street-level windows, varied architectural details, and accessible street furniture directly correlated with increased pedestrian presence and vibrancy. Similar findings from observational studies in Copenhagen and Boston highlight the role of "soft edges" such as transitional spaces and seating areas, in encouraging longer stays and greater social engagement (Gehl and Svarre, 2013).

In addition to walkability and aesthetics, community engagement is significantly affected by the physical characteristics of streets. Carmona (2010) and Mehta (2009) showed that features such as pavement quality, lighting, and spatial permeability shape how people interact with their environment and with each other. These physical factors directly influence the levels of social trust, informal surveillance, and civic participation, all of which are key indicators of livability.

Collectively, these physical street design elements—walkability, spatial enclosure, connectivity, traffic-calming, and active frontages—do not merely enhance day-to-day experiences and serve as powerful drivers of urban growth. Well-designed streets attract residents and businesses, increase land value, and encourage higher-density mixed-use development (Montgomery, 1998; Talen, 2002; Litman, 2015). They support both horizontal and vertical urban expansion, aligning growth with sustainability principles and long-term urban resilience. As Boeing (2017) notes, street-level forms fundamentally shape urban accessibility and adaptability, making the physical qualities of streets central to both livability and strategic growth.

4.2.1 Key Theoretical Concepts

4.2.1.1 Urban Livability

Urban livability is a multifaceted concept encompassing the quality of life in urban areas, focusing on factors contributing to residents' well-being and satisfaction. According to Marans and Stimson (2011), urban livability involves the perception of the quality of life in a city or region based on the availability and accessibility of resources and amenities and the extent to which the built and natural environments contribute to the well-being of residents. This includes access to green spaces, efficient public transportation, cultural amenities, safety, and social inclusion. Ewing and Handy (2009) emphasized the role of urban design in fostering livable communities, highlighting the importance of walkability, mixed land use, and public spaces. Furthermore, Pacione (2003) discusses urban quality of life, which is heavily related to urban livability, and how it is measured using both objective and subjective indicators.

Within this broader framework, the concept of livable streets has emerged as a critical element in shaping urban livability. Livable streets prioritize human-scale design, safety, accessibility, and social interaction, transforming streets from mere transport corridors into vibrant public spaces (Appleyard, 1981; Mehta, 2014). These streets support diverse activities such as walking, sitting, cycling, and gathering, and contribute to a stronger sense of place and community identity (Gehl, 2010). By encouraging social engagement, promoting active transportation, and improving environmental quality, livable streets play a vital role in enhancing both everyday life and long-term urban sustainability (Forsyth and Krizek, 2011). This concept is crucial for this study, as the goal is to identify and enhance the physical factors that contribute to the livability of urban streets, with direct implications for urban growth and its redevelopment.

4.2.1.2 Urban Growth

Urban growth refers to the quantitative and qualitative expansion of urban areas, propelled by various drivers, including demographic increases, economic development,

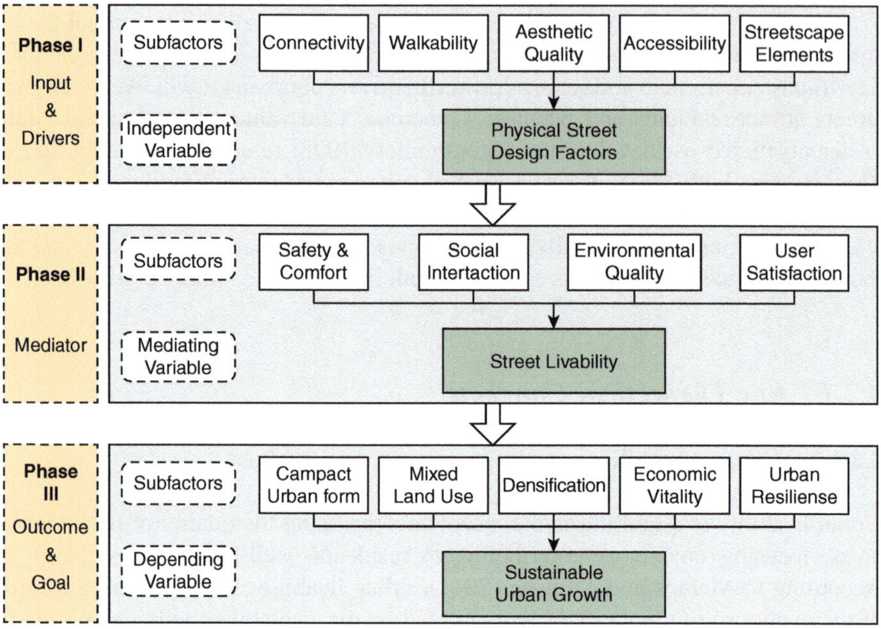

Fig. 4.1 Theoretical framework.

infrastructure investment, and evolving land use patterns (Angel et al., 2011; UN-Habitat, 2020). Urban expansion occurs through two principal modes: horizontal growth, commonly identified as urban sprawl, and vertical densification, which entails intensification of development within established urban footprints (Neuman, 2005). Urban growth is not solely a matter of physical expansion; it also entails significant shifts in governance frameworks, societal configurations, and environmental systems (Seto et al., 2011).

Within contemporary urban planning discourse, the concept of urban growth is closely linked to sustainability goals, emphasizing the value of compact, connected, and inclusive urban forms (OECD, 2012; Jenks and Jones, 2010). Physical design factors at the street level—such as connectivity, accessibility, walkability, and integration with the built environment—play a crucial role in shaping the direction and quality of urban growth (Marshall, 2005; Cervero and Kockelman, 1997). Well-designed streets that facilitate active modes of transport, mixed-use development, and social interaction are essential components of sustainable urban growth, as they optimize land use efficiency and foster the creation of vibrant and livable public spaces (Gehl, 2010; Dempsey et al., 2012).

This study conceptualizes urban growth not only because of demographic and economic drivers but also as an ongoing process deeply influenced by the physical characteristics of urban street design. By analyzing how specific street-level design features either facilitate or impede urban transformation, this study aims to elucidate the significance of street interventions in fostering urban environments that are livable, resilient, and capable of adapting to future challenges.

The theoretical framework of this study examines how physical street design impacts urban growth and livability (Fig. 4.1). It highlights how well-designed streets enhance mobility, accessibility, and social well-being, which in turn support and drive sustainable urban growth. Streets are conceptualized not just as transportation corridors but as vital public spaces that facilitate community interaction, environmental quality, and user satisfaction.

Structured in three phases, the framework positions physical street design factors—such as connectivity, walkability, aesthetic quality, and streetscape elements—as key inputs. These influence street livability, which acts as a mediating factor encompassing safety, comfort, social interaction, and environmental quality. The final phase connects this livability to broader urban outcomes, including compact form, mixed land use, densification, economic vitality, and urban resilience—essential components of sustainable urban growth. Drawing on recent literature and empirical studies, this model offers a cohesive lens for understanding and improving the role of street design in shaping resilient and people-centered cities.

4.3 Research Methodology

This study used a sequential mixed-methods approach to explore how visual and physical factors affect the livability of streets. The research unfolded in five phases, starting with detailed field observations of street features, followed by surveys to gather user perception. Expert evaluations and resident focus groups provided professional and community insights. Finally, all the data were combined to identify the key design improvements. This approach ensured a comprehensive understanding by blending objective assessments with subjective experiences of the participants.

The case studies selected for this research are two prominent arterial streets located in the central core of Sari City, Iran: Enqelab and Qaran Streets. Sari, the capital of Mazandaran Province in northern Iran, is a historically rich and socio-economically dynamic city that blends traditional and modern urban fabrics. The arterial streets in the central core of Sari city, characterized by dynamic social and economic activities and suitable accessibility, were selected for this study (Fig. 4.1). The selection of these streets was based on their importance in terms of street landscape within the central core and the historical fabric of Sari city, their multifunctional characteristics, high citizen presence, and significant daily traffic volume. Additionally, their accessibility and convenience for data collection by researchers contributed to their selection. Enqelab Street is one of Sari's principal transportation axes and forms part of the city's historical urban fabric, with its origins tracing back to the early twentieth century and reflecting layers of Sari's architectural and infrastructural development over time.

It is characterized by a concentration of administrative (governmental) and administrative-commercial functions. The street is approximately 600 m long and 20.90 m wide. The pedestrian paths on each side are 4 m wide, and the roadway is 12.90 m wide.

Qaran Street is the main street in the heart of Sari city center. The two sides of this historic street are lined with market and commercial buildings. This street, approximately 800 m in length, stretches linearly from east to west. Its width is 20 m, with pedestrian paths on each side measuring 4 m wide, and the roadway is 12 m wide.

Initially, a thorough physical assessment of both streets was conducted using direct field observations. The evaluation focused on the key visual and physical elements essential for street livability. The harmony of architectural styles was assessed by comparing the consistency and compatibility of building designs along the street, considering factors such as building height, materials, color schemes, and architectural details, to determine how well they created a cohesive streetscape. The attractiveness of building façades was evaluated based on factors such as façade maintenance, presence of decorative elements, cleanliness, and overall visual appeal. The quality of street lighting was measured by observing the type, distribution, and intensity of the lighting fixtures, as well as their effectiveness in providing uniform illumination without glare or dark spots. Physical characteristics, such as the adequacy of pathway widths, were measured by physically recording sidewalk dimensions and assessing whether they allowed comfortable pedestrian flow and accessibility. The availability of seating was noted by counting the number and distribution of benches and other resting spots along streets.

Alongside the physical assessment, this study conducted a public survey ($n = 78$) of residents' attachment and satisfaction to capture the social and perceptual aspects of urban livability on Qaran and Enqelab Streets. A total of 42 males and 36 females, aged 18 to 65—including residents, shop owners, and visitors—were recruited through convenience and purposive sampling. Convenience sampling involved selecting readily available participants, while purposive sampling targeted individuals based on specific roles or characteristics relevant to the study. All participants were surveyed in person at various times.

The survey examined how economic factors, neighborhood identity, security, and satisfaction with municipal services influence residents' emotional and functional ties to their neighborhoods. These insights complement physical evaluation by highlighting the human experience and sense of belonging. Additionally, the findings shed light on urban growth patterns in Sari's city center, revealing residents' preferences and challenges related to physical changes, housing, and mobility, thus informing sustainable urban growth strategies for the future. This integrated approach links street-level livability to broader urban development dynamics.

The questionnaire consisted of closed- and multiple-choice questions organized into four thematic sections: (1) place attachment, (2) key challenges, (3) satisfaction with urban and municipal services, and (4) housing and mobility preferences. Most items were recorded as binary (yes/no) responses—for example, satisfaction indicators or support for pedestrianization—whereas place attachment factors and key challenges were captured through multiple-choice items, allowing participants to select the most relevant option. The survey design aimed to balance clarity and brevity to suit on-site administration while ensuring sufficient depth for a comparative analysis.

Furthermore, an expert panel (n = 35) consisting of urban professionals—including architects, urban designers, and planners—conducted photo audits of 32 street images (16 per street). These professionals were selected through purposive sampling based on their experience in urban design, with at least five years of professional practice or an academic background in the field. The group included 20 males and 15 females, ranging in age from 30 to 60 years. During the audits, they coded positive visual elements (e.g., architectural detailing, continuity of building frontages) and negative visual elements (e.g., obstructive signage, poor material transitions), providing structured evaluations of each street's physical and aesthetic characteristics.

In addition, a focus group was conducted with ten residents (five men and five women, aged 25 to 55 years) who were regular users of the selected streets. Participants were recruited through local community networks and neighborhood organizations in the area. The focus group discussions aimed to identify residents' top visual complaints, most needed physical improvements, and preferences regarding future street enhancements. This combination of expert assessments and user insights added depth and local relevance to the evaluation of the livability factors.

Ultimately, the findings were synthesized using a priority matrix to identify high-impact visual upgrades (such as façade treatments and improved lighting), critical physical modifications (including accessible seating, wider pathways, and shade structures), and other feasible design interventions. This approach ensured that the study outcomes directly informed actionable street design strategies aimed at enhancing urban livability and supporting sustainable urban growth.

4.4 Results

A comprehensive physical evaluation of Qaran and Enqelab Streets revealed the key visual and functional elements that shape urban livability in the central core of Sari (Fig. 4.2). Although both streets serve as vital public corridors, their contrasting physical and infrastructural conditions underscore the need for integrated urban design interventions. The assessment, based on direct field observations, focused on six critical criteria: harmony of architectural style, attractiveness of building façades, quality of street lighting, adequacy of pathway widths, and seating availability.

Qaran Street exhibited relatively strong architectural coherence, demonstrated by consistent building heights, materials, and façade treatments that contributed to a unified and legible streetscape. This conclusion is based on direct field observations and systematic photo audits of the street's buildings, where over 70% of structures were found to maintain height consistency within a narrow range of approximately 3.5 to 4 m. Additionally, material analysis showed a predominance of traditional brick and plaster finishes applied in a harmonious manner across façades (Fig. 4.4 (i)). Expert evaluations of 32 street images highlighted positive visual elements such as detailed architectural features and the continuity of building frontages, reinforcing the quantitative findings (Fig. 4.4 (j)). Resident surveys and focus groups further

Fig. 4.2 Location of studied streets in Sari City (Source: The authors).

supported this, with participants describing Qaran Street as historically significant and a central commercial hub with a strong visual identity (Fig 4.4 (h)). These multiple lines of evidence—from statistical measurements to qualitative feedback—provide a robust basis for recognizing Qaran Street's architectural coherence and its urban importance.

In contrast, Enqelab Street showed greater variation in building styles, color schemes, and material usage, especially in older segments, where façade neglect and visual clutter were prominent (Fig. 4.3 (b)). Although Qaran benefited from recent signage regulations and generally well-maintained façades (Fig. 4.4 (f)), both streets displayed informal storefront extensions and signage clutter, underscoring the tension between economic survival strategies and the need for urban order (Fig. 4.3 (k)). Design guidelines aimed at harmonizing façade treatments while respecting local identity and informal practices would be beneficial.

Lighting infrastructure further reflects differences in municipal investment and commercial development. Qaran Street, benefiting from the presence of modern commercial buildings, such as shopping centers, has enhanced street illumination and improved nighttime lighting conditions (Fig. 4.4 (b)). These commercial centers contribute to uniform and effective fixture placement along the street, reducing dark zones and promoting safety and visibility. However, localized dark pockets still persist in some commercial hotspots, indicating the need for ongoing maintenance. In contrast, Enqelab Street featured older and deteriorated lighting systems with dim or nonfunctional fixtures (Fig. 4.3 (d)). These deficiencies raise safety

Fig. 4.3 Evaluation of physical-visual factors affecting the vitality of Enghelab Street in Sari City through direct observations: (a) Administrative/office land uses dominate the street; (b) Stark contrast between old and new architectural styles; (c) Insufficient vegetation along the street; (d) Inadequate nighttime lighting; (e) Public green spaces; (f) Inappropriate and insufficient street furniture; (g) Attractive façades line the street edges; (h) Historic buildings with unique architecture; (i) Vehicle-pedestrian traffic conflict; (j) Proper maintenance and cleanliness; (k) Appropriate signage and traffic signs; (l) Proper paving/surface quality (Source: Photo taken by the authors).

concerns, especially for women and older adults, and reinforce the need for equitable infrastructure maintenance.

Pedestrian infrastructure on both streets met the minimum Iranian urban code for sidewalk width—ranging between 2.5 and 4 meters—but this regulatory compliance failed to ensure walkability. Qaran's sidewalks are frequently obstructed by vendors, informal stalls, and poorly placed signage, causing severe congestion near market zones. Although less commercially dense, Enqelab was hampered by uneven paving (Fig. 4.3 (l)) and sporadic encroachments that undermined pedestrian mobility (Fig. 4.3 (i)). Critically, neither street offered universal access: the absence of ramps at crossings and entrances violated the 2018 Disability Rights Act, revealing a disconnect between policy and practice in the built environment. This shortfall disproportionately impacts older adults and people with disabilities, diminishing their independence and dignity in public space use.

A significant shared issue was the shortage and poor distribution of seats. Qaran Street lacked formal seating entirely (Fig. 4.4 (l)), whereas Enqelab featured a limited number of benches near the municipal park (Fig. 4.3 (f)). These are underutilized due to poor accessibility, absence of shade, and weak integration with pedestrian flows, suggesting a reactive rather than systematic approach to street furniture planning. This inadequacy contradicts international livability standards, such as New York City's benchmark of ten benches per kilometer, and

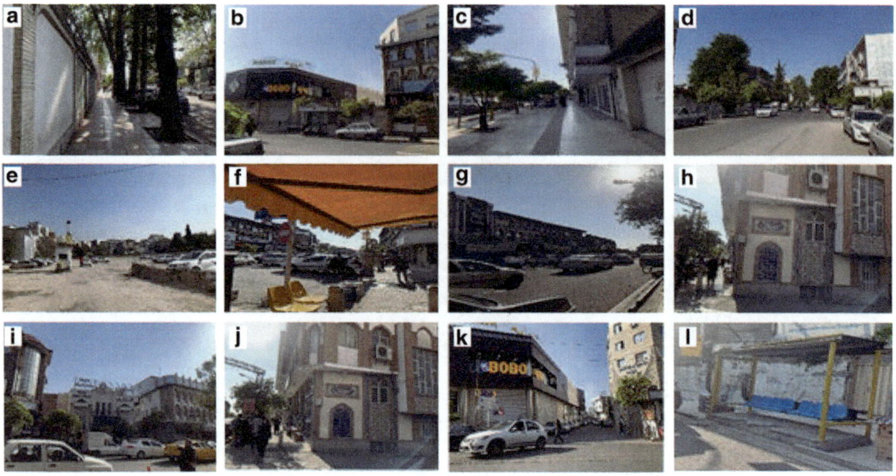

Fig. 4.4 Evaluation of some factors affecting the vitality of Qaran Street in Sari City through direct observations: (a) Proper sidewalk width; (b) Existence of modern commercial centers (Elahieh Shopping center); (c) Proper and uniform sidewalk paving; (d) Relatively suitable vegetation cover; (e) Parking layout; (f) Proper signage placement; (g) Motor vehicle traffic; (h) Existence of identity-defining elements featuring unique architectural design; (i) Varied and diverse architectural styles; (j) Key physical elements; (k) Conflict between vehicle and pedestrian traffic; (l) Climate appropriate furniture (Source: Photo taken by the authors).

was a recurring concern in resident feedback, especially among older users who reported having "nowhere to rest." A connected seating network designed with accessibility and shade in mind is essential for improving comfort and inclusivity in urban corridors.

Additionally, to assess the vitality of the studied streets, factors influencing street vitality were observed. Direct observations, conducted by photographing the studied streets, revealed physical factors on Enghelab Street that demonstrated notable aspects, including the contrast between old and new architectural styles (Fig. 4.3 (b)), insufficient vegetation along the street (Fig. 4.3 (c)), inadequate nighttime lighting (Fig. 4.3 (d)), the presence of public green spaces (Fig. 4.3 (e)), and attractive façades along the street (Fig. 4.3 (g)). Positive physical attributes of Qaran Street include good vehicular access, a significant number of pedestrians, appropriate sidewalk width (Fig. 4.4 (a)), and well-located parking spaces (Fig. 4.4 (e)). The shading and tree canopies, as well as sun exposure on the building façades, were also satisfactory (Fig. 4.4 (d)). However, as in other corridors, vegetation and urban furniture require improvement and maintenance. The presence of distinctive landmarks, human scale, and prominent physical elements are also notable features of this street. Among its outstanding characteristics are attractive façades, diverse architectural styles, and the quality and historical value of the buildings along the street (Figs. 4.4 (h)) and 4.4 (i)). Field observations also indicate that the studied streets, in many cases, do not meet optimal physical quality standards. Some physical factors, such as lighting, signage, seating, adequate path width, and parking space, were more favorable on Qaran Street (Figs. 4.4 (b), 4.4 (f), 4.4 (l), 4.4 (a), and 4.4 (e)), while vegetation, unique buildings with distinctive architecture, and public green spaces are more desirable on the Enghelab Street (Figs. 4.3 (c), 4.3 (h), and 4.3 (e)).

Table 4.1 highlights that residents of Qaran and Enghelab Streets are primarily attached to their neighborhoods due to economic factors and identity-defining features, with Qaran Street emphasizing economic considerations and Enghelab Street focusing more on identity. Security issues and inadequate urban services are the most pressing challenges in both areas, with Enghelab Street reporting a higher incidence of security concerns than in other areas. Satisfaction with residence and employment is higher on Qaran Street, but dissatisfaction with municipal services is widespread on both streets. Additionally, most residents in both areas prefer detached houses to apartments.

Notably, a higher proportion of residents on Enghelab Street desire physical and visual changes to their environment, and parking difficulties are more commonly reported there. Support for pedestrianization, however, is stronger in Qaran Street, reflecting differences in urban mobility preferences between the two neighborhoods. These findings underscore the complex interplay of social, economic, and environmental factors that shape residents' attachment and satisfaction in the studied streets (Table 4.1).

Based on the expert panel's photo audit, Qaran Street demonstrated stronger overall livability, with well-maintained greenery, consistent architectural quality, and accessible pedestrian amenities. The panel noted that Qaran's human-scale façades and shaded seating areas contribute positively to street comfort and visual appeal. However, some inactive ground floors and minor monotony were observed in certain segments (Table 4.2).

In contrast, Enghelab Street benefits from the presence of a pocket park and significant official land uses, such as the old municipality building and governance offices, which enhance the civic identity and attract daily foot traffic. The panel observed that while the pocket park provides valuable green space and social opportunities, Enghelab's main sidewalk areas often lack greenery and seating, and blank walls and security barriers characterize some stretches. Both streets present opportunities for nature-based solutions, such as green walls and expanded tree planting, to further enhance their livability and environmental performance.

The focus group of ten regular street users (five men and five women, aged 25 to 55) provided insights consistent with field observations, addressing lighting, pedestrian safety, and street conditions. Participants noted that the street lighting quality varied, with some parts better lit than others. Several participants mentioned that daily electricity outages, caused by energy-saving strategies, led to streetlights being off for at least two hours each day, increasing safety concerns. In some spots, the lack of lighting made the streets feel particularly unsafe after dark. A woman, 33, from Qaran Street said:

Table 4.1 Results of questionnaire on residents' attachment and satisfaction levels in studied streets *(n = 78; percentages represent selected responses per item).*

Category	Factor/Indicator	Qaran St. (%)	Enghelab St. (%)	Measurement Type
Place Attachment Factors	Neighborhood relations/ Memories	23.7	19.8	Single-choice (multiple options)
	Proximity to work/Education	5.3	4.2	Single-choice (multiple options)
	Environment/Street character	12.5	15.6	Single-choice (multiple options)
	Peace & comfort	19.2	16.3	Single-choice (multiple options)
	Economic factors	25.4	21.8	Single-choice (multiple options)
	Identity-defining features	13.5	22.3	Single-choice (multiple options)
	No attachment	0.4	0	Single-choice (multiple options)
Key Challenges Faced	Security issues	35.6	43.6	Single-choice (multiple options)
	Inadequate urban services	33.4	29.4	Single-choice (multiple options)
	Poor sanitation/Health conditions	13.3	17.2	Single-choice (multiple options)
	Lack of recreational opportunities	17.7	9.8	Single-choice (multiple options)
Satisfaction & Urban Development	Satisfaction with residence/ Employment	79.5 (Yes)	61.2 (Yes)	Yes/No
	Desire for physical/Visual changes	43.6 (Yes)	71.4 (Yes)	Yes/No
	Parking difficulties	70.3 (Yes)	83.7 (Yes)	Yes/No
	Support for pedestrianization	65.6 (Yes)	40.3 (Yes)	Yes/No
	Importance of knowing neighbors	69.7 (Yes)	55.4 (Yes)	Yes/No
	Perception of street safety	57.7 (Yes)	33.3 (Yes)	Yes/No
	Approval of nearby commercial spaces	88.3 (Yes)	71.9 (Yes)	Yes/No
	Accessibility for vulnerable groups	15.2 (Yes)	14.4 (Yes)	Yes/No
Municipal Service Satisfaction	Satisfied	19.1	14.2	Multiple choice (3 options)
	Moderate	31.3	26.7	Multiple choice (3 options)
	Dissatisfied	49.6	59.1	Multiple choice (3 options)
Housing Preference	Detached Houses (Villas)	92.4	95.2	Preferred option
	Apartments	7.6	4.8	Preferred option

Table 4.2 Results from the comparative qualitative analysis of streets by experts.

Element/Category	Qaran St. Positive Factors	Qaran St. Negative Factors	Enghelab St. Positive Factors	Enghelab St. Negative Factors
Architectural Details	Human-scale façades, articulated building details.	Some inactive ground floors, minor monotony.	Historic and administrative buildings add gravitas and identity.	Some blank walls, lack of façade articulation in newer structures.
Greenery & Landscaping	Well-maintained trees, consistent planting, shade.	Limited seasonal variety in some sections.	Pocket park adds greenery and gathering space.	Sparse greenery along sidewalks, some neglected planting.
Seating & Rest Areas	Shaded, comfortable seating, well-placed benches.	Occasional lack of seating in busy areas.	Seating in pocket park, social seating near offices.	General shortage of benches along main sidewalk.
Lighting	Warm, pedestrian-scale, evenly distributed lighting.	Some dark spots at intersections.	Lighting in the park area, illuminated administrative buildings.	Insufficient or harsh lighting in some sidewalk sections.
Pavement & Transitions	Smooth, accessible, cohesive surfaces.	Minor unevenness at crossings.	Improved paving in the park and around official buildings.	Uneven, mismatched, or slippery surfaces elsewhere.
Signage & Wayfinding	Clear, context-sensitive signage.	Occasional visual clutter from excessive signage.	Official signage and wayfinding near government buildings.	Obstructive or excessive signage, visual clutter.
Street-Building Interface	Active shopfronts, visible entrances.	Some inactive edges, a lack of permeability in stretches.	Frequent public access points at government/municipal offices.	Inactive edges, security barriers, lack of permeability in some areas.
Public Art & Placemaking	Local art, murals, and unique features.	Some generic or neglected spaces.	Art installations in and near the pocket park.	Lack of placemaking in administrative zones.
People Presence	Active street life, diverse users, visible interaction.	Less activity at night.	High pedestrian flow due to offices and public services.	Crowding and congestion at peak times, and unwelcoming at night.
Functional Diversity	Mix of uses: retail, services, housing.	Some monofunctional stretches.	Concentration of administrative, governance, and service functions.	Inactive at certain times, lack of amenities in some parts.
Safety & Security	Good lighting, clear sightlines, and active uses.	Isolated areas at night.	Security presence near official buildings, safety in the park.	Some dark, isolated, or unsafe-feeling areas.

(continued)

Table 4.2 (continued)

Element/Category	Qaran St. Positive Factors	Qaran St. Negative Factors	Enghelab St. Positive Factors	Enghelab St. Negative Factors
Accessibility	Barrier-free routes, curb ramps, intuitive crossings.	Occasional obstacles or poor connectivity.	Accessible paths in the park and around government offices.	Security barriers, poor accessibility for the disabled in some sections.
Aesthetic Cohesion	Consistent materials, harmonious colors, attention to detail.	Minor visual clutter in commercial areas.	Cohesive design in the park and around historic/official buildings.	Mismatched materials, neglected maintenance elsewhere.
Nature-Based Solutions	Opportunities for green walls, bioswales, tree planting.	Limited space for large new green interventions.	Pocket park as a model NBS, potential for green corridors.	Lack of continuous green infrastructure, limited greening around offices.

> *The lights are often off for a couple of hours every day because of energy saving. This makes walking at night risky, especially in commercial areas.*

A man, 42, on Enqelab Street added:

> *Power cuts mean some streetlights do not work regularly, which really worries women and older people after dark. There are spots with no lighting at all that feel unsafe.*

Traffic congestion was also raised as a safety issue, especially on Qaran Street, where dense commercial and clinical activities increased pedestrian-vehicle conflicts. A man, 50, agreed:

> *Heavy traffic near clinics and markets on Qaran reduces pedestrian safety. Drivers don't always stop, and sidewalks are crowded.*

A woman, 29, noted:

> *"Enghelab Street is so busy with cars and shops that it is hard to cross safely. The traffic makes walking stressful and dangerous."*

Regarding sidewalks, vendors and signage obstruct pedestrian paths, particularly on Qaran Street, while Enqelab's sidewalks suffer from uneven paving and poor accessibility. A woman, 29, on Qaran Street, stated:

> *Sidewalks get blocked by stalls and signs, forcing people into the street where cars pass.*

A man, 50, on Enqelab Street, explained:

> *Uneven sidewalks and lack of ramps make walking difficult, especially for disabled people.*

The availability of seating was limited. Qaran Street had no formal seating, while Enqelab Street had a few benches that were poorly accessible and lacked shade. A woman, 55, from Qaran Street said:

> *There is nowhere to rest on Qaran. Most seating is in front of fast-food shops, not really for public use. Older people really need places to sit in the shade.*

A man, 38, on Enqelab Street, observed:

> *"The benches are few and uncomfortable, so most people don't use them."*

Finally, vegetation and street furniture were insufficient on both streets, contributing to an unpleasant environment. A woman, 40, mentioned:

> *More trees and greenery would make both streets nicer and provide needed shade.*

A man, 35, added:

> *Better maintenance of benches, lighting, and signs would improve the overall experience for pedestrians.*

These comments highlight the need to improve electricity reliability, manage traffic better, enhance pedestrian facilities, increase accessible seating, and upgrade streetscape elements to boost safety and comfort along both the streets. The comprehensive evaluation of Qaran and Enqelab Streets highlighted the key priorities for enhancing both the visual and physical quality. High-impact visual upgrades include improving façade treatments to achieve architectural harmony and upgrading street lighting to create safer, more inviting environments. From a physical perspective, critical modifications, such as adding accessible seating with shade, widening pedestrian pathways, and introducing shade structures, were identified to improve comfort and mobility. These prioritized actions form the basis for targeted urban design strategies aimed at boosting overall livability and supporting sustainable urban growth in the city center.

4.5 Discussion and Implications

This study demonstrates that the specific visual and physical elements of urban streets play a critical role in shaping the livability and functionality of these spaces, particularly in the context of a growing city like Sari. The contrasting conditions of Qaran and Enqelab Streets reveal how differences in infrastructure quality, design, and maintenance directly affect not only the user experience but also broader patterns of urban vitality and development.

Qaran Street benefits from more coherent architectural features, shaded pedestrian zones, and human-scale design elements that promote comfort, walkability, and visual cohesion. These qualities support sustained social interactions and economic activities. In contrast, Enqelab Street suffers from inconsistent architectural treatments, outdated or ineffective lighting, and sidewalk conditions that limit the mobility of pedestrians. These issues reduce comfort, compromize accessibility, and negatively affect safety perceptions, particularly among women, older adults, and people with disabilities.

The findings make it clear that livability is not merely a matter of aesthetics; it is fundamental to how public space functions. Inadequate lighting, blocked pathways, a lack of shaded seating, and poorly maintained sidewalks may seem like minor issues, but they collectively contribute to a degraded urban experience. These deficiencies discourage walking, restrict public life, and act as barriers to inclusive growth. More importantly, they send a signal that public space is not a priority, which can erode trust in urban authorities and the planning systems.

Survey data confirm that a combination of identity, economic opportunity, and physical quality shapes residents' emotional and functional ties to their neighborhoods. While economic factors were especially important for residents of Qaran Street, those on Enqelab Street placed greater emphasis on identity-defining features. However, both areas showed low levels of satisfaction with municipal services and a strong desire for physical and visual improvements to the same. This indicates a clear gap between residents' values and what is currently being delivered through local governance and planning efforts.

A critical concern highlighted in this study is the lack of accessibility. Despite regulatory requirements for universal access, neither street provided barrier-free routes or adequate support for people with mobility impairment. The poor state of pedestrian infrastructure not only undermines inclusivity but also limits participation in public life, particularly among older adults and individuals with disabilities.

These shortcomings reveal a disconnect between urban design policies and their actual implementation, indicating deeper issues of inequality and exclusion in the urban environment.

The expert panel and resident focus groups identified several practical interventions that could significantly improve street livability. These include better lighting systems, enhanced façade treatments, shaded and accessible seating, and removal of physical obstructions from pedestrian pathways. Importantly, these are not expensive interventions; they are small-scale, high-impact improvements that could yield immediate benefits. However, such upgrades must be context sensitive. While Qaran Street would benefit most from managing sidewalk clutter and increasing vegetation, Enqelab Street requires lighting improvements, sidewalk repairs, and activation of the currently underutilized building edges.

This study has broader implications for urban planning policies. Livability criteria should be integrated into zoning regulations, infrastructure investment decisions, and public space management strategies. Urban development cannot focus solely on increasing density or traffic capacity; it must also prioritize human experience, public comfort, and inclusivity. Incorporating resident feedback into planning processes through community consultations or participatory design methods can help ensure that physical improvements align with real user needs.

In conclusion, creating more livable streets is not an optional or decorative pursuit; it is a foundational strategy for cities facing the pressures of urban growth, climate change, and shifting social needs. Evidence from Sari shows that thoughtful, inclusive, and well-maintained street designs can directly support healthier, more equitable, and more sustainable urban growth.

4.5.1 Implications for Transformative Growth

This study supports a broader and more integrated understanding of Transformative Growth as a concept that moves beyond the traditional opposition of ecological modernization and degrowth. Rather than limiting urban change to incremental environmental efficiency or potentially impractical reductions in growth, Transformative Growth envisions cities thriving within planetary boundaries through resilience, ecological innovation, and active citizen engagement. It expands the notion of urban productivity beyond economic output to include the ecological functioning of natural systems, the social vitality of inclusive communities, and advancements in technology and design. This comprehensive framework offers a pragmatic yet ambitious pathway for reimagining cities that are socially inclusive, ecologically sound, and politically relevant.

The conditions and contrasts observed between Qaran and Enqelab Streets in Sari illustrate these principles clearly. Qaran Street's coherent architectural style, enhanced pedestrian infrastructure, and improved lighting linked to commercial developments exemplify how focused urban improvements can simultaneously enhance ecological, social, and economic productivity. These improvements contribute to safer, livelier, and more accessible public spaces, underscoring the intertwined nature of physical environment quality and social inclusivity emphasized by Transformative Growth.

Conversely, Enqelab Street's fragmented façades, inconsistent lighting, limited accessibility, and scarce seating reveal persistent challenges that hinder inclusive and sustainable urban development. These shortcomings highlight the necessity of systemic, community-oriented interventions that prioritize equitable access and social engagement, which are foundational elements of transformative urban futures.

Recommendations from this study—including upgrades to lighting systems, provision of shaded, accessible seating, improved pedestrian routes, and introduction of nature-based solutions—reflect practical, context-sensitive actions aligned with the goals of Transformative Growth. Such measures not only enhance urban livability but also align urban design and policy with resilience and sustainability, illustrating how local-scale interventions contribute to broader, systemic transformation.

In conclusion, the findings present Transformative Growth as a valuable framework for guiding cities like Sari toward futures that balance economic viability, ecological integrity, and social equity, establishing a foundation for transformative, inclusive, and sustainable urban development.

4.6 Concluding Remarks

This study highlights the critical role of both visual and physical factors in shaping street livability in the central core of Sari. Through a detailed examination of Qaran and Enqelab Streets, the findings revealed how architectural consistency, pedestrian infrastructure, lighting, and amenities influenced residents' attachment and satisfaction. The diverse challenges identified emphasize the need for targeted urban-design interventions that address accessibility, safety, and comfort.

Importantly, these improvements are essential not only for enhancing the immediate quality of life but also for supporting sustainable urban growth within the city's historic and dynamic center. Integrating community preferences with objective assessments provides a valuable foundation for planning streets that foster social cohesion and accommodate future development. This approach offers a pathway for creating vibrant, inclusive urban environments that reflect the evolving needs of Sari's population.

References

Angel S, Sheppard SC, Civco DL (2011) The dimensions of global urban expansion: Estimates and projections for all countries, 2000–2050. Prog in Plan 75(2):53–107. https://doi.org/10.1016/j.progress.2011.04.001

Appleyard D (1981) Livable streets. University of California Press, California, USA

Ashihara Y (1983) The aesthetics of the urban landscape. University of Tokyo Press, Tokyo, Japan

Boeing G (2017) Street network structure and urban growth: Evidence from 100 US cities. Urban Stud 54(10):2342–2363. https://doi.org/10.1177/0042098016633278

Carmona M (2010) Contemporary public space: Critique and classification, part one: Critique. J of Urban Des 15(1):123–148. https://doi.org/10.1080/13574800903435651

Carmona M (2019). Public places, urban spaces: The dimensions of urban design, 3rd edn. Routledge, New York, USA
Cervero R, Kockelman K (1997) Travel demand and the 3Ds: Density, diversity, and design. Transp Res Part D: Transp and Environ 2(3):199–219. https://doi.org/10.1016/S1361-9209(97)00009-6
Dempsey N, Brown C, Bramley G (2012) The key to sustainable urban development in UK cities? The influence of density on social sustainability. Prog in Plan 77(3):89–141. https://doi.org/10.1016/j.progress.2012.01.001
Ewing R, Cervero R (2010) Travel and the built environment. J of the Am Plan Assoc 76(3):265–294. https://doi.org/10.1080/01944361003766766
Ewing R, Clemente O (2013) Measuring urban design: Metrics for livable places. Island Press, DC
Ewing R, Handy S (2009) Measuring the unmeasurable: Urban design qualities related to walkability. J of Urban Des 14(1):65–84. https://doi.org/10.1080/13574800802451155
Ewing R, Handy S, Brownson R, Clemente O, Winston E (2006) Identifying and measuring urban design qualities related to walkability. J of Phys Activity and Health 3(s1):S223–S240. https://doi.org/10.1123/jpah.3.s1.s223
Forsyth A (2015) What is a walkable place? The walkability debate in urban design. Urban Des International 20(4):274–292. https://doi.org/10.1057/udi.2015.10
Forsyth A, Krizek KJ (2011) Urban design: Is there a distinctive view from the bicycle? J of Urban Des 16(4):531–549. https://doi.org/10.1080/13574809.2011.586239
Gehl J (2010) Cities for people. Island Press, Washington, DC, USA
Gehl J, Koch N (2011) Life between buildings: Using public space. Island Press, Washington, DC, USA
Gehl J, Svarre B (2013) How to study public life. Island Press, DC
Jacobs J (1961) The death and life of great American cities. Random House, New York, USA
Jenks M, Jones C (2010) Dimensions of the sustainable city. Springer, Dordrecht
Kasraian D, et al (2021) Preference and perception of urban street design: A stated-preference study in Toronto. J of Urban Plan and Dev 147(4), 04021034. https://doi.org/10.1061/(ASCE)UP.1943-5444.0000731
Litman T (2015) Evaluating transportation land use impacts. Victoria Transp Policy Inst. www.vtpi.org/landuse.pdf
Lynch K (1960) The image of the city. MIT Press, Cambridge, Massachusetts, USA
Marshall, S. (2005). Streets and patterns. Spon Press, London, UK
Marans RW, Stimson RJ (2011) Investigating quality of urban life: Theory, methods, and empirical research. Springer, Dordrecht
Mehta V (2009) Look closely and listen carefully: Urban design and social interaction on streets. J of Urban Des 14(1):29–64. https://doi.org/10.1080/13574800802459125
Mehta V (2014) The street: A quintessential social public space. Routledge, London, UK
Millard-Ball A, Barrington-Leigh C (2015) Accessibility and urban form: A quantitative analysis of street networks and transit in US cities. Urban Stud 52(7):1294–1309. https://doi.org/10.1177/0042098014526186
Montgomery C (1998) Making a city: Urbanity, vitality and urban design. J of Urban Des 3(1):93–116. https://doi.org/10.1080/13574809808724492
Montgomery C (2013) Happy city: Transforming our lives through urban design. Farrar, Straus and Giroux.
National Association of City Transportation Officials (2013) Urban street design guide. Island Press, Washington, DC, USA
Neuman M (2005) The compact city fallacy. J of Plan Educ and Res 25(1):11–26. https://doi.org/10.1177/0739456X04270466
OECD. (2012). Compact city policies: A comparative assessment. OECD Publishing, Paris, France
Pacione M (2003) Urban environmental quality and human well-being—a social geographical perspective. Landsc and Urban Plan 65(1–2):19–30. https://doi.org/10.1016/S0169-2046(02)00187-6

Seto KC, Güneralp B, Hutyra LR (2011) Global forecasts of urban expansion to 2030 and direct impacts on biodiversity and carbon pools. Proc of the Natl Acad of Sci 109(40):16083–16088. https://doi.org/10.1073/pnas.1211658109

Southworth M (2014) Designing the walkable city. J of Urban Plan and Dev 140(4), 04014032. https://doi.org/10.1061/(ASCE)UP.1943-5444.0000196

Speck J (2012) Walkable city: How downtown can save America, one step at a time. Farrar, Straus and Giroux.

Stevens Q (2017) The ludic city: Exploring the potential of public spaces. Routledge, London, UK

Talen E (2002) Measuring urbanism: Issues in smart growth research. J of Urban Des 7(3):195–215. https://doi.org/10.1080/1357480022000030306

UN-Habitat (2020) World cities report 2020: The value of sustainable urbanisation. United Nations Human Settlements Programme.

Part 2
Resilience

Resilience, under a *transformative growth* lens, moves beyond recovery to proactive adaptation. It embeds climate readiness, social equity, and ecological capacity into the city's core systems. This ensures urban areas thrive under change rather than merely survive it.

Chapter 5
Planning Cooling Structures for Climate-Resilient Cities: A Configurational Approach to Optimizing Urban Green Spaces in Tehran, Iran

Hadi Soltanifard, Sima Pourhashemi, and Mahdi Boroughani

Abstract The rapid pace of urbanization globally has led to the expansion of unplanned surfaces, intensifying urban heat, increasing climate risks, and exacerbating vulnerabilities in cities. While urban areas significantly contribute to these challenges, they also have the potential to enhance resilience to climate change and related stressors through effective planning and adaptation strategies. In this context, Nature-Based Solutions (NBS) have emerged as a powerful approach in urban planning, design, and management to mitigate urban heat, combat climate change, and improve urban quality of life. Despite the urgent need to integrate NBS into urban planning, many urban areas still struggle to effectively implement strategies that enhance climate resilience while mitigating heat emissions. To fill the gap, this study explores cooling structures in Tehran, Iran, by adopting an integrated planning approach to optimize the cooling effects of urban green spaces (UGS).

The research aims to assess how UGS can be strategically planned and adapted to the city's existing spatial configuration to function as an effective NBS. To this end, the study employs space syntax theory to analyze urban form and structure, and landscape ecology principles to model UGS spatial patterns. Several indices are used to evaluate urban spatial structure, including global integration, connectivity, and spatial depth. UGS spatial patterns are quantified using landscape metrics such as LPI, MPS, TE, ED, MSI, MPI, and MNN. The findings reveal a lack of alignment between Tehran's existing spatial configuration and UGS spatial patterns and distribution via a combination of spatial and statistical analyses. This misalignment has resulted in suboptimal cooling performance, limiting the effectiveness of UGS in mitigating urban heat.

H. Soltanifard (✉)
Department of Environment, Hakim Sabzevari University, Sabzevar, Iran
e-mail: H.soltanifard@hsu.ac.ir; Hsoltanifard@gmail.com

S. Pourhashemi • M. Boroughani
Research Center for Geoscience and Social Studies, Hakim Sabzevari University, Sabzevar, Iran

A. Cheshmehzangi et al. (eds.), *Transforming Cities*, Urban Sustainability,
https://doi.org/10.1007/978-981-95-7378-3_5

To enhance cooling efficiency, the study proposes the development of a connected ecological network that integrates UGS with urban spatial configurations to improve spatial connectivity and ecological functionality. By adopting this framework, cities can strengthen the synergy between physical-spatial urban layouts and the biophysical characteristics of UGS, creating more resilient urban environments. This approach not only enhances cooling performance but also supports broader climate adaptation efforts, helping cities reduce their exposure to climate risks while fostering greener, smarter, and more inclusive communities.

Keywords Cooling structures • Nature Based Solutions (NBS) • Urban Green Spaces (UGS) • Space Syntax • Landscape ecology • Climate resilience • Tehran

5.1 Introduction

Today, Urban Green Spaces (UGS) have become an integral component of human settlements. They not only enhance the visual quality of urban landscapes but also provide various ecosystem services, boosting social interaction and bringing economic benefits. In the last decade, the cooling effect of green spaces on urban areas has received increasing attention from researchers and urban planners as a nature-based solution(Yan et al., 2021), the importance of the cooling effect of UGS warrants investigation from two main perspectives. Globally, rising concerns about intensifying global challenges, such as greenhouse gas emissions, global warming, and climate change, along with their negative impacts on the quality of life in urban areas, have led UGS planning and development to become an effective and sustainable solution for urban planners, managers, and policymakers (Wong et al., 2024). Locally, over 56% of the world's population, approximately 4.7 billion people, currently resides in urban areas, and according to a United Nations report from 2023, this number is projected to add another 2.5 billion people to urban areas by 2050 (United Nations 2018; United Nations Department for Economic and Social Affairs 2023).

This upward trend in population growth will, in most cases, lead to unplanned, rapid urban development, accompanied by changes in land use, fragmented landscapes, and an increase in impervious urban surfaces (Yao et al., 2022). Such changes will cause imbalances in local flows, increased heat absorption by the replaced surfaces, and ultimately elevate the temperature of the built environment relative to the surrounding environment, a phenomenon known as the urban heat island (UHI) effect (Li et al., 2019). The UHI effect typically raises the temperature in urban areas by 1.0–6.0°C (Dimoudi et al., 2013).

However, studies focused on American cities have reported temperature increases ranging from 0.5 to 4.0°C during the daytime and from 1.0 to 2.5°C at nighttime, compared to nearby rural areas (USGCRP et al., 2017). In addition to adverse environmental effects, such as declining air quality, reduced thermal comfort, and increased energy demand (Ghribi and Dahech, 2023), UHIs are also recognized as a threat to public health, exposing residents to an increased risk of mortality (Singh et al., 2020). Although a growing body of literature has been presented to find effective solutions to combat heat, UGSs remain a strategic approach to addressing the challenges of urbanization on both global and local scales. Consequently, urban planners and policymakers continuously seek effective methods to develop and integrate UGSs into the urban context.

Early research has proven that UGS directly contributes to lowering urban heat by creating cool islands through shading, evapotranspiration, and air flow regulation (Zhou et al., 2011; Wang and Ren, 2025). Consequently, these works highlighted the importance of planning, implementing, and developing UGSs as a key strategy for alleviating urban heat (Farhadi et al., 2025). This approach is effective only if it takes into account climate considerations, the water needed for irrigation, and the available space for green space development; otherwise, urban managers may face significant challenges in implementing these plans(Coutts et al., 2013; Zhao et al., 2023). In response to these challenges, urban ecologists have proposed alternative strategies.

From a landscape ecology point of view, studies indicated that optimizing existing UGSs through interventions in their spatial patterns can enhance the cooling effects of urban areas. The suggested strategies include reducing heterogeneity, strengthening connectivity, building networks between existing green patches, and improving spatial patterns (Masoudi et al., 2019; Zhou et al., 2025). The improved spatial patterns of UGSs through applying shape complexity, cohesion, aggregation, size, and density directly and quantitatively influence the cooling performance at the city scale, with clustered, larger, and more cohesive/compact green patches delivering stronger heat reductions (Myint et al., 2015; Kowe et al., 2021).

Nevertheless, the cooling potential of UGSs is strongly influenced by the interplay between their spatial patterns and the broader urban morphological context. This context, referred to as city layout, encompasses buildings, streets, urban infrastructure, urban blocks, and land uses such as UGS. In terms of urban morphology, this complex context is characterized by urban form and the relationships between its parts. However, addressing urban form is often viewed as an abstract study that focuses solely on physical attributes, neglecting the relationships between components and the existing spatial logic (Kostof and Castillo, 1992; Karimi, 2023). Therefore, to gain a deeper understanding of the city's form and structure, space syntax theory introduces the concept of spatial configuration.

This theory posits that spatial configuration encompasses the network of relationships formed by interconnected urban open spaces. This network shapes the structure of the entire city and embodies the urban form (Hillier and Hanson, 1984; Hillier, 2007). While traditional morphology relies on studying the city's physical evidence, the concept of spatial configuration through quantitative analysis encompasses not only geometric features but also complex non-geometric aspects, processes (social, cultural, and economic), relationships between components, and structural patterns (Karimi, 2023). The following figure shows the concept of spatial configuration (Fig. 5.1).

In general, UGSs are planned based on the attributes of urban morphology, growing in harmony with its spatial configuration, and forming accordingly. While the concept

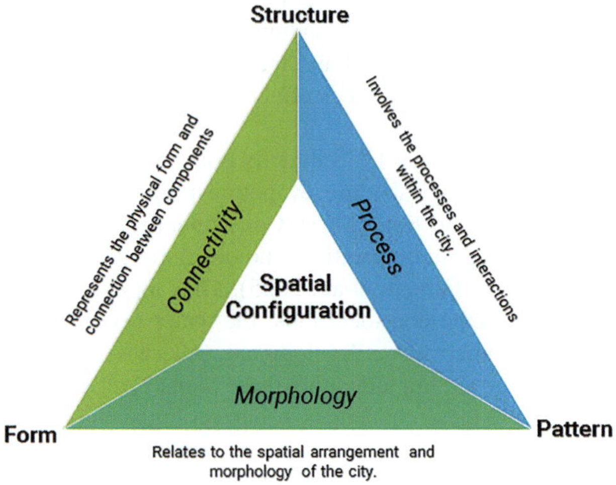

Fig. 5.1 The conceptual framework of the urban spatial configuration, terms, and components (Source: The authors).

of configuration has been extensively studied within landscape ecology, it is viewed as a multidimensional and complex idea in modern urban studies. Nevertheless, the influence of urban configuration on the spatial patterns of UGSs and the ecosystem services they provide, such as the cooling effect, remains a topic of debate. Therefore, adopting a configuration approach to analyzing this relationship reveals its underlying spatial logic, thereby leading to a deeper understanding of the spatial patterns of UGS.

In this study, the relationship between spatial patterns of UGS and urban spatial configuration was examined in the Tehran urban area, Iran. The current work aimed to investigate how urban spatial configuration interacts with UGS characteristics and their combined influence on the cooling effect within the urban environment. The study applied the space syntax approach and the landscape ecology knowledge to analyze urban spatial configuration and the spatial patterns of UGS, respectively. The results of our study provide new insights into the features of urban morphology and their impact on the cooling efficiency of UGS. Additionally, it provides an integrated approach to planning the cooling structure in urban areas, enabling urban planners and policymakers to optimize the cooling effect of UGS in these areas.

5.2 Background Knowledge

Although the concept of configuration has been utilized in various studies within landscape ecology, it has received less attention regarding urban morphology, particularly in assessing its impacts on the spatial patterns of green spaces and their ecosystem services. For instance, a growing body of literature has shown that landscape metrics, including patch size (CA and LPI), edge (TE), fractal dimension (MPFD), cohesion

(COHESION), aggregation index (AI), and mean shape index (MSI), significantly mediate cooling efficiency(Soltanifard et al., 2025b). In this regard, the assessment of COHESION values indicated that cohesive green patches have cooling ranges that are 1.2–1.5 times greater than those of fragmented configurations. Additionally, green patches with MSI > 1.4 enhanced the cooling intensity of urban parks by 18–22% due to the edge effect (Jiang et al., 2021; Gong et al., 2023; Li et al., 2023).

At a glance, the concept of spatial configuration in urban morphology studies has primarily focused on indicators such as building layout, sky view factor (SVF), land use mix, and anthropogenic heat. For instance, research has demonstrated that urban heat islands in areas with tall, dense buildings are typically warmer than in open areas of similar height due to a heat-trapping issue (Liang et al., 2020; Xu et al., 2024). However, some studies indicate that by managing the relationship between building height and coverage, urban planners can effectively enhance the SVF to mitigate UHI effects in densely populated areas(Yuan and Chen, 2011).

There are a few studies that applied the space syntax approach to quantify the spatial configuration for assessing UHI in urban areas. The relevant research found that integrated cores exhibit a more consistent cooling effect due to a negative correlation between the integration index, land surface temperature (LST), and UHI. These studies emphasize that enhancing street network configuration emerges as a more effective strategy for mitigating UHI effects in mid-rise buildings compared to other morphological factors (Soltanifard and Aliabadi, 2019; Nasehi et al., 2022; Wang et al., 2023; Güller and Toy, 2024).

Innovative works have recently integrated space syntax theory and landscape metrics to present a framework for optimizing UGS cooling. For instance, Wang et al. (2022) proposed a combination of morphological spatial pattern analysis (MSPA) and space syntax to optimize green infrastructure networks. The results showed that the cooling effect of UGSs was amplified by 1.8 times through an increase in syntax choice value to greater than 1.5 and the aggregation index (AI) (Wang et al., 2022). Additionally, Lou et al. (2024) explored a model for green network resilience in their work by integrating circuit theory with space syntax. In this work, the results indicate that NDVI values (> 0.6) and patch areas (> 50 ha) can be connected by a network with high-choice syntactic corridors (integration > 2.5) to enhance green resilience (Luo et al., 2024). In another study, the results indicated that the cooling efficiency in urban parks was 1.2–1.5°C higher compared to deeper spaces, where urban parks are located within 400m of high-intensity streets (Long et al., 2023).

5.3 Materials and Methods

5.3.1 Study Area

The capital of Iran is Tehran, situated on the southern slopes of the Alborz Mountains, with altitudes ranging from 1,050 m in the south to 1,800 m in the north (Fig. 5.2). Covering an area of 730 km², with a total population of 8.7 million in the city and

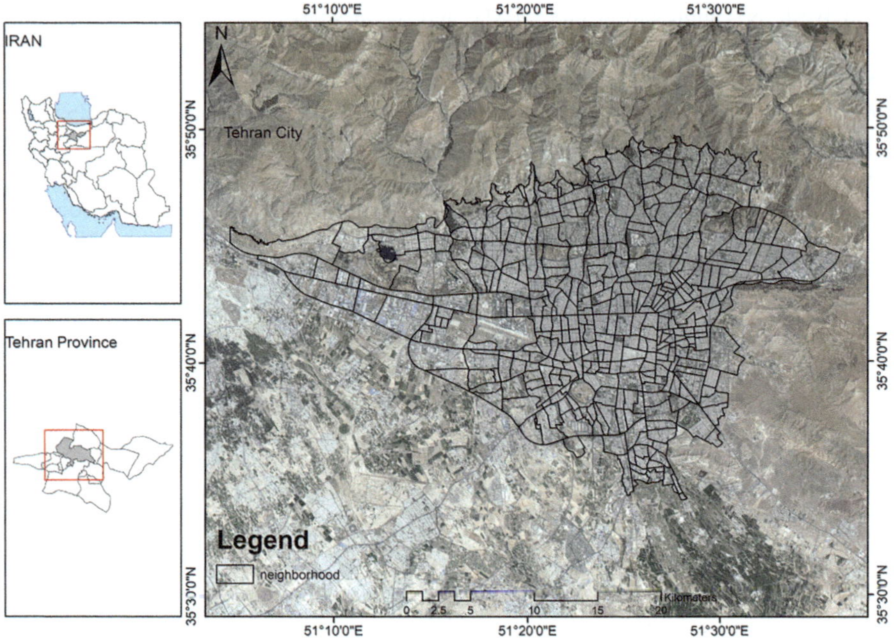

Fig. 5.2 Location of the study area.

over 15 million in the metropolitan area, Tehran is estimated to be the most densely populated and populous urban region in Iran (Zargari et al., 2025).

Administratively, Tehran is divided into 22 municipal districts and over 350 formally designated neighborhoods. Districts 4, 5, and 15 are among the most densely populated, reflecting patterns of residential expansion and socio-economic diversity. This city experienced rapid and largely unplanned expansion from 1985 to 2019, driven by a significant increase in population growth, which led to the replacement of vegetation with concrete. As a result, the built-up area grew by approximately 88%. These changes have primarily been rooted in economic centralization, rural-urban migration, and suburban sprawl since 1986 (Rahimzadeh and Jafarpour Ghalehteimouri, 2024). However, the transformation of the city has given rise to a wide range of environmental issues in recent decades. Related to the scope of this study, urban sprawl development has caused an uneven distribution of UGS, where the city's per capita is very low (~4.5m² per person), and many low-income districts fall well below this level. Conversely, northern and central neighborhoods have far higher green provision (Zargari et al., 2025). Furthermore, the replacement of natural vegetation with urban surfaces, dense high-rise blocks, and narrow streets has raised concerns about urban heat issues and the need for strategies to cool urban areas.

5.3.2 Data Acquisition

In our study, we need the Normalized Difference Vegetation Index (NDVI) to model the spatial patterns of UGS. To extract the NDVI value, the Landsat 8 satellite image in 2021 was used. The source of the imagery is the United States Geological Survey's (USGS) Earth Explorer online platform (https://earthobservatory.nasa.gov/). Moreover, the land use map of Tehran (2020) was utilized for identifying urban parks and their distributions. Additionally, this study employed axial line modeling to analyze the urban spatial configuration. The land use map was used to examine the road network of Tehran and draw the axial lines.

5.3.3 Method

The study of the relationship between the amount of UGS and relevant spatial patterns includes the following steps:

1. This index reflects the quantity, density, and health of the vegetation, which can have a direct effect on the cooling performance of green spaces. This index ranges from −1 and +1, where the higher the number, the denser (quantitatively) and healthier (qualitatively) vegetation will be, which is also a significant improvement of the capacity of UGS to lower land surface temperatures and mitigate the impact of the heat island (Qiu et al., 2017; Na et al., 2024).
2. Besides the vegetation index, the spatial patterns of the green spaces could significantly determine their total cooling characteristics. Landscape ecology describes these patterns as being commonly defined in two broad terms: spatial composition and spatial configuration. Spatial composition encompasses characteristics such as quantity, area, distance, and variety of green patches. The arrangement or configuration of these patches, concerning distribution, form, intricacy, connectivity, and edges, is, however, referred to as spatial configuration. In this study, the Largest Patch Index (LPI) and Mean Patch Size (MPS) were selected as compositional indicators to assess the cooling capacity of green spaces. LPI is the measure of the ratio between the largest green patch and a definite area, e.g., a neighborhood. The more dominant green spaces and the greater the LPI, the more likely it is to cool urban environments. Such impacts are primarily facilitated by mechanisms such as evapotranspiration, shading, and improved air movement through larger green spaces (Wang and Ren, 2025). MPS determines the mean size of single patches of green space within a specified study zone, which is an urban neighborhood. It can be utilized as an effective indicator for comparative evaluation of various units of study regarding their green space form and composition. Evapotranspiration, shading, and airflow are processes controlled by the MPS index, given the size

of a green patch, which thus affect cooling in green spaces. Consequently, the higher the value of this index, the greater and more widespread the vegetation cover, resulting in a stronger cooling effect in green spaces (Zhou et al., 2023).

Among the configuration criteria, the mean proximity index (MPI) was used to assess the degree of connectivity of green patches. This index, in addition to distance, also considers the size of the patches and provides a realistic view of how green patches are distributed or aggregated (McGarigal et al., 2002). Thus, it can demonstrate how the arrangement of green spaces influences ecosystem services, especially UGS cooling, and their effectiveness in mitigating heat. Generally, higher MPI values indicate that green patches are larger and more closely spaced, suggesting a well-connected green space network that typically enhances cooling benefits. Conversely, lower MPI values suggest isolated, small, and widely dispersed patches that are less effective in ecological and climatic performance (Feng and Myint, 2016; Soltanifard et al., 2025a).

The Mean Nearest Neighbor (MNN) index is commonly employed to quantify the average proximity between adjacent green patches, thereby reflecting their degree of spatial isolation or connectivity (McGarigal et al., 2002). Although the direct influence of MNN on urban cooling remains a subject of ongoing debate, elevated MNN values generally indicate that green patches are dispersed and isolated, which restricts their cooling effects to limited, localized zones. Conversely, lower MNN values imply that patches are situated closer together and exhibit higher connectivity, which can enhance ecological linkages and facilitate interactions that result in more pronounced and extensive cooling effects (Zhou et al., 2011).

Edge Density (ED) is another spatial metric utilized to assess the configuration of green spaces, such as those in Tehran. This index quantifies the total length of patch edges within a unit area (e.g., meters per hectare), serving as an indicator of the complexity, fragmentation, and spatial dispersion of green patches, factors that affect their interaction with surrounding environments. Typically, lower ED values correspond to larger, more contiguous, and simpler-shaped patches, which can increase shading potential and strengthen cooling capacity. In contrast, higher ED values correspond to smaller, more elongated patches, which often correlate with diminished cooling effectiveness (Rakoto et al., 2021).

Similarly, Total Edge (TE) measures the cumulative length of all green patch boundaries, representing the complexity of patch shape (McGarigal et al., 2002). Patches with more intricate or numerous edges tend to exhibit increased TE values. Functionally, longer edge lengths may promote greater heat exchange between vegetation and the atmosphere, contributing to environmental cooling. However, excessive increases in TE often signal heightened fragmentation, which can undermine the overall cooling efficiency of green spaces. Conversely, patches characterized by lower TE values and simpler morphologies tend to maintain more consistent cooling performance.

The MSI is employed to evaluate the average geometric complexity of green patches and their associated cooling effects. Higher MSI values denote more irregular and complex patch boundaries, whereas lower values suggest more compact and regularly shaped patches (Zhang et al., 2023). The utility of MSI for assessing landscape structure and cooling potential remains contentious. Generally, lower MSI values, indicative of simpler shapes with larger core areas, are associated with enhanced cooling stability and efficiency in larger patches. Nonetheless, the impact of MSI appears context-dependent and scale-sensitive, with its influence on cooling varying according to patch size and vegetation density (Masoudi and Tan, 2019). Table 5.1 lists the applied landscape metrics.

After selecting the metrics, the spatial patterns were quantified in Fragstate and represented through spatial analysis in ArcGIS. The spatial maps were prepared for each metric using the spatial analysis tool in ArcGIS.

To illustrate the spatial configuration of Tehran, axial line analysis was applied. The axial lines were prepared by drawing the axis of the existing street networks and analyzed in the UCL depth map. The results were imported into ArcGIS for further investigation. The indicators used in this study are: connectivity, global integration (Rn), and depth value. In space syntax theory, the connectivity index indicates the number of connections a space has with other urban areas within an interconnected urban system (Hillier, 2007). This index forms the foundation of an urban spatial configuration. Connectivity may have an indirect relationship with the intensity of the UHI effect, influenced by factors such as urban land use, activities, morphology, and land cover patterns(Soltanifard and Aliabadi, 2019; Güller and Toy, 2024). Additionally, integration and mean depth are two crucial spatial parameters. These indicate the effects of urban form and configuration on movement patterns and spatial hierarchy, which are associated with such environmental phenomena as UHI. In space syntax, integration refers to the degree of accessibility, or centrality, of an area within the totality of the urban space (Jiang and Claramunt, 2002; Hillier and Hanson, 1984; Hillier et al., 1993). Although further investigation is necessary to determine the impact of these indices on environmental heat, it appears that integration is a robust indicator of where activities and use are concentrated in urban cores, where environmental heat may be heightened. Conversely, the average depth index is inversely related to integration; thus, there is a reduction in the density of land uses and activities, which could result in a decrease in environmental heat (Soltanifard and Aliabadi, 2019; Soltanifard et al., 2022). In brief, the space syntax indices are presented in Table 5.2.

The relationship between variables was explored using Pearson's correlation. This section provides a deeper understanding of the relationships between all spatial configuration indicators and landscape metrics.

5.4 Results

5.4.1 UGSs' Spatial Patterns Analysis

The distribution of urban greenery in the Tehran urban area was depicted in this study by extracting the NDVI value (Fig. 5.3).

The findings indicate that the northern and northwest parts of the research area have an abundance of urban vegetation cover. Although fragmented patches of

Table 5.1 The landscape metrics utilized in this study.

Index	Description	Unit	Range	Equation
Normalized Difference Vegetation Index (NDVI)	It is a useful index for quantifying the health, change, and density of vegetation using sensor data.	None	$-1 \leq NDVI \leq +1$	$R_{NIR} - R_{RED} / R_{NIR} + R_{RED}$ NDVI uses the NIR and red channels in its formula.
The largest patch index (LPI)	It is calculated by the percentage of a landscape that is covered by the largest patch in the landscape.	%	1 < LPI ≤ 100	$\frac{\max(aij)}{A}(100)$ aij = area (m^2) of patch ij. A = total landscape area (m^2)
Mean patch size(MPS)	It computes the average size of the patches in a specific type	m^2	MPS > 0	$\frac{A}{n}$ MPS equals the total landscape area (m^2) divided by the total number of patches.
Mean proximity index (MPI)	MPI is a measure of the average distance of a green space to other green spaces in the landscape. This metric expresses the degree of isolation and fragmentation of a patch in the landscape.	None	MPI ≥ 0	$\sum_{i=1}^{n} \frac{a_i}{h_{ij}^2}$ MPI equals the sum of patch area (m2) divided by the nearest edge-to-edge distance squared (m2) between the patch and the focal patch of all patches of the corresponding patch type.
Mean nearest neighbor (MNN)	This metric measures the average distance to the nearest neighborhood patches to identify the degree of patch isolation.	m	MNN > 0	$\frac{\sum_{i=1}^{n} hij}{n}$ The mean nearest neighbor distance at the landscape level is the mean of the class nearest neighbor distances.
Edge density (ED)	ED equals the sum of the perimeters of all patches in the landscape.	Meter/hectare	ED ≥ 0	$\frac{\sum_{k1}^{m} e_{ik}}{A}(1000)$ eik = total length (m) of edge in landscape involving patch type (class) i; includes landscape boundary and background segments involving patch type i. A = total landscape area (m^2).

Table 5.1 (continued)

Index	Description	Unit	Range	Equation
Total edge (TE)	TE calculations include all edges on the landscape, including boundary edges.	m	TE ≥ 0	$\sum_{k=1}^{m} e_{ik}$ eik = total length (m) of edge in landscape involving patch type (class) i; includes landscape boundary and background segments involving patch type i.
Mean shape index (MSI)	It is measured by calculating the average perimeter-to-area ratio for a specific patch type (class) or for all patches in the landscape.	None	MSI ≥ 1	$\frac{\sum_{i=1}^{m}\sum_{j=1}^{n}\left(\frac{0.25\,pij}{\sqrt{aij}}\right)}{N}$ MSI is equal to 1 when all patches are circular (for polygons), and it increases with increasing irregularity of patch shape.

Sources: McGarigal and Turner, 2001; McGarigal et al., 2002; Havens et al., 2018.

Table 5.2 The space syntax indices.

Index	Description	Unit	Range	Equation
Connectivity	Measures the number of direct connections a node has to other nodes. This implies how well a space is connected to other spaces in the city.	None	C > 0	$C_I = K_I$ C_i shows the value of the connection for space i, and K is the number of directly connected spaces with i.
Mean Depth Value (MD)	Depth is a value to represent the distance of all the axial lines from an axial line, and the mean depth is equivalent to the average of these distances divided by the number of axial lines or nodes of the system.	None	MD > 0	$MD = \frac{L}{N-1}$ L is the distance and N is the number of the node or axial lines.
Global Integration (Integ_Rn)	It describes a value by quantifying the relative accessibility of a spatial element to all others within a defined radius or network. This value is used to measure spatial configuration.	None	Int - Rn > 0	$\frac{\mathbf{k}\left\{\log_2\left(\frac{\mathbf{k}+2}{3}\right)-1\right\}+1 / \mathbf{k}-1}{\mathbf{MD}-1}$ K is the total number of axial lines in an urban system, and MD represents the mean depth value.

Sources: Hillier and Hanson, 1984; Asami et al., 2003; Li et al., 2016; van Nes et al., 2021.

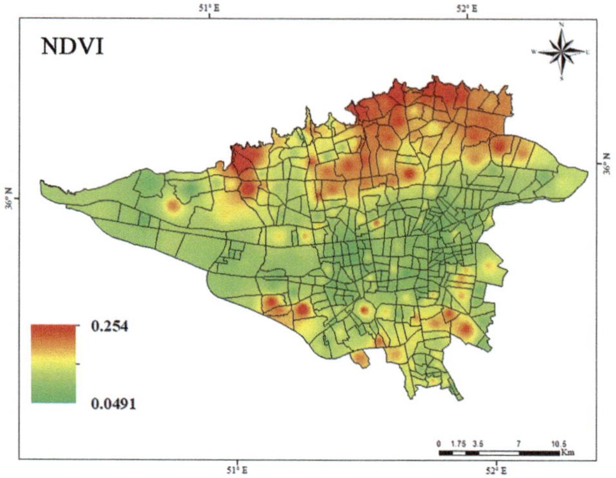

Fig. 5.3 The map of the NDVI index in the study area.

urban vegetation are also visible in southern Tehran, a high to very highly connected vegetation index has mainly concentrated in the north of Tehran. In contrast, the central and western parts of the city have recorded the lowest NDVI values. According to the obtained vegetation cover index, the values of the metrics mentioned in this study were calculated, and hence, maps of each metric were created based on the obtained values (Fig. 5.4).

The findings show that the largest green patches (LPI = 98.79) in Tehran are concentrated in the city's northern areas, indicating the UGS with the highest landscape connectivity. Additionally, this part of the city recorded the highest values for the average size of green patches (MPS = 12.98). In contrast, more fragmented and scattered patterns are observed among the smaller green patches in other areas, particularly in the southern and western parts of the city. Furthermore, the highest MPI values (115.74) are observed in the northern neighborhoods of Tehran. The accumulation of high values for this index indicates that the green spaces in these neighborhoods, which are larger and more closely spaced, have a strong ability to connect and form a cohesive green network. Through synergy, they can improve the ecosystem functions of green spaces in reducing heat. The assessment of the MNN metric indicates maximum values in the central parts of the study area. This issue reveals that the green patches are distant from their nearest neighbors, making them more isolated than those in the northern and, to some extent, southern parts of the city. This concentration of small and fragmented green patches in the city center has resulted in the highest recorded edge density values (ED = 1317.6) in these areas, while in some northern parts of Tehran, the total length of the edges shows significant values due to the larger size and extent of the green patches (TE = 1772.95). Although the spatial map of TE indicates that the spatial structure of existing UGSs is heterogeneous and discontinuous,

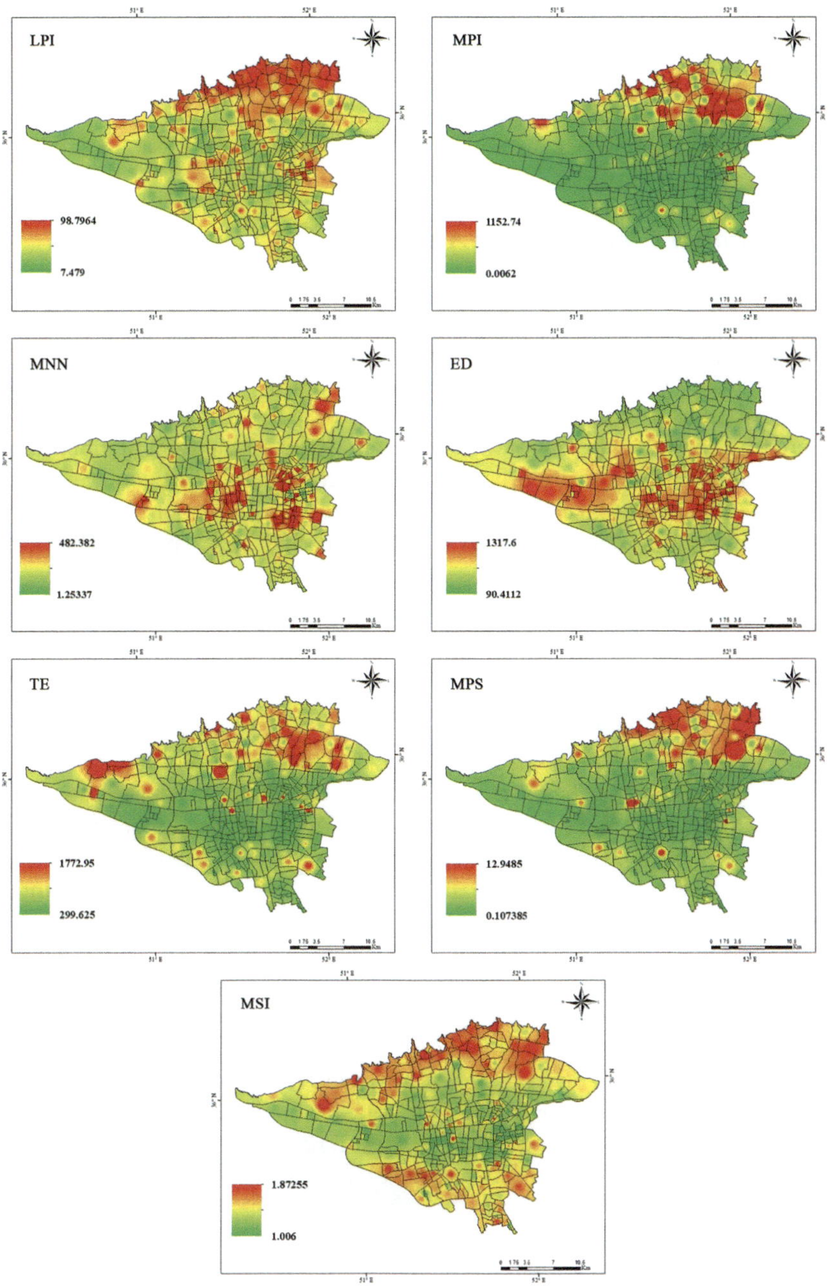

Fig. 5.4 The spatial analysis of the landscape metrics.

this part of the city holds significant value compared to other areas in the city. In terms of shape, the evaluation shows the highest values of MSI (1.87) in the northern parts of the city. Accordingly, high values of MSI signify more complex shapes of UGSs compared to other areas. While an increase in more regular shapes suggests greater cooling efficiency of UGS, this relationship in MSI alone cannot account for the rise in such ecosystem services.

5.4.2 Spatial Configuration Analysis

The findings of spatial configuration analyses of Tehran's street network, using mean depth, global integration, and connectivity indices, are displayed in Fig. 5.5. For each index, the blue spots indicate the fewest values, while the red spots display the greatest number of values.

Here, the connectivity value was calculated based on the total number of node connections within the entire urban network. Accordingly, the center and eastern parts of Tehran represent high values of connectivity, where urban streets have the strongest spatial connections to other streets and open spaces. This feature

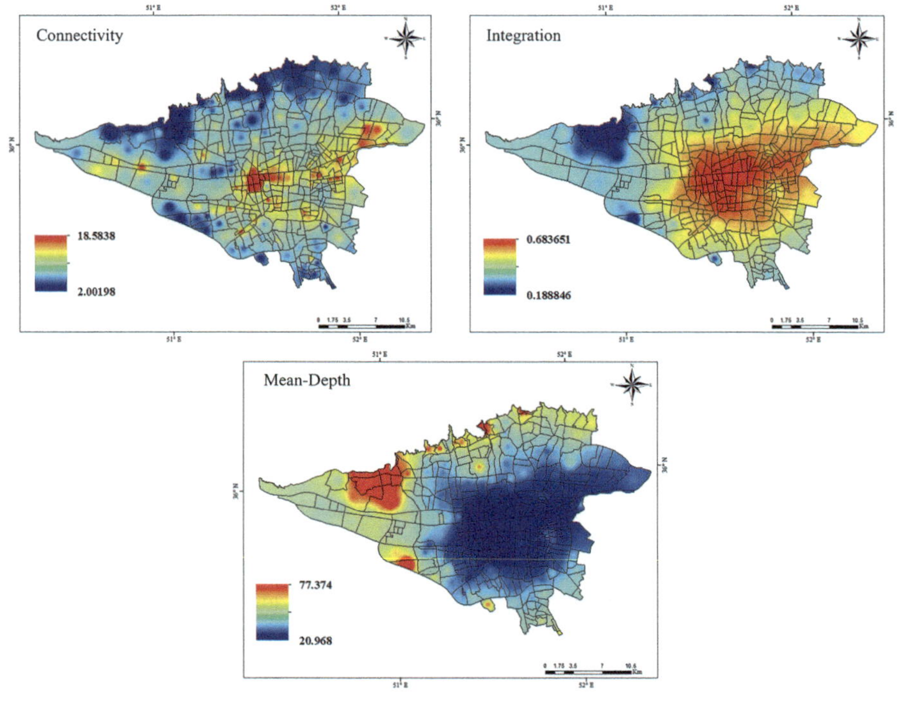

Fig. 5.5 Spatial configuration of the study area.

is spatially presented as fragmented red spots on the map. Similarly, global integration values reveal that these locations have built the highest integration values compared to other parts, as measured by the total number of direction changes from all streets to all others. In the case of Tehran, the most integrated streets are situated in the center (0.68), which is considered the densest and busiest part of the city and is mostly known for its concentration of commercial, administrative, and educational uses. In contrast, streets with the lowest connectivity and integration values have shaped more isolated spaces in Tehran. These points, primarily located in the north, northwest, and south of Tehran, exhibit a weaker structural connection with the city's main structure, indicating that they are situated at a greater spatial depth.

5.4.3 Correlation Analysis

In this study, Pearson correlation analysis was employed to investigate the associations between spatial configuration variables, NDVI, and the spatial patterns of UGS. The findings are presented in Table 5.3.

The overall findings demonstrate statistically significant correlations at the 0.01 significance level. Specifically, the relationship between the Normalized Difference Vegetation Index (NDVI) and spatial configuration indices exhibits relatively strong correlations, whereas the associations involving UGS spatial patterns are comparatively weaker. Structurally, the analysis identifies a negative correlation between NDVI and spatial configuration metrics related to connectivity and integration. This issue suggests that urban areas with higher connectivity and more integrated cores tend to have lower NDVI values. In other words, the urban spaces that are pivotal in shaping Tehran's spatial structure, by directing movement flows and organizing activities, generally feature reduced green space density. Conversely, locations that are more spatially segregated or peripheral display a greater concentration of urban vegetation.

Further examination of the interplay between the spatial patterns of UGS and spatial configuration indices reveals distinctive differences regarding the size, shape, proximity, distance, and dispersion of green patches. The results indicate that larger green patches, characterized by more regular geometric shapes and shorter distances to neighbors, are typically found in spatially segregated areas relative to the urban configuration. In contrast, areas with higher levels of spatial connectivity and integration often contain fragmented green patches with more irregular shapes and greater spatial dispersion. This trend is also reflected in edge density metrics, where highly connected and integrated urban zones exhibit increased edge density, thereby contributing to heightened spatial heterogeneity within green space structures. Moreover, a positive correlation is observed between the TE length and spatial depth, indicating a direct link between the extent of spatial segregation and the cumulative perimeter length of green patches.

Table 5.3 Pearson correlation coefficients between spatial configuration indices, NDVI, and spatial patterns.

Variable		NDVI	MPI	MNN	MPS	TE	ED	MSI	LPI
Connectivity	Pearson Correlation	−.550**	−.232**	.149**	−.349**	−.206**	.293**	−.364**	−.174**
	Sig. (2-tailed)	.000	.000	.004	.000	.000	.000	.000	.001
Integration	Pearson Correlation	−.517**	−.266**	.154**	−.359**	−.273**	.293**	−.391**	−.197**
	Sig. (2-tailed)	.000	.000	.003	.000	.000	.000	.000	.000
Depth Map	Pearson Correlation	.523**	.272**	-.153**	.363**	.279**	−.294**	.391**	.196**
	Sig. (2-tailed)	.000	.000	.003	.000	.000	.000	.000	.000

**. Correlation is significant at the 0.01 level (2-tailed).

5.5 Discussion and Implications

Cities represent complex interactions among geography, climate, physical factors, environment, economic conditions, social dynamics, and urban policy and management. Collectively, these factors shape the city's form and morphology, providing a distinct framework for other elements influencing movement, land-use distribution, and the urban landscape. However, researchers, planners, and urban designers have consistently required a common language to describe and effectively compare the characteristics of cities, enabling an accurate representation of these complex relationships. In recent decades, space syntax theory has sought to establish a common language through indicators such as connectivity, integration, and spatial depth. This language, known as spatial configuration, affects urban form and growth, shapes spatial structure, and influences both the movement of people and their perception of urban space (Hillier, 2007).

In landscape ecology, it is well established that the attributes of UGS, such as area, size, shape, complexity, and connectivity, significantly affect their cooling effects. For example, urban areas with greater green cover generally contribute more effectively to mitigating the urban heat island (UHI) effect. Larger and more contiguous green patches typically yield more substantial cooling benefits compared to smaller and fragmented ones (Masoudi and Tan, 2019; Sun et al., 2020). Additionally, among green spaces of comparable size and coverage, a higher Mean Proximity Index (MPI) indicates that green patches are more clustered and closer to each other, forming well-connected networks that typically provide more substantial cooling. This spatial clustering facilitates synergistic cooling through enhanced ecological connectivity. Conversely, lower MPI values indicate more dispersed and isolated green spaces, which limit the formation of effective green networks and reduce the overall cooling performance of UGS. Regarding shape, it has been observed that green patches with irregular or complex edges may exhibit different thermal performance compared to those with more regular and smooth edges (Li et al., 2023).

However, the cooling efficiency of UGS behaves differently with respect to Edge Density (ED) and TE. It has been found that green spaces with higher ED tend to have weaker cooling effects, primarily due to their fragmented structure disrupting heat exchange and air circulation. In contrast, an increase in TE, the total boundary length of green patches, enhances interaction with the surrounding environment, thereby improving heat exchange and natural ventilation, which consequently increases cooling capacity (Li and Zhou, 2019; Wang et al., 2025).

The empirical and theoretical findings presented in this chapter highlight the critical role of Tehran's spatial configuration in shaping the spatial patterns of urban green spaces and enhancing their cooling performance to mitigate adverse thermal conditions. The study emphasizes that the cooling potential of UGS depends not only on size, shape, edge, connectivity, and vegetation density but also on urban spatial configuration, particularly indicators of connectivity, spatial depth, and integration. The results underline a critical yet often overlooked aspect of urban climate adaptation: the spatial configuration of green spaces within the urban fabric. By integrating space syntax analysis with landscape ecological metrics, this research provides deeper insight into how the city's structural logic influences UGS cooling performance. Notably, the findings reveal a negative correlation between NDVI and both integration and connectivity indices, suggesting that more integrated and highly connected urban areas in Tehran, typically central commercial and administrative zones, exhibit lower vegetation density.

Furthermore, the quantified spatial patterns demonstrate that the expected cooling performance of UGS is significantly constrained by spatial configuration. Specifically, areas characterized by high connectivity and strong spatial integration gain fewer ecological benefits from green spaces, primarily due to their function as high-traffic zones with concentrated, high-demand land uses, which exacerbate UHI effects and air pollution amid high building density. These areas prioritize accessibility and movement flow but offer limited ecological functions, often resulting in fragmented and thermally inefficient green patches. In contrast, areas with higher mean spatial depth—that is, more spatially segregated zones—typically contain larger, more cohesive, and richer green patches exhibiting stronger cooling effects. Therefore, these areas have an urgent need for the ecological and climatic benefits provided by UGS.

One effective approach in urban planning and policymaking is to enhance the cooling performance of existing green spaces by intervening in the city's spatial configuration and improving the spatial patterns of UGS. Such breakthroughs can be achieved through strengthening green corridors, modifying patch shapes and vegetation types, and strategically managing their distribution to optimize the cooling effects of UGS within integrated urban environments.

5.5.1 Planning Cooling Structure for Transformative Growth

Planning for cooling structures in cities necessitates a re-evaluation of traditional views on urban sustainability and a shift toward newer concepts of urban sustainability and resilience. This approach requires achieving a holistic view of the current urban crises, encompassing socio-economic, infrastructure, unmanaged pollution, energy, water, and climate challenges (Næss and Vogel, 2012; Ernst et al., 2016). Although the realization of concepts such as urban sustainability and resilience requires constant change, complex interdependencies play a role, and there is little opportunity for learning through trial and error (Woolthuis et al., 2013). Each implemented solution requires significant investments and carries high risks; the diversity of challenges, scales, and levels of intervention has made the need for a codified intellectual and practical framework inevitable to achieve urban sustainability goals(UN-Habitat, 2022). In this regard, the concept of transformative growth offers an integrated framework for advancing urban resilience in ways that go beyond conventional sustainability models. Unlike modern approaches to urban ecology, which often focus on the development of

UGSs, transformative growth emphasizes resilience, ecological innovation, and citizen participation as key pillars of urban futures. By incorporating ecological efficiency, the concept redefines social well-being and innovation in design and technology, aligning resilience with contemporary debates about sustainable urban development.

In this chapter, the planning of cooling structures through UGS in Tehran illustrate how the concept of transformative growth can be operationalized. By aligning UGS planning with urban spatial configuration, expected ecological processes and functions in UGS can be canalized in optimizing microclimates regulation, UHIs mitigation, and enhancing ecosystem services. From this perspective, UGS operates as living, productive assets that grow in accordance with the city's structure, transforming and evolving with the city's changes. Socially, the planning of UGS through integrated spaces provides an ecological network contributing to healthier, more inclusive communities by improving thermal comfort and quality of life in vulnerable neighborhoods. Methodologically, the integration of space syntax and landscape metrics represents design innovation that enables a better balance between urban form and ecological performance. By integrating cooling structures with urban spatial configuration, the configurational approach transforms the nature of planning from a technical intervention to a transformative growth. Consequently, the planning framework will shape and develop an urban future based on strengthening ecological productivity by enhancing natural systems, fostering social productivity through equitable public spaces, and advancing technological productivity through innovative planning tools. Although Tehran is considered a case study in this study, the proposed framework can provide valuable lessons for other cities, especially in hot and dry regions. The results of this study can be essential for developing cooling strategies to support a climate-resilient urban future and increase resilience and transformation.

5.6 Concluding Remarks

The research findings reveal that the spatial patterns and distribution of UGS in Tehran are not aligned with the city's spatial configuration, indicating a contradiction between ecological needs and the city's spatial structure. This spatial misalignment between urban structure and ecological performance reveals a structural tension within Tehran's urban form: areas critical for movement and socio-economic activities underperform ecologically, while zones with better ecological functions remain spatially isolated and underutilized in terms of connectivity. Resolving this paradox is essential to enhancing the city's resilience to heat stress. To translate these insights into planning actions, the following strategies are recommended:

- Reintegrate Green Spaces into the Spatial Core
  Urban planners should prioritize the reintegration of UGS into highly integrated zones by creating small-scale, yet strategically located, pocket parks, green

plazas, and vegetated pedestrian corridors. This would enhance vegetation presence in the spatial core without disrupting its high accessibility function.

- Leverage Connectivity for Green Corridors
The space syntax connectivity index should be used to identify well-connected but under-vegetated routes. These can be transformed into green corridors or linear parks, which not only improve ecological connectivity but also promote active mobility and reduce heat exposure along transit paths.
- Optimize Depth and Edge Structures in Peripheral Zones
Areas with high spatial depth and larger green patches (as identified in the northern parts of Tehran) should be maintained and enhanced as thermal buffer zones. Urban planning should avoid over-integrating these areas to preserve their cooling function, while carefully managing their edge structure to improve microclimate regulation.
- Balance Spatial Accessibility with Ecological Function
A dual-index framework combining Integration (for human flow) and Edge Density/TE (for cooling function) should be adopted in park design. Parks in medium-integration zones should be designed to maximize both social access and environmental performance through compact shapes, diversified vegetation, and semi-permeable edges.

5.6.1 Recommendations for Optimization

The research findings highlight the importance of integrating two perspectives, the space syntax approach and landscape spatial patterns, to optimize cooling in existing and planned green spaces, a key objective of this study. Accordingly, recommendations are provided to support this effort. These recommendations encompass a broad range, from revising current comprehensive plans to adapt them for climate change mitigation, to establishing an interdisciplinary framework that incorporates spatial planning experts and landscape ecologists. Emphasized in this framework are ongoing monitoring of plans and programs within the current structure, as well as modifications to maximize cooling effects, combat heat emission, and address climate change. These recommendations:

- Spatial-Structural Alignment of UGS Networks
Developing a UGS master plan aligned with the city's spatial configuration is highly recommended. This enables the application of space syntax-based simulations and modeling to explore how green network designs (e.g., the amalgamation of smaller patches or the placement of connector greenways) can increase not only spatial integration but also ecological cooling.
- Incorporate Space Syntax in UGS Impact Assessments
The current guidelines for UGS evaluation tools, which serve as standards, should be revised to reflect indicators of spatial configuration (Rn, MD, and Connectivity) as well as ecological measures. With this more integrative

framework, it is possible to predict better and optimize the cooling capacity of planned or existing green spaces.

- Restructure Inner-City Urban Fabric
  In densely populated regions with fragmented vegetation, such as the central neighborhoods of Tehran, the process of urban redevelopment (e.g., construction of pedestrian routes and reconfiguration of plazas) should be accompanied by elements of green spaces, as determined by accessibility analysis based on space syntax. This procedure also means that such spaces will remain the functional center, as well as having an elevated environmental value.
- Dynamic Monitoring and Simulation
  It is suggested that real-life future-stimulated urban growth scenarios be implemented using space syntax modeling within a GIS setup, and rendering the effects on UGS spatial performance. It will enable adaptive planning for urban expansion and climate change.
- Institutionalize a Syntax-Ecology Design Practice
  It is highly recommended to promote interdisciplinary working of space syntax specialists, landscape ecologists, and urban designers. Urban planning Institutions in Tehran need to adopt a well-structured syntax-ecology pattern that can be employed to guide future developments and the regeneration of existing urban spaces.

References

Asami Y, Kubat AS, Kitagawa K, Iida S (2003) Introducing the third dimension on Space Syntax: Application on the historical Istanbul. In: Hanson J (ed) Proceedings of the 4th International Space Syntax Symposium. Space Syntax Laboratory, London, pp 48.1–48.8

Coutts AM, Tapper NJ, Beringer J, et al (2013) Watering our cities: The capacity for water sensitive urban design to support urban cooling and improve human thermal comfort in the Australian context. Prog in Phys Geogr: Earth and Geogr 37(1):2–28

Dimoudi A, Kantzioura A, Zoras S, et al (2013) Investigation of urban microclimate parameters in an urban center. Energy Build 64:1–9

Ernst L, de Graaf-Van Dinther RE, Peek G-J, Loorbach DA (2016) Sustainable urban transformation and sustainability transitions; conceptual framework and case study. J of Clean Prod 112(4):2988–2999

Farhadi R, Soltanifard H, Alizadeh B (2025) Enhancing climate resilience through urban landscape design: Strategies for green infrastructure and adaptation. In: He B, Wang Y, Cheshmehzangi A (eds) Urban climate and urban design. Springer, Singapore, pp 117–128

Feng X, Myint SW (2016) Exploring the effect of neighboring land cover pattern on land surface temperature of central building objects. Build and Environ 95:346–354. https://doi.org/10.1016/j.buildenv.2015.09.019

Ghribi M, Dahech S (2023) The impact of the Urban Heat Island on the sensation of thermal comfort and electricity consumption in Sfax in Central-Eastern Tunisia during the hot season. Energies 16(2), 911

Gong Y, Li X, Liu H, Li Y (2023) The spatial pattern and mechanism of thermal environment in urban blocks from the perspective of green space fractal. Build 13(3), 574

Güller C, Toy S (2024) The impacts of urban morphology on urban heat islands in housing areas: The case of Erzurum, Turkey. Sustain 16(2), 791

Havens KJ, Hershner C, Rudnicky T, et al (2018) Virginia Wetland Condition Assessment Tool (WetCAT): A model for management. In: Rheinhardt JL, Sohngen MCK (eds) Wetland and stream rapid assessments. Elsevier, Amsterdam, pp. 135–149

Hillier B (2007) Space is the machine: A configurational theory of architecture. Space syntax. UCL University Press, London

Hillier B, Hanson J (1984) The social logic of space. Cambridge University Press, New York

Hillier, B., Penn, A., Hanson, J., et al. (1993) Natural movement-or, configuration and attraction in urban pedestrian movement. Environ and Plan B: Urban Anal and City Sci 20(1):29–66

Jiang B, Claramunt C (2002) A comparison study on space syntax as a computer model of space. In: Proceedings of the 3rd International Space Syntax Symposium, Atlanta, pp 30.1–30.11

Jiang Y, Huang J, Shi T, Wang H (2021) Interaction of urban rivers and green space morphology to mitigate the urban heat island effect: Case-based comparative analysis. Int J Environ Res and Public Health 18(21), 11404

Karimi K (2023) The configurational structures of social spaces: Space syntax and urban morphology in the context of analytical, evidence-based design. Land 12(11), 2084

Kostof S, Castillo G (1992) The city assembled: The elements of urban form through history, First Nort. Little Brown and Company, Boston

Kowe P, Mutanga O, Odindi J, Dube T (2021) Effect of landscape pattern and spatial configuration of vegetation patches on urban warming and cooling in Harare metropolitan city, Zimbabwe. GISci & Remote Sens 58(2):261–280

Li D, Liao W, Rigden AJ, et al (2019) Urban heat island: Aerodynamics or imperviousness? Sci Adv 5(4), eaau4299

Li D, Yan X, Yu Y, et al (2016) The analysis of pingyao ancient town street spaces and view spots reachability by space syntax. J of Data Anal and Inf Process 4(4), 177

Li X, Zhou W (2019) Optimizing urban greenspace spatial pattern to mitigate urban heat island effects: Extending understanding from local to the city scale. Urban Forestry & Urban Green 41:255–263

Li Y, Ren C, Ho JY, Shi Y (2023) Landscape metrics in assessing how the configuration of urban green spaces affects their cooling effect: A systematic review of empirical studies. Landsc and Urban Plan 239, 104842

Liang Z, Wu S, Wang Y, et al (2020) The relationship between urban form and heat island intensity along the urban development gradients. Sci of the Total Environ 708, 135011

Long Y, Qin J, Wu Y, Wang K (2023) Analysis of Urban Park accessibility based on space syntax: take the urban area of Changsha City as an example. Land 12(5), 1061

Luo Y, Zhou Y, Li B, et al (2024) Constructing a future green space ecological network based on multi-scenario urban expansion: A case study of Chengdu, Sichuan, China. For 15:1818

Masoudi M, Tan, PY (2019) Multi-year comparison of the effects of spatial pattern of urban green spaces on urban land surface temperature. Lands and Urban Plan 184:44–58

Masoudi, M., Tan, P. Y., & Liew, S. C. (2019). Multi-city comparison of the relationships between spatial pattern and cooling effect of urban green spaces in four major Asian cities. Ecol Indic 98:200–213

McGarigal K, Cushman SA, Neel MC, Ene E (2002) FRAGSTATS: Spatial pattern analysis program for categorical maps. Computer Software Program University of Massachusetts. www.umass.edu/landeco/research/fragstats/ fragstats.html

McGarigal K, Turner MG (2001) Landscape Metrics for Categorical Map Patterns

Myint SW, Zheng B, Talen E, et al (2015) Does the spatial arrangement of urban landscape matter? Examples of urban warming and cooling in Phoenix and Las Vegas. Ecosyst Health and Sustain 1(4):1–15

Na N, Lou D, Xu D, et al (2024) Measuring the cooling effects of green cover on urban heat island effects using Landsat satellite imagery. Int J of Digit Earth 17(1), 2358867

Næss P, Vogel N (2012) Sustainable urban development and the multi-level transition perspective. Environ Innov and Soc Transit 4:36–50

Nasehi S, Yavari A, Salehi E, Emmanuel R (2022) Role of local climate zone and space syntax on land surface temperature (case study: Tehran). Urban Clim 45, 101245

Qiu GY, Zou Z, Li X, et al (2017) Experimental studies on the effects of green space and evapotranspiration on urban heat island in a subtropical megacity in China. Habitat Int 68:30–42

Rahimzadeh A, Jafarpour GK (2024) Identify successful performance of urban creativity indicators dimensions between 22 regions of Tehran Metropolitan: A comparative study. City and Built Environ 2(5)

Rakoto PY, Deilami K, Hurley J, et al (2021) Revisiting the cooling effects of urban greening: Planning implications of vegetation types and spatial configuration. Urban Forestry & Urban Green 64, 127266

Singh N, Singh S, Mall RK (2020) Urban ecology and human health: Implications of urban heat island, air pollution and climate change nexus. In: Douglas I, James P (eds) Urban ecology: Emerging patterns and social ecological systems. Elsevier, Amsterdam, pp 317–333

Soltanifard H, Aliabadi K (2019) Impact of urban spatial configuration on land surface temperature and urban heat islands: A case study of Mashhad, Iran. Theor and Appl Climatol 137:2889–2903

Soltanifard H, Kashki A, Karami M (2022) Analysis of spatially varying relationships between urban environment factors and land surface temperature in Mashhad city, Iran. The Egypt J of Remote Sens and Space Sci 25(4):987–999

Soltanifard H, Kashki A, Zandi R, Amani-Beni M (2025a) Investigating efficient urban planning strategies for reducing land surface temperature of the neighborhoods in Tehran, Iran. PLoS One 20(4), e0313417

Soltanifard H, Mehrfrooz O, LeBrasseur R (2025b) A configurational analysis of urban form and its influence on urban green space spatial patterns: A case study of Sabzevar, Iran. Environ, Dev and Sustain 1–29

Sun X, Tan X, Chen K, et al (2020) Quantifying landscape-metrics impacts on urban green-spaces and water-bodies cooling effect: The study of Nanjing, China. Urban Forestry & Urban Green 55, 126838

UN-Habitat (2022) World cities report 2022: Envisaging the future of cities. UN

United Nations (2018) World Urbanization Prospects: The 2018 Revision

United Nations Department for Economic and Social Affairs (2023) World population prospects 2022: Summary of results. UN. https://www.un.org/development/desa/pd/content/World-Population-Prospects-2022. Accessed on 30 July 2023

USGCRP W, DJ F, DW H, et al (2017) USGCRP, 2017: Climate Science Special Report: Fourth National Climate Assessment, Volume I (Washington, DC, USA: US Global Change Research Program)

van Nes A, Yamu C, van Nes A, Yamu C (2021) Analysing linear spatial relationships: The measures of connectivity, integration, and choice. In: Introduction to Space Syntax in Urban Studies. Springer, Cham, pp 35–86

Wang F, Chen J, Tong S, et al (2022) Construction and optimization of green infrastructure network based on space syntax: A case study of suining county, jiangsu province. Sustain 14(13), 7732

Wang L, Wang W, Tang F, Xu H (2025) Optimizing urban park cooling effects requires balancing morphological design and landscape structure. Sci Rep 15, 15435

Wang X, Yao X, Shao H, et al (2023) Land use quality assessment and exploration of the driving forces based on location: A case study in Luohe City, China. Land 12(1), 257

Wang Y, Ren Y (2025) Cooling efficiency of urban green spaces across functional zones: mitigating heat island effects through spatial configuration. Sustain 17(5), 2275

Wong GKL, Ma ATH, Cheung LTO, et al (2024) Visiting urban green space as a climate-change adaptation strategy: Exploring push factors in a push--pull framework. Clim Risk Manag 43, 100589

Woolthuis RK, Hooimeijer F, Bossink B, et al (2013). Institutional entrepreneurship in sustainable urban development: Dutch successes as inspiration for transformation. J of Clean Prod 50:91–100

Xu C, Wang W, Zhu H (2024) Spatial gradient differences in the cooling island effect and influencing factors of urban park green spaces in Beijing. Build 14(5), 1206

Yan L, Jia W, Zhao S (2021) The cooling effect of urban green spaces in metacities: A case study of Beijing, China's capital. Remote Sens 13(22), 4601

Yao L, Sun S, Song C, et al (2022) Recognizing surface urban heat 'island' effect and its urbanization association in terms of intensity, footprint, and capacity: A case study with multi-dimensional analysis in Northern China. J of Clean Prod 372, 133720

Yuan C, Chen L (2011) Mitigating urban heat island effects in high-density cities based on sky view factor and urban morphological understanding: a study of Hong Kong. Archit Sci Rev 54(4):305–315

Zargari M, Mofidi A, Entezari A, Baaghideh M (2025) Climatology of Tehran surface heat Island: A satellite-based spatial analysis. Sci Rep 15, 10567

Zhang Y, Wang Y, Ding N, Yang X (2023) Assessing the contributions of urban green space Indices and spatial structure in mitigating urban thermal environment. Remote Sens, 15(9), 2414

Zhao J, Meili N, Zhao X, Fatichi S (2023) Urban vegetation cooling potential during heatwaves depends on background climate. Environ Res Lett 18(1), 14035

Zhou W, Huang G, Cadenasso ML (2011) Does spatial configuration matter? Understanding the effects of land cover pattern on land surface temperature in urban landscapes. Lands and Urban Plann 102(1):54–63. https://doi.org/10.1016/j.landurbplan.2011.03.009

Zhou W, Yu W, Zhang Z, et al (2023) How can urban green spaces be planned to mitigate urban heat island effect under different climatic backgrounds? A threshold-based perspective. Sci of the Total Environ 890, 164422

Zhou W, Yu Y, Zhang S, et al (2025) Methods for quantifying the cooling effect of urban green spaces using remote sensing: A comparative study. Lands and Urban Plan 256, 105289

Chapter 6
Resilient Urban Planning to Enhance the Architectural, Urban and Landscape Identity of the City: The Example of Dunkirk

Vassilia Pirierros and Huan Zou

Abstract Faced with decline and pollution, the Port-City of Dunkirk responded at the end of the twentieth century to a very degraded collective imagination. In the century of sustainable development, the Urban Community of Dunkirk (CUD) has proved to be a forerunner. Urban policies aimed at improving the quality of life of inhabitants and increasing the city's attractiveness have been put in place, and the objective of transforming the city is set out in the Local Urban Plan (PLU), which deals today jointly with issues of urban planning, housing, transport and heritage.

After contextualizing the research regarding urban and environmental challenges and identifying the resources of the territory, spatial-configuration tools are adopted to model a *projected* image of the urban core considering the urban planning strategies undertaken. The impact of recent development projects is examined in terms of (i) the creation or reinforcement of urban identities, (ii) mobility and sociability hubs, and (iii) landscape and water. Finally, an evaluation of the influence of the projects on the representations of different districts will enable us to identify their shortcomings and merits.

The choices made in Dunkirk to cope with the disruptions of the twentieth century, and a fragile environment define the basis of resilient urban planning that mitigates past, present and future shocks and tensions, and whose assets are not just ecological. Projects tend to compensate for urban fractures and enhance architectural, urban and landscape heritage. The promotion of an urban narrative aims to overcome the stigma and transition towards a post-industrial coastal city to encourage new practices and mental representations. The city of Dunkirk plays a central role in the cohesion and adaptation of a Port-Beach-City territory. These three components coexist and interact in the reclaiming of the waterfront.

V. Pirierros (✉)
School of Architecture, Tianjin University, Tianjin, China
e-mail: vassilia_pirierros@tju.edu.cn

H. Zou
School of Architecture, Tsinghua University, Beijing, China

A. Cheshmehzangi et al. (eds.), *Transforming Cities*, Urban Sustainability,
https://doi.org/10.1007/978-981-95-7378-3_6

Keywords Dunkirk (France) • Urban renewal • Landscape • Urban and architectural heritage • Resilient urban planning • Image of the city

6.1 Introduction

Dunkirk is known nationally as France's third-largest port, but regionally as the most beautiful beach in the North. In between, the city is attempting to reclaim its seafront. As a land reclaimed from the North Sea during the Middle Ages and located at the crossroads between England and the Netherlands, the site is strategic for the State: a stronghold in the seventeenth century, a commercial port in the nineteenth century, post-war reconstruction, an industrial port in the twentieth century. The city experienced environmental, economic and social changes, including rapid demographic and urban growth followed by a steady decline. In response, the Dunkirk Urban Community (CUD) asserted its local power and specificity by being a pioneer in urban planning and sustainable development.

Research has already been carried out, proving that Port Cities are resilient. Regardless of the form of their governance, they have the capacity to persist in their functions and to face future challenges (Hein and Schubert, 2021). Furthermore, the urban identity and in particular the *image projected* by the city is a representation at the service of a public urban development strategy (Faye and Lacour, 2020). In this respect, the CUD is implementing proactive urban policies and is determined to overcome the *inherited* image of the declining Factory Town. Its "urban planner mayor"[1] has stated his ambition to improve the quality of life and attractiveness of the conurbation and thus influence the *perceived* image.

The city of the twenty-first century is characterized by its response to the challenge of pollution, its adaptation to climate change and the increase in the presence of nature in the city. In addition to these ecological considerations, this study further examines the transformation of the city's image through the resorption of urban fractures, the reconversion of industrial wastelands and the reclamation of the seafront. What urban planning tools are local authorities using to rehabilitate the urban core and what urban projects are being implemented? What influence do they have on urban morphology and what is their impact in terms of mental representations? First, the context of Dunkirk, its challenges and its resources are explored, and then the local policy framework, the stakes of resilience and the objective of enhancing the image of the city are studied. The analysis of projects carried out in the city centre over the past 25 years is broken down into three parts: (i) districts and icons, (ii) mobility and polarities, and (iii) landscape and water, and their shortcomings and merits are highlighted. Finally, a resilient urban planning model in the twenty-first century is deduced from the example of Dunkirk.

[1] Patrice Vergriete is a graduate of the Ecole Polytechnique and the Ponts et Chaussées engineering school. From 2000 to 2008, he headed the Dunkirk Urban Planning Agency (AGUR) and in 2013 submitted a doctoral thesis on housing policies. He was elected Mayor of Dunkirk in 2014 and became Chairman of the Dunkirk Urban Community (CUD).

This study can serve as a reference for towns in decline, factory towns, port cities or, more broadly, stigmatized communities wishing to rethink their territory and promote new prospects. A political project and resilient urban planning make it possible to compensate for fragility and project a *rehabilitated* image of the city. It operates at architectural, urban and landscape levels through the reinterpretation of the built identity, the implementation of new modes of transport, the regeneration of convivial spaces and the emphasizing of the natural environment.

6.2 The Dunkirk Context, Challenges, and Resources

6.2.1 A Strategic Territory

At the crossroads between England and the Netherlands, the port of Dunkirk, like the ports of Hamburg, Rotterdam, and Antwerp, lies on the North Sea shipping route. Nevertheless, the geographical position of the city is unusual since it is not located at the estuary of a large river but directly on the seafront and the urban area extends along the coast.

The Dunkirk maritime plain stretches from Calais to West Flanders in Belgium with an average width of 12 km and a maximum of 20 km inland. Polderization in the Aa estuary began in the Middle Ages (Desoutter, 2023) and became systematic between the nineth and twelfth centuries. The purpose of draining coastal marshes was to create arable land through a process of successive dyking and desalination along small watercourses. In the twelfth century, wateringues, a network of dikes, siphons, pumps and canals, were built to control sea intrusion and irrigate the land with fresh water. The result is this flat territory, most of which is less than 5 m above sea level and known as Blootland (bare land in Dutch).

In the seventeenth century, Dunkirk was a strategic and coveted location. Dunkirk became definitively French in 1662, and Louis XIV turned it into the largest military port in the kingdom and a powerful trading city. Vauban was entrusted with charge of the military protection of the city: he reinforced the existing citadel built by the English, designed new fortifications and created a naval arsenal for the construction and maintenance of warships. Vauban also took charge of the planning of the city: he designed a simple and functional urban grid around a royal square and a lower town outside the first enclosure.

6.2.2 The Disruptions of the Nineteenth and Twentieth Centuries

The city underwent a metamorphosis in the second half of the nineteenth century, with the industrial revolution and the arrival of the railway in 1848. To the east, the seaside resort of Malo-les-Bains was established on dunes which were unsuitable for cultivation. With sea bathing in vogue, its location on the seafront was an attractive one. Its casino and holiday villas made it a prestigious resort during the Belle Époque. At the same time, to the north of the town, on the western foreshore along the channel leading to the open sea, a commercial port was developed as part of the Freycinet plan to modernize national infrastructure. An initial circular dock, built in the 1830s, was replaced in 1880s by four parallel docks that can accommodate ships and enable the efficient transport of goods. A lock was commissioned to control the water level.

In the twentieth century, the city experienced profound changes. At the end of the Second World War, 90% of the city was devastated and destroyed. The fortifications, whose demolition had begun in the 1930s, were definitively erased from the cityscape generating areas of fringes and fractures as highlighted by the mapping (Fig. 6.1). In the 1950s, under the direction of the Ministry of Reconstruction and Urban Planning (MRU), the city was rebuilt and acquired a modern image. An important place was given to the automobile, and the tramway was dismantled in 1952.

In 1956, the French government decided to establish a waterfront steelwork in Dunkirk with its "feet in the water." The industrial port extends 17 km west along the coastline with the digging of a maritime basin reclaimed from the sea and protected by the Braek dike. The Autonomous Port of Dunkirk (PAD) is now the third-largest port in France. Demographic growth followed industrial development. The rapid transformations of the urban area were reminiscent of the urban-industrial areas of mining or steel cities, and the city became something of a Factory Town

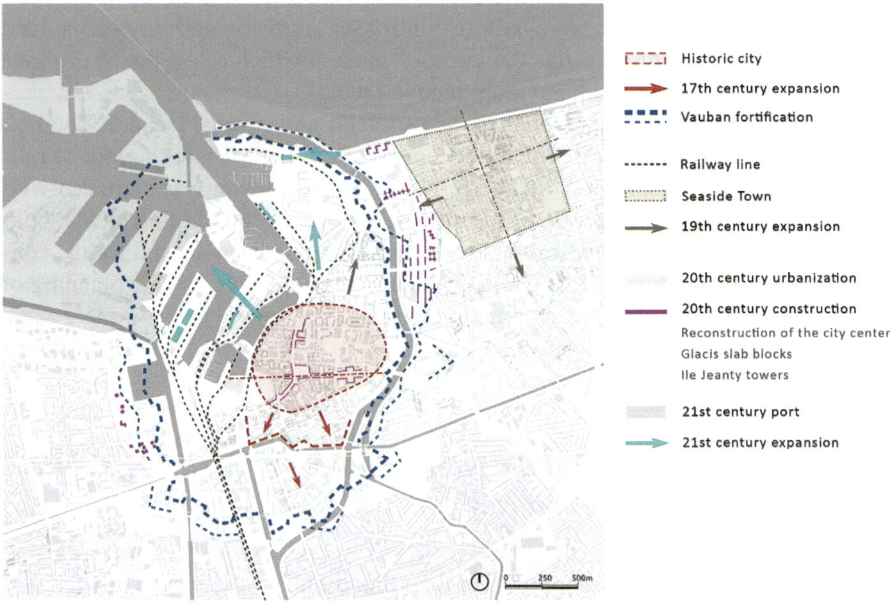

Fig. 6.1 Spatial and temporal development of the urban core (Source: Map drawn by the authors).

(Beaurain and Chembessi, 2018). However, the 1973 oil crisis had a severe impact on industrial dynamics and local growth. The decline of industry was followed by a socio-economic and urban decline in the 80s and 90s, culminating in the closure of the shipyards in 1989.

6.2.3 A Threatened and Threatening Environment

From the Netherlands to Calais, the maritime plain is submersible (Climate Central, n.d.). The CUD has taken on the responsibility of managing water resources and preventing flooding (GEMAPI)[2] and created in 2016, together with six other intermunicipal authorities in the Aa Delta, an Intermunicipal Institution of Wateringues. Two Coastal Risk Prevention Plans (PPRL)[3] regulate urban development in the most exposed areas. Moreover, the Autonomous Port of Dunkirk (PAD) is home to a nuclear power plant and 14 major risk establishments (classified as Seveso[4]). Five Technological Risk Prevention Plans (PPRT)[5] have been drawn up to protect populations from accidents. The oil industry's past also poses decontamination problems and uncertainties weigh on the conversion of the sites (Hauser, 2019). In terms of air quality, local industry is responsible for emissions of particulates and chemical pollutants, including nitrogen oxides (NOx) and sulphur dioxide (SO2). Finally, the Port of Dunkirk is responsible for 20% of French industrial CO_2 emissions, and a "DKarbonation" program was launched in 2023 to initiate a net zero emissions transition for the industrial cluster.

6.3 The Stakes of Resilience

The destructions during the Second World War, the closure of the shipyards, the demographic decline and the fractures in the territory are all historical traumas and structural tensions inherited from the past. In addition, industrial presence, pollution and climate change are significant risk factors for the population. Lastly, a territorial diagnosis carried out by the CUD in 2022, noted a fragile social and economic

[2] The 2014 law on the Modernisation of Territorial Public Action and the Assertion of Metropolises (MAPTAM) entrusts Management of Aquatic Ecosystems and Flood Prevention (GEMAPI) to local authorities to take over from the départements.

[3] Article L562-1 of the French Environment Code provides for Natural Risk Prevention Plans (PPRN), and within this framework two Coastal Risk Prevention Plans (PPRL) were prescribed in the Nord region in 2011, one of which covers the communes of Dunkerque and Bray-Dunes.

[4] The Seveso Directive is a common European policy on the prevention of major industrial hazards, introduced following the accidental release of dioxin in 1976 in the municipality of Seveso in Italy.

[5] In accordance with the French Risks Act of 2003, Technological Risk Prevention Plans (PPRT) are set up by the government in association with local authorities and in consultation with local residents, to control urban development in the vicinity of high-risk industrial facilities.

context and a worrying state of health of the inhabitants. These vulnerabilities lead to early awareness and the implementation of appropriate planning, and the term "resilient urban planning" is proposed to investigate the case of Dunkirk.

6.3.1 A Political Ambition

The city of Dunkirk is a pioneer: in 1968, the elected representatives of the municipalities decided to join forces within the Dunkirk Urban Community[6] (CUD) and to pursue a concerted policy. In 1972, the community set up its own urban planning agency, the Urban Planning Agency of the Dunkirk Region (AGUR), to implement a master plan. From then on, the city became aware of the need to respond to environmental and air quality issues. In 1996, the city signed the Aalborg Charter and received a European Sustainable Cities Award. It launched its Local Agenda 21 in 2002, well before Grenelle law I[7] and was awarded the Sustainable Development Ribbon in 2005. Finally, in 2009, the CUD joined the WHO's French Healthy Cities Network.

In accordance with the French urban planning code[8] and the national sustainable development strategy initiated in the 2000s, the CUD has successively implemented several Local Urban Plans (PLU), the latest of which was approved in 2024. This plan is part of an intercommunal approach and takes the place of the Housing and Transport programs to become PLUiHD (CUD, 2024). The Sustainable Development and Planning Project (PADD) is the policy document that defines the strategic vision of the territory (social, environmental, economic and cultural). The CUD promotes health-promoting urban planning[9] and "an attractive agglomeration where life is good."

[6] The system of urban communities was created in 1966 and the first four urban communities (Bordeaux, Lille, Lyon and Strasbourg) were imposed by the State. Dunkirk broke new ground in 1968, when a local initiative created the Dunkirk Urban Community (CUD). Today, it brings together 17 communes and 192,000 inhabitants (2021) over an area of 300 km².

[7] Through the Grenelle I and II environmental laws of 2009 and 2010, the French government is promoting the widespread use of greenhouse gas emission assessments and Territorial Climate Air Energy Plans (PCAET).

[8] In France, the Urban Planning Code dates to 1954 and was substantially amended by the law on Solidarity and Urban Renewal (SRU) of 13 December 2000. The aim of this law is to ensure greater consistency between town planning policies, housing policies and transport policies, with a view to promoting social diversity and sustainable development. It introduces, in place of the previous measures, the Territorial Coherence Schemes (SCoT), the Sustainable Development and Planning Project (PADD) and the Local Urban Planning Plan (PLU).

[9] Health is defined in the 1946 preamble to the Constitution of the World Health Organisation (WHO) as "a state of complete physical, mental and social well-being and not merely the absence of disease or infirmity". Health depends on individual factors (age, gender, behaviour, etc.), environmental factors (quality of air, water, noise, etc.), social and economic factors (access to housing, employment, education, etc.) and political factors (urban areas, housing, transport, etc.).

The urban planning code in France requires several assessments of the PLU: on the one hand, an analysis of the results after six years, focusing on sustainable development objectives or land use, on the other hand, environmental assessments after nine years at the latest. Evaluations of the Housing and Transport programs are also required by law. This research further introduces a spatial evaluation of public urban planning projects and the result on the representations and attractiveness of the city.

6.3.2 The Inherited Image of The City

The study looks at the morphological and architectural form and focuses on the urban core, bounded from west to east, between the railway station and Malo les Bains and, from north to south, between the seafront and the Furnes canal. To model an image of the city, the configuration tools established by Kevin Lynch (edge, district, landmark, path and node) (Lynch, 1969) are adopted. However, the nodes element is replaced with a sociability hub. The conceptual framework proposed by Benoit Faye and Claude Lacour to define the city's identity enables us to distinguish the *inherited* image (representation of an effective historical trajectory), the *projected* image (representation in the service of a public urban development strategy) and the *perceived* image (individual and social representation of the lived space).

At the turn of the twenty-first century (Fig. 6.2), the city is cut off from the seafront by the port to the north at dock 3, and from the western part of the city by the railway line. The districts of the city centre and of Malo les Bains have an identifiable urban coherence. The urban skyline is marked by the Reuze tower built in the 1970s, and three historical monuments: the Church of Saint Eloi, the belfry and the town hall. In the seaside resort, the Notre Dame Church on Place Turenne in the centre of the district and the first-aid station on the beach are landmarks. The historic Place Royale, where the statue of Jean Bart stands, is the main meeting place, followed by the train station and the Place du Kursaal on the site of the former casino. The roads follow the local logic of the urban networks, and connections between districts depend on crossing the waterways. This urban fragmentation, the curvature of the routes and the complexity of the crossroads do not facilitate the imageability of the city.

Now that this first *inherited* image has been mapped, the urban planning projects launched by the CUD and the urban planning tools deployed in the twenty-first century are examined. In the following sections, the evolution of the city of Dunkirk will be broken down under the prisms of (i) the creation or reinforcement of urban identities, (ii) mobility and sociability hubs, and (iii) landscape and water. The impact of the projects on the configurations of six sectors (City Centre, Seaside Town, Docks, Grand Large, Glacis and Gare/Ile Jeanty) are studied to identify their shortcomings and merits. Finally, a *projected* image of the city is modeled.

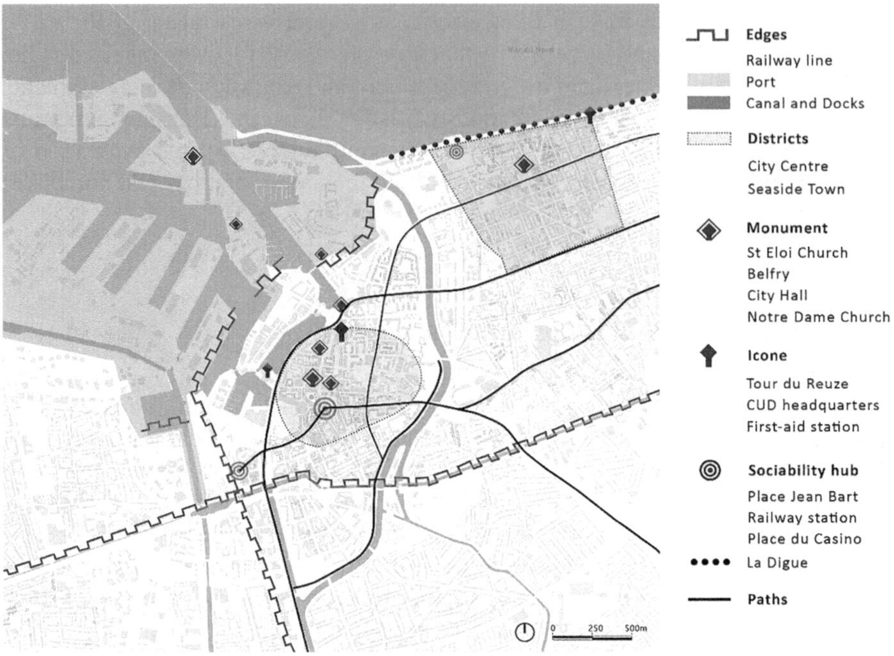

Fig. 6.2 Inherited image of the urban core at the dawn of the twenty-first century (Source: Map drawn by the authors).

6.4 Districts and Icons

6.4.1 The Urban Coherence and Historical Landmarks

In Dunkirk, the belfry of the church of St Eloi hosts the first tourist office, encouraging visitors to begin their discovery of the town with a panoramic view towards the sea. (Fig. 6.3). A second tourist office was opened in the Kursaal on La digue de Malo-les-bains. This choice shows the visitors' interest and demonstrates the strong visibility of these two districts. In the collective imagination and in physical reality these two districts have a clear and coherent structure.

The seaside town of Malo-les-bains, created at the end of the nineteenth century, is organized along the beach and around the Notre Dame church, which overlooks Place Turenne and its kiosk. From this centre, two main roads running East-West and North-South have structured the expansion of the town and formed geometric blocks of houses. The seafront is built in rows of Belle Epoque villas, with the Casino at the far end. The architectural typomorphology characteristic of the district is the villa malouine, a brick townhouse on the street, with a half-basement and two to three floors. Families were distinguished by the ornamentation on their façades. The destruction of World War II did not damage this urban fabric and houses of the same type were rebuilt, confirming the urban identity of the seaside resort (Fig. 6.4).

Fig. 6.3 View from the Belfry towards the sea (Source: Photo taken by the authors).

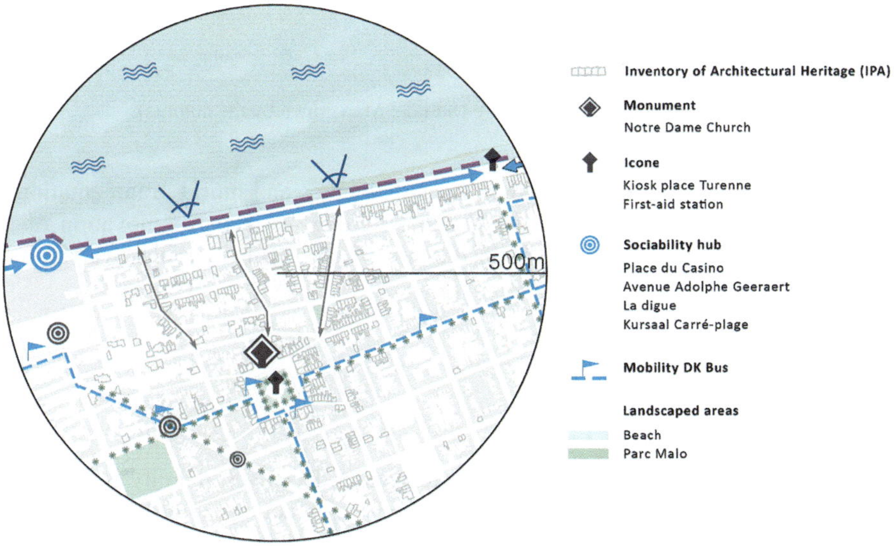

Fig. 6.4 Projected image of the seaside town (Source: Map drawn by the authors).

On the other hand, bold choices were made for the reconstruction of the city centre. Théodore Leveau's urban plan was adopted in 1948. The urban planner maintained most of the roads and decided that the three surviving "towers" would continue to overlook the town. However, he was not into regionalism and followed the national ideal of modernity. He opted for half the previous density, with large green spaces, sunny housing and block architecture (closed blocks, open blocks and oriented blocks) as shown on the master plan (Fig. 6.5). Chief architect Jean Niermans coordinated the design of the blocks and ensured that the architectural language was both coherent and modern, specifically using brick, continuous ground

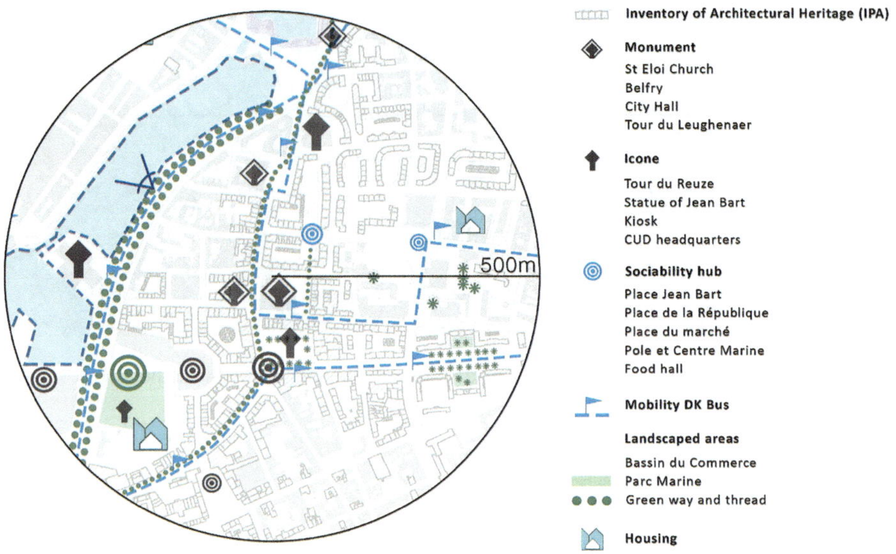

Fig. 6.5 Projected image of the city centre (Source: Map drawn by the authors).

floors, flat roofs, concrete window frames, balconies and mural ornamentation. A renovation of the façades combined with an energy-efficiency renovation of the housing was launched in 2018 with the aim of enhancing this post-war heritage (Cerema, 2020).

6.4.2 New Districts

The railway lines define the town's western edge. However, the city is expanding towards the port on the north side. During the reconstruction period, the citadel was first converted. Following the closure of the shipyards in 1989, the Neptune project[10] aimed to reclaim the industrial-port brownfield. Two Concerted Development Zones[11] (ZAC), the ZAC of the Docks and the ZAC of the Grand Large. In the 1990s, the construction of the flagship Université Côte d'Opale project and the Pôle Marine shopping centre completed the Bassin de la Marine. In a second phase, the city will take over Mole I, where the Sugar Hall is located, then the embankment

[10] A joint Dunkerque-Neptune syndicate has been set up by the CUD, the city and the PAD to consider the future of the 180 hectares of docks, quaysides and industrial buildings.

[11] Zones d'Aménagement Concertées (ZAC) were set up in 1985 to provide a legal, financial and technical framework to facilitate development projects by public authorities in conjunction with private developers.

up to Radoub dry dock n°5, part of Mole II with the construction of a Euraénergie innovation park and the Ile Jeanty canal (Husser and Raison, 2015).

The regeneration of the Grand Large wasteland was launched in 2005, with the iconic building of FRAC (Regional Contemporary Art Fund), the Gable buildings and the Guy Debeyre high school. The project makes a radical break with the industrial past and is creating a mixed housing neighborhood around a central park. The city is now continuing its appropriation with plans for a marina and a new Museum of Fine Arts, which the mayor has announced will be "a strong architectural gesture in this rather magical part of the Dunkirk territory" (Vergriete, 2024) but questions remain about the appropriateness of building denser developments along the waterfront and the need for a new symbolic element in this sector.

For its part, the Glacis district is located on an vacant site between the historic city and the seaside town, at the foot of the old Vauban fortifications, which were levelled after the war. This segment of the city was initially occupied by barracks for the war victims and reconstruction workers. In the 1960s, social housing was built in this area: a repetition of slab blocks, inspired by Théodore Leveau's plan, is organized around Avenue de la Libération. In 2022, an EcoQuartier initiative[12] was launched to promote a natural city model (AGUR, 2024). Nevertheless, it is limited to opening up towards the canal and a connection to Ziegler Park and does not propose any structural modification of the lawns between the towers (Fig. 6.6).

Lastly, the Ile Jeanty district is the subject of a regeneration project as part of the New National Urban Renewal Programme[13] (NPNRU). The «Housing» Programme of Orientation and Action[14] (POA) stipulates that the NPNRU is a lever for rebalancing the urban area by tackling the neighborhoods where urban and social problems are concentrated. To break the isolation and connect the sector to the city centre, a bridge and a pedestrian walkway are planned over the canal and the railway tracks. Some apartment towers will be demolished, and new individual and collective housing will increase the density of the area along the Bassin de la Batellerie. These developments will improve the comfort of residents. However, charismatic architecture would be needed here to enhance the neighborhood's identity, give it a singularity and ensure its long-term integration.

[12] The EcoQuartier initiative is a national label launched in 2012 and now structured in four stages (planned, under construction, delivered, experienced and confirmed). It enables local authorities to benefit from government support in meeting the challenges of sustainable cities (sobriety, inclusion, value creation, resilience).

[13] The New National Urban Renewal Programme (NPNRU) 2014–2024 continues the objectives of the National Urban Renewal Programme (PNRU) launched in 2003, and provides for the in-depth transformation of more than 450 priority urban policy neighborhoods, with a strong focus on housing and public facilities, to promote diversity in these areas.

[14] The 2014 Law on Access to Housing and Renewed Urbanism (ALUR) establishes that PLU can take the place of housing or transport and transport policies. A Program of Orientation and Action (POA) includes all the measures or information needed to implement them.

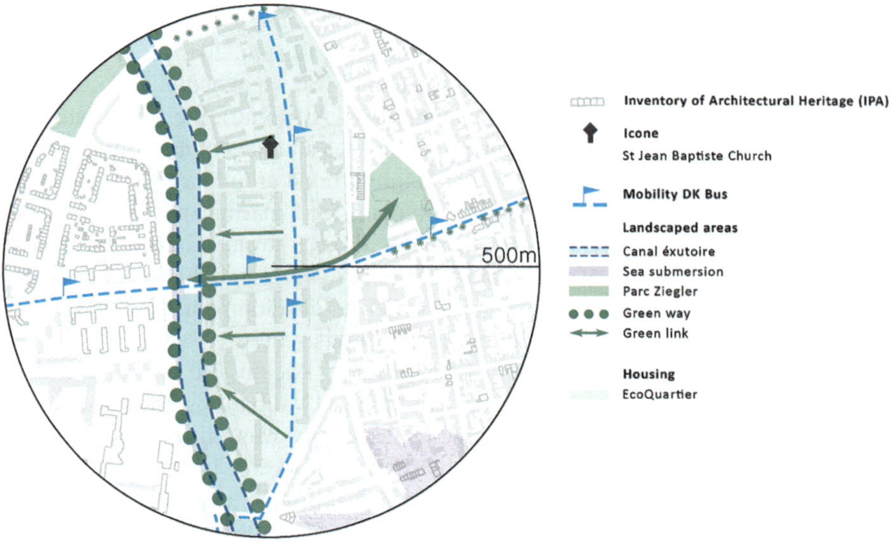

Fig. 6.6 Projected image of Les Glacis (Source: Map drawn by the authors).

6.4.3 Urban Identities

Urban identity is the subject of a dichotomy between two types: *emblem or stigma* (Galland, 1993) and four sectors studied respond to this classification. Working-class neighborhoods are socially stigmatized. However, the Glacis neighborhood is in the process of urban integration. The challenge of the Ile Jeanty project will be to provide a morphological substance that will positively influence social practices and mental representations. By contrast, the historic districts benefit from positive recognition as *an emblem*. They are particularly promoted by the city, which is reclaiming its past, and particularly the architecture of the reconstruction. Thus, guides, flyers and architectural discovery tours are offered for a reinterpretation of local specificities (Jebrak, 2010). Within the PLUiHD, a «Heritage » Orientation for Planning and Programming (OAP)[15] sets out a strategy to reveal and preserve a Heritage of Local Interest[16] listed in the

[15] The Orientation for Planning and Programming (OAP) are operational urban planning measures that appeared in France in 2010 with the Grenelle II law and are codified in article L.151-6 of the Urban Planning Code.

[16] In accordance with article L.151-19 of the Urban Planning Code, an Inventory of Architectural Heritage (IPA) is used to identify a certain number of buildings which, in addition to the historic monuments protected by the State, are testimony to the history and memory of the place.

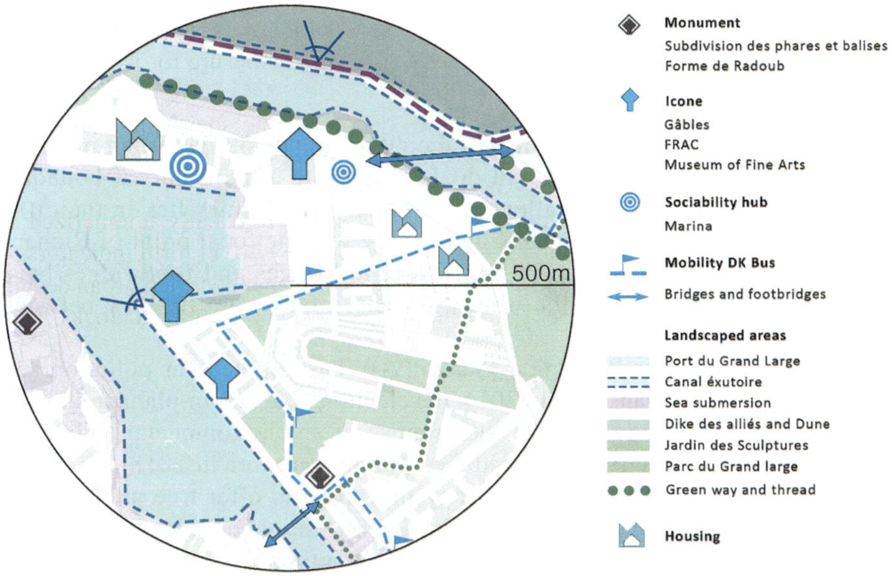

Fig. 6.7 Projected image of the Grand Large (Source: Map drawn by the authors).

Inventory of Architectural Heritage (IPA), and the CUD could qualify for recognition as a Remarkable Heritage Site.[17]

Finally, the Grand Large and the docks are unique, a symbol of industrial decline and the flagship of the city. A strategy for reclaiming the urban identity here consists of building architectural icons. The Grand Large has lent itself to this representational strategy. The result is nevertheless mixed. While the FRAC designed by Lacaton Vassal conveys the local history and should act as a structuring landmark in space, Nicolas Michelin's Gâbles are more successful in fixing the collective imagination and crystallizing the identity of the place (Prévot and Leclercq, 2010). Both are visible faces of the new district and symbolic architectures. They are enough to shift the public perception of the former industrial wasteland by reinterpreting the hangar and chimneys in a contemporary language despite a loose urban plan and an ordinary architectural background (Fig. 6.7). Can however this single piece of architecture become tomorrow's heritage? Heritage is not just about architectural achievement, as the promotion of monuments might suggest. Today, it is essential to create an urban landscape and ensure integration into the site and into the culture. The question then arises in terms of legacy and the collective narrative left to future generations.

[17] The Remarkable Heritage Sites programme was created in 2016. It aims to conserve, restore, rehabilitate or enhance towns, villages or districts that are of public interest from a historical, architectural, archaeological, artistic or landscape point of view.

6.5 Mobilities and Polarities

6.5.1 *Public Transport*

The integration of each segment into the city depends on the urban mobility network. However, the canals and docks constitute a multitude of boundaries to be crossed, and the challenge for the city of Dunkirk lies in the difficulty of creating an urban mesh. Connectivity was the focal point of Richard Rogers' Neptune master plan in 1991. Several bridges and footbridges have already been built to cross the waterways and improve traffic flow between the embankments.

Since 2002, the Air Protection Plan (PPA)[18] has targeted car pollution and aims to decarbonize transport. In 2014, the election of a "urban-planner mayor," Patriece Vergriete, made possible the promise of an environmental and social mobility policy. The city introduced free public transport in 2018 and prides itself on having become "the largest city in Europe to offer free public transport for all, with no conditions" and of having increased bus ridership by 135% from 2018 to 2024. This strategy of "urban marketing" is deliberately used in the city's communication and promotional activities. As a result, it has given Dunkirk internationally renown and free public transport is a source of pride for its inhabitants.

Beyond campaigns, this proactive policy has involved an overhaul of the intercommunal public transport network and the creation of a multimodal transport hub at the train station. The mobility project is an opportunity for a vast public realm improvement and a new urban conviviality (Huré, 2020). The square in front of the train station has been redesigned and the canopy of the former bus station has been incorporated into a bicycle parking area with a stainless steel "mirror." This urban design extends into the town centre with bus shelters and rain protection punctuations.

In conclusion, this transport operation has a very positive impact on the *perceived* image of the city including young people and tourists, whose mobility is greatly facilitated and immediately beneficial. However, a waterbus service could be introduced to increase connectivity on the north side of the city and encourage urban regeneration projects in the docks via the waterways. On another note, the environmental assessment is nuanced because greenhouse gas emissions and pollution have not been reduced despite the use of bio-gas or hydrogen-powered buses, and the impact on public health is not guaranteed since users take the bus on short trips to the detriment of walking and cycling (Héran, 2020).

[18] The Air Protection Plans (PPA) have been compulsory since 1996 for towns with more than 250,000 inhabitants.

6.5.2 *Soft Mobilities*

Within the latest PLUiHD, the «Transport» Programme of Orientation and Action (POA) emphasizes walking and cycling as modes of transit. The promotion of soft mobility requires improving the quality of public spaces, particularly the continuity of pathways and accessibility for all. A Road and Public Space Accessibility Plan (PAVE) was drawn up in 2019 and it is estimated that 55% of pavements are accessible to people with reduced mobility.

Soft mobility contributes to the quality of life and well-being of its inhabitants. It is frequently associated with the greening of the city. In addition, the canals are commonly used for walks. Greenways are thus being rehabilitated along the banks of the Canal éxutoire (outfall canal) and on the Quai des Hollandais along the Bassin du Commerce. In total, a 410 km network of cycle paths has been created across a large area from the Belgian border to Gravelines, and the CUD is pursuing the goal of generating a comprehensive mesh.

In the urban core, a 2 km-long "green thread" project was launched in 2023 to link the train station to the beach across the city centre and the Grand Large. With a central rainwater collection ditch and the planting of trees, it aims to green the city core. Plans to pedestrianise the centre are also underway. Yet the role of public squares remains unclear. Only Place Jean Bart is valued and appropriated by residents. The squares around the town hall, the theater, FRAC and the Kursaal seem devoid of meaning and use and do not constitute places for gathering or conviviality.

6.5.3 *Sociability Hub*

The sociability hubs embody places such as educational establishments, cultural facilities and commercial spaces that generate mobility, exert a functional and social attractiveness and have a high visibility. In the 1990s, urban planners allocated strategic spaces to shopping malls. Thus, the marine centre occupied part of the park designed by Théodore Leveau along the axis of rue Sainte Barbe and on the site of the naval arsenal. Today, the Docks de la Marine project aims to enhance this green lung and revitalize this entrance to the city centre. New apartment blocks are arranged in a radial pattern around the bandstand, and the shops on the ground floor open on Boulevard Alexandre III.

The ZAC du cœur d'agglomération pole théâtre project, set up in 2011, initially enabled the renovation of the theater and the library and densification through the creation of housing. North of Rue Sainte Barbes, Théodore Leveau's block plan was not followed, and a large empty space serves as a car park and market square. The construction of a food hall on Cours François-Bart aims to promote local products and should become a new tourist attraction. This modest and elegant architecture has found its place in this urban environment. Tourists and Dunkirk residents quickly took over the spaces and terraces.

In March 2018, downtown Dunkirk was designated a beneficiary of the «Action Cœur de ville»[19] programme, which was subsequently extended to 2023–2026 and expanded beyond the inner city. The objective is to revitalize the city centre and attract commercial spaces that have been located in the suburbs. Thus, near the train station, and as part of the urban renewal of Ile Jeanty, a leisure centre dedicated to games and entertainment (performance and sports hall, go-karting, bowling, arcade games, casino, etc.) is planned. But a closed building will probably not be enough, from a symbolic and spatial standpoint, to repair this fragmented territory. As the mapping shows (Fig. 6.8), the district needs better-structured and better-connected public spaces.

6.6 Landscapes and Reclamation of Water

6.6.1 City and Nature

The PLUiHD and the PADD focus on health and environmental issues, which have been identified as major concerns for the region. The «Landscape and Biodiversity» Orientation for Planning and Programming (OAP) aims to define development intentions that contribute directly to improving health.

Landscape issues are determined throughout the territory to preserve the characteristic views and skylines of maritime Flanders. A green and blue grid is composed of a mesh of natural coastal elements, whether they result from indigenous factors or human action (dune environments, hydrographic networks, wetlands, woods, etc.). The natural heartlands and ecological corridors are being restored to create a living ecological network across the whole territory. The Canal éxutoire and the Ile Jeanty canal are two ecological corridors that cross the city and connect the plain to the seafront. A pocket of biodiversity has formed on the Quatre Écluses peninsula and is known to be host to cormorants. In the urban area, green spaces are few and spread out. To make up for this lack, the CUD has set a greening target that should also help to fight against heat islands, reduce soil impermeability and promote places of tranquillity.

The parks and gardens are being strengthened as nature relays (Parc Marine and Parc Ziegler) and the planting of trees is widespread (Mail de la Rue des Fusiliers Marins). Nevertheless, the situation on the seafront with the wind and sea spray limits the possibilities for replanting vegetation, particularly in the Moles area. The introduction of flora into the city is limited, but nature is not limited to greening. Water forms a constitutive component of Dunkirk's spatial identity.

[19] The « Action Cœur de ville » programme was launched at the end of 2017 by the Ministry of Territorial Cohesion to revitalize the centres of medium-sized towns and develop a new and attractive housing supply.

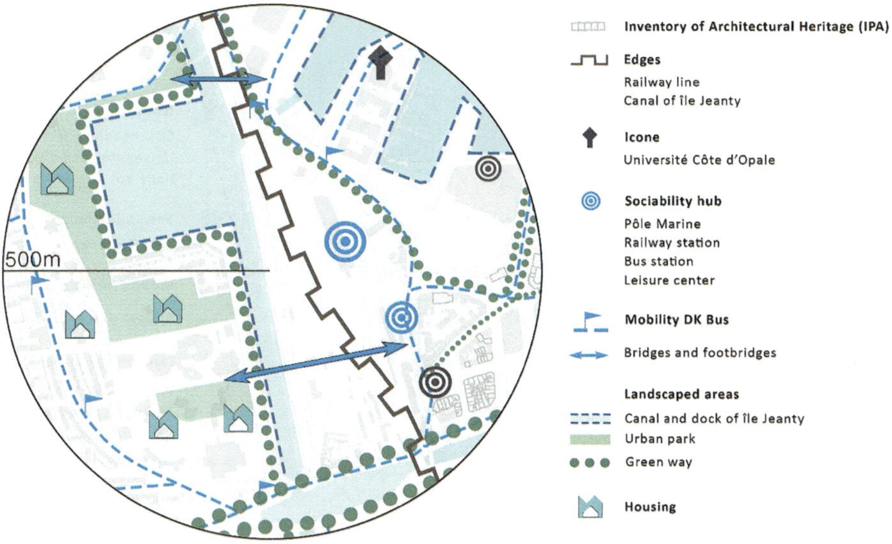

Fig. 6.8 Projected image of the train station/Ile Jeanty (Source: Map drawn by the authors).

6.6.2 The Ambivalence of Water

Water is part of Dunkirk's heritage: the sea, the harbor docks and the network of wateringues, canals and locks shape the landscape. The inhabitants have nevertheless an ambivalent relationship with water and the sea-level rise is a cause for concern. The areas at risk of marine submersion are identified in the PLUiHD and the «Landscape and Biodiversity» OAP keeps the maintenance of the hydrographic network a priority for the protection of the territory. Sand has been added to the Beach every year since 2019. The reinforcement of a dune ridge below the digue des Alliés is a nature-based solution that serves as a defense against the possibility of a large wave causing the sea to enter inland via the Canal éxutoire.

The PADD aims to develop the coastline through tourism and to consider water an essential asset. La digue de Malo-les-Bains is a combination of a seawall, a promenade and a landmark. Stabilization and enhancement works were carried out from 2020 to 2023. The public facilities are associated with a charter for terraces and commercial furniture. The ambition to increase the resort's tourist potential is displayed under the slogan "the most beautiful beach in the North."

The Grand Large port and the Bassin du Commerce host yachts. The reintegration of these water surfaces brings added value to the landscape, but the image of the Freycinet docks remains devalued (Fig. 6.9). Following the Danish model, water-based sports and leisure activities such as water-skiing or equipped floating pontoons could activate the dock waterbodies, especially as the water level is stabilized by the locks.

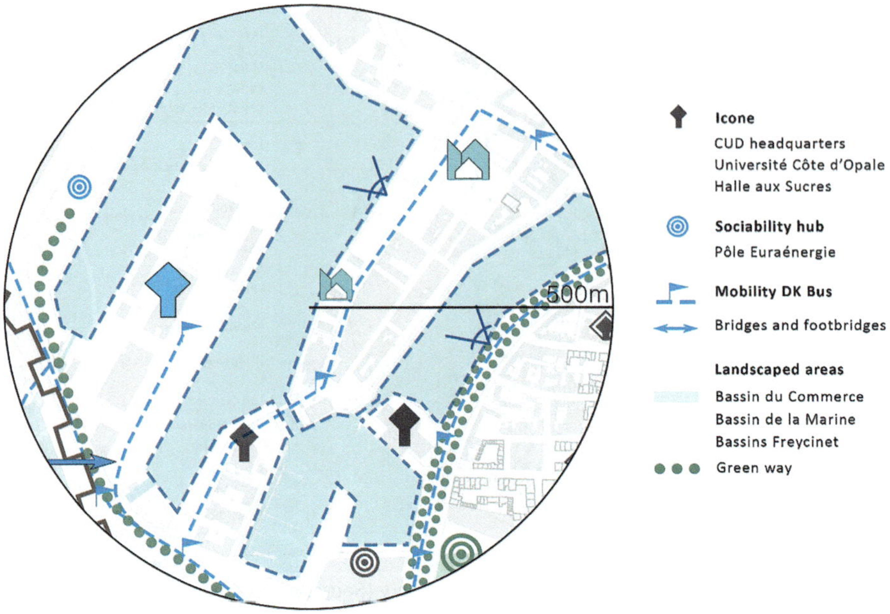

Fig. 6.9 Projected image of the Docks (Source: Map drawn by the authors).

In addition, the reclamation of the waterfront requires a port city project coordinated with the Autonomous Port of Dunkirk (PAD) and the development of port-related activities such as strengthening the fishing industry or cruise tourism by relocating piers close to the city (Lavaud-Letilleul, 2001). It also appears necessary that the Île Jeanty project preserves the boatmen's livelihoods and supports the local history of inland waterway transport. Finally, in 2024, the people of Dunkirk took a great interest in the renovation of the famous three-masted Duchesse Anne which will be unveiled to the public at the international Tall Ships Races 2025. In this regard, the creation of a training shipyard open to the public at the Radoub dry docks, which is already listed as a historical monument, could revive the shipbuilding industry and the history of the site.

6.6.3 The Resilient City

CUD urban planning projects employ a wide range of urban planning tools available under French law and have a significant impact on the urban form. Centralized urban governance at intercommunal level, the choice of a sustainable holistic approach (economic, social and urban) and the coordination of transport, housing and environmental policies foster the emergence of political and visionary urban planning. The study of sectors further demonstrates that urban interventions involve

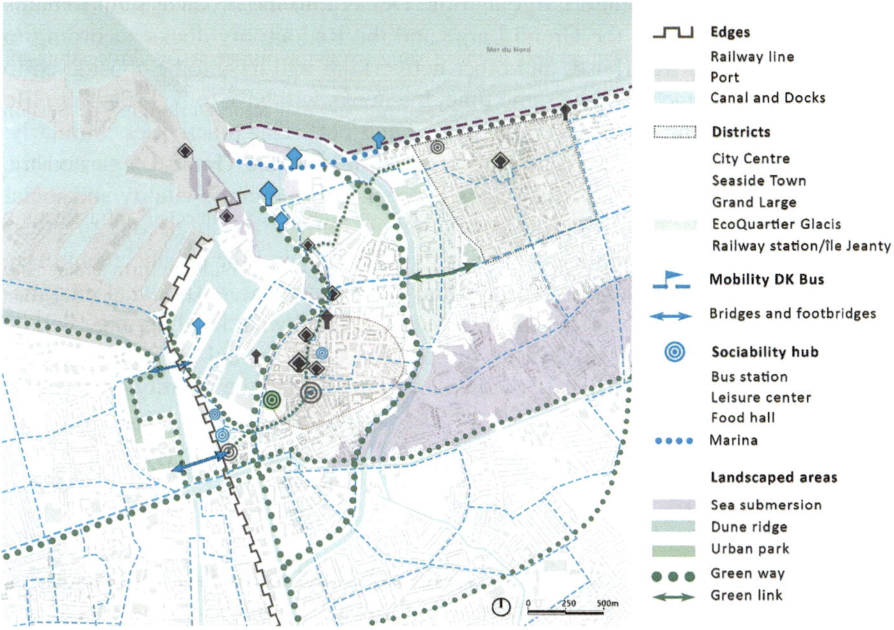

Fig. 6.10 Projected image of a resilient urban core in the twenty-first century (Source: Map drawn by the authors).

acting on three levers—spatial, symbolic, and societal—and that a good balance of these three parameters ensures the attractiveness of districts. The projected image of the city (Fig. 6.10) displays the architectural, urban, and landscape interactions that shape a "resilient urban identity."

This model highlights the following virtues: (1) The inhabited area is protected from marine submersion by a mixed natural and technical system (sand dune against the dike work); (2) The ecological restoration of the green and blue mesh contributes to enhance landscape heritage; (3) The edges of the urban area are being pushed north towards the sea over the port and west beyond the railway lines; (4) New neighborhoods are being introduced in neglected urban areas: Projects such as the nature city at the Glacis or the urban renewal of Ile Jeanty are intended to reduce fractures; (5) New sociability hubs revitalize the heart of the urban area; (6) The free public transport and greenways are vectors for the redesigning of public realm; (7) The promotion of tourism in the seaside resort and the rebuilt city centre reveals the potential of these architectural features; and (8) The construction of iconic buildings is generating a new urban identity for the Grand Large on the wasteland.

Besides, certain limitations must be noted: from an ecological and climatic point of view, the network of wateringues will have to drain larger volumes of water to contain the risk of flooding, and the introduction of plants in a maritime environment requires a specific ecosystem. Moreover, will the port edges be pushed back even further, and to what extent can the city take over the industrial wasteland?

The integration of the Grand Large and the Docks into the urban system remains fragile. A bridge linking the Grand Large and the Radoub dry docks according to R. Rogers' project, waterbuses and other networking will have to be considered to cross the water surfaces. At the same time, the docks will have to be brought to life through the development of leisure activities and port-related activities. Similarly, the revitalization of the city centre, the friendliness of the Grand Large and the regeneration of working-class districts will depend on economic vitality and social cohesion initiatives. The city promotes its image, but studies are needed to assess the impact on tourism and the economic spin-offs. Beyond the enhancement of an urban landscape and the improvement of the city's *projected image*, what imaginations are conveyed, what narratives are constructed, and what local heritage will be transmitted? Sociological surveys on the practices and mental representations of places will be necessary to clarify the *perceived* image and the identity of the city.

6.7 Conclusion

The city of Dunkirk is renowned for having committed itself early on to a sustainable development model, and the free public transport framework has won strong adherence from the population and international visibility. The ambition to make the agglomeration more attractive has led it to pursue a development policy in favor of well-being and proactive urban renewals that enable us to better understand how the image of the city of the twenty-first century is created. Mental representations are based on defined spatial-configuration elements such as edges, districts, landmarks and sociability hubs, but also additional factors that contribute to the quality of life, such as mobility and landscaped spaces.

This study demonstrates that an urban rehabilitation policy compensates for fragility, encourages the emergence of a new perception of the city and enhances local architectural, urban and landscape heritage. The example of the city of Dunkirk enables us to identify the characteristics of a resilient urban planning model, namely: (i) the repair of past shocks and tensions, such as the shocks of war or industrial decline through architectural, urban and landscape reinterpretation, (ii) the revitalization and attractiveness in the present to overcome urban and demographic vulnerabilities, and (iii) the adaptation of the territory to future natural and technological risks, in this case marine submersion and industrial accidents.

This example could provide inspiration for towns in decline, factory towns, port cities or, more broadly, stigmatized communities. Although, Dunkirk relies on the national legal framework of French urban Law (PLUiHD, NPNRU, ZAC, IPA …), the success of this model is above all the result of a proactive urban policy that has been implemented in a coordinated approach at intercommunal level, with a continuous focus on sustainable development and over the long term for 50 years. To avoid economic, social and urban decline, resilient urban planning involves spatial, symbolic, and societal fields. The writing of a new narrative requires strong communication and local political leadership.

According to the concept of Transformative Growth introduced in this book, resilient urban planning enables Dunkirk to move forward beyond the race of ecological modernization or the idea of degrowth. The prosperity of the city depends to a greater extent on achieving a balance of environmental, economic and social components of its territory. Urban productivity is not just about producing economic value through tourism or commerce, but also about enhancing nature, including communities and improving well-being in the city. In doing so, pragmatic yet ambitious policy affects space, usability and conviviality of the city and provides a pathway for desirable growth, transformation of urban identity and attractiveness.

This is how the "urban planner mayor" of Dunkirk is promoting the transformation of the industrial city into a tourism-oriented city and reversing its image from *stigma to emblem*. The new urban identity, as a product of the *projected image* in a development strategy and the *perceived image* by the inhabitants, results not only from the new face of the city but also from the collective imaginations conveyed. Transformative Growth has influenced the city's image and value creation consist in improving quality of life and strengthening the relationship between men and their environment. Between the third-largest port in France and the most beautiful beach in the North, the city of Dunkirk asserts its central place in the cohesion of the territory and ensures the synergies of a Port-Beach-City with a view to reclaiming the seafront.

Abbreviations

AGUR	Agence d'Urbanisme de la Région Dunkerquoise	Urban Planning Agency of the Dunkirk Region
CUD	Communauté Urbaine de Dunkerque	Dunkirk Urban Community
GEMAPI	Gestion des Milieux Aquatiques et de Prévention des Inondations	Management of Aquatic Ecosystems and Flood Prevention
IPA	Inventaire du Patrimoine Architectural	Inventory of Architectural Heritage
NPNRU	Nouveau Programme National de Renouvellement Urbain	New National Urban Renewal Programme
OAP	Orientation d'Aménagement et de Programmation	Orientation for Planning and Programming
ODD	Objectifs de Développement Durables	Sustainable Development Goals
PAD	Port Autonome de Dunkerque	Autonomous Port of Dunkirk
PAVE	Plan de mise en Accessibilité de la Voirie et des Espaces publics	Road and Public Space Accessibility Plan
PCAET	Plan Climat Air Energie Territorial	Territorial Climate Air Energy Plan
POA	Programme d'Orientations et d'Actions	Programme of Orientation and Action
PPA	Plan de Protection de l'Atmosphère	Air Protection Plan
PADD	Projet d'Aménagement et de Développement Durable	Sustainable Development and Planning Project

AGUR	Agence d'Urbanisme de la Région Dunkerquoise	Urban Planning Agency of the Dunkirk Region
PPR	Plan de Préventions des Risques	Risk Prevention Plan
PPRT	Plan de Préventions des Risques Technologiques	Technological Risks Prevention Plan
PPRN	Plan de Préventions des Risques Naturels	Natural Risk Prevention Plan
PPRL	Plan de Préventions des Risques Littoraux	Coastal Risk Prevention Plan
PLU	Plan Local d'Urbanisme	Local Urban Planning Plan
PLUiHD	Plan Local d'Urbanisme intercommunal Habitat Déplacement	Intercommunal Local Urban Planning Plan Housing Transport
ZAC	Zone d'Aménagement Concertée	Concerted Development Zone

References

AGUR (2024, février) Ecoquartier des glacis, ici on sème le quartier de demain (Le cahier de l'AGUR, No. 30). https://www.agur-dunkerque.org/blog/ecoquartier-des-glacis-ici-on-seme-le-quartier-de-demain-billet-8582.html

Beaurain C, Chembessi CD (2018) Les villes-usines face à la problématique environnementale: Réflexions à partir de l'exemple de l'agglomération dunkerquoise. Revue Géographique de l'Est 58(3–4). https://doi.org/10.4000/rge.8512

Cerema (2020, octobre 27) Dunkerque: Analyse comparée des méthodes d'approche de la rénovation et de la mise en valeur du patrimoine reconstruit des centres-villes. Centre d'études et d'expertise sur les risques, l'environnement, la mobilité et l'aménagement. https://doc.cerema.fr/doc/SYRACUSE/600916/dunkerque-analyse-des-methodes-d-approche-de-la-renovation-et-de-la-mise-en-valeur-du-patrimoine-rec

Climate Central (n.d.) Coastal risk screening tool. https://coastal.climatecentral.org/mapview/9/2.7943/50.8178/f13c00df31864056489103b8f16efca86acdad5711b49d857be84e092c63b6f4

Communauté Urbaine de Dunkerque (2024) Plan local d'urbanisme intercommunal Habitat Déplacement (PLUiHD) approuvé par délibération du 26 avril 2024 et ses pièces constitutives. https://www.communaute-urbaine-dunkerque.fr/pluihd

Desoutter S (2023, octobre 12) La plaine dunkerquoise, un ancien estuaire polderisé (Nord). Institut national de recherches archéologiques préventives (INRAP). https://www.inrap.fr/la-plaine-dunkerquoise-un-ancien-estuaire-polderise-nord-17540

Faye B, Lacour C (2020, décembre) Identités urbaines, question d'images. Revue d'Économie Régionale et Urbaine (5):803–827. https://doi.org/10.3917/reru.205.0803

Galland B (1993) Les identités urbaines: Cultures, sous-cultures et déviances. Convention romande de 3e cycle de sociologie (3). https://www.artfactories.net/IMG/pdf/identite_urbaine_blaise_galland.pdf

Hauser S (2019) The interplay of economic development and environmental protection: Dunkirk and the search for balance. PORTUSplus 8(Special Issue: Governance in Port City Regions). https://www.portusplus.org/index.php/pp/article/view/176

Hein C, Schubert, D (2021) Resilience and path dependence: A comparative study of the port cities of London, Hamburg, and Philadelphia. J of Urban Hist 47(2):389–419. https://doi.org/10.1177/0096144220925098

Héran F (2020) Gratuité des transports publics et cohérence des politiques de déplacements urbains. Transp Urbains 136(1):28–34. https://doi.org/10.3917/turb.136.0028

Huré M (2020, juin) La gratuité des transports ou le retour du politique dans le pilotage de l'action publique: L'exemple de Dunkerque. Transp Urbains 136(1):18–22. https://doi.org/10.3917/turb.136.0018

Husser C, Raison S (2015, novembre) Dunkerque: La reconversion de terrains portuaires. Annales des Mines – Réalités Industrielles (4):66–71. https://doi.org/10.3917/rindu1.154.0066
Jebrak Y (2010) La reconstruction et la résilience urbaine: L'évolution du paysage urbain. Master's thesis, Université du Québec à Montréal. https://archipel.uqam.ca/3124/
Lavaud-Letilleul V (2001) Dunkerque-Neptune: Dix années d'une expérience de waterfront à la française. In: Manneville P (ed) Actes du 124ème Congrès national des sociétés savantes. Section d'histoire moderne et contemporaine: Milieu, littoral et estuaires. Éditions du CTHS, pp 271–283. https://www.persee.fr/doc/acths_0071-8440_2001_act_124_1_5941
Lynch KA (1969) L'image de la cité (trans: Renau C). Dunod, Malakoff, France
Prévot M, Leclercq C (2010) De la prolifération des études au projet sans auteur: Essai sur les projets urbains dunkerquois et sur la ville visible et invisible. Géocarrefour 85(4). https://doi.org/10.4000/geocarrefour.8151
Vergriete P (2024, juin 14) Le nouveau musée des Beaux-Arts a trouvé sa place (Interview by C. TV). https://www.corsairetv.fr/actualites/Le-nouveau-musee-des-Be%20aux-Arts-a-trouve-sa-place-14-juin-2024-836bf345fc.html

Chapter 7
Solar-Optimized Urban Form: Reimagining the Sustainable Block for Energy Resilience and Social Connectivity

Shadi Galehdar and Mahmoud Ghalehnoee

Abstract Solar energy, one of the most abundant and accessible renewable resources, has become increasingly central to the vision of sustainable urban futures. Yet, much of the existing research has been confined to its morphological potential—optimizing building orientation, roof geometry, and shading—while paying limited attention to the social dimensions that condition its acceptance and long-term success. This study addresses this gap by reimagining solar integration through the lens of the Community Block (Oikonomou, 2015), an urban design framework that foregrounds the interactions between buildings, streets, open spaces, and shared social life.

Adopting a mixed-method approach that combines urban morphological analysis with participatory design case studies, the study evaluates how seven key urban design qualities—flexibility, enclosure, adaptability, ownership, diversity, sense of belonging, and permeability—mediate the relationship between energy efficiency and community needs. The findings reveal that participatory-designed blocks not only optimize solar performance but also foster stronger resident engagement in energy stewardship, reinforce networks of collaboration, and cultivate shared spaces that support social cohesion.

By advancing a solar-optimized urban form that unites technical efficiency with socio-cultural vitality, the chapter contributes both a theoretical framework and practical insights for planners, designers, and policymakers seeking to align energy resilience with inclusive and connected urban living.

Keywords Solar Energy • Urban Form • Social Connectivity • Community Block

---

S. Galehdar
Faculty of Architecture and Urban Planning, Art University of Isfahan, Isfahan, Iran

M. Ghalehnoee (✉)
Professor of Urban Design, Department of Urban Planning, Faculty of Architecture and Urban Planning, Art University of Isfahan, Isfahan, Iran
e-mail: m.ghalehnoee@aui.ac.ir

A. Cheshmehzangi et al. (eds.), *Transforming Cities*, Urban Sustainability,
https://doi.org/10.1007/978-981-95-7378-3_7

7.1 Introduction

Urban areas account for approximately 75% of primary energy consumption and 60% of greenhouse gas emissions (Shi et al., 2021). Under current climate projections, the cooling demand for buildings is expected to increase significantly 70% by 2030 and 115% by 2050 compared to 2017 levels (Mohajeri et al., 2019). Despite the urgency of these trends and notable advances in solar technology, much of the research and practice remains focused on technical optimization, often overlooking the multifactorial influences—social, ecological, and governance-related—that shape success at the neighborhood scale. No comprehensive methodology yet holistically assesses how renewable energy, particularly solar power, can be meaningfully integrated into urban form to simultaneously enhance energy resilience and social connectivity.

A central approach to sustainable urbanism is the study of urban form, which intersects with diverse social, economic, environmental, and ecological parameters. Moreover, urban morphology can alter the amount of shading and the level of desirable or undesirable solar radiation in urban areas (Perera et al., 2021). In recent years, numerous studies have integrated solar energy considerations into building design, urban planning, and urban design as a combined approach (Shi et al., 2021).

With the accelerating transition to zero-energy neighborhoods and buildings, the deployment of renewables has become essential for the future of sustainable cities. Policy frameworks reinforce this momentum—such as the European Union's directive mandating on-site energy generation for all new buildings after 2019 (Zeng, 2011). The Community Block concept (Oikonomou, 2015) aligns with sustainable urban futures, combining buildings, blocks, streets, and open spaces into integrative frameworks fostering resident interaction and participatory design. This study investigates how solar infrastructure can be embedded into these blocks to optimize energy use and strengthen social networks, recognizing that such integration requires balancing technical morphology with socio-cultural dynamics.

Current strategies for sustainable urban development increasingly emphasize the importance of interventions at the local and micro scales (Pakzad and Salari, 2018). However, the implementation of optimal solar solutions demands a multi-scalar perspective—addressing city, regional, neighborhood, and building levels (Manni et al., 2023). This chapter therefore, adopts a holistic approach, employing both urban morphological analysis and participatory design case studies to explore how key design qualities—flexibility, enclosure, adaptability, ownership, diversity, sense of belonging, and permeability—mediate the integration of solar energy and social resilience.

To anchor this investigation within the evolving research landscape, Fig. 7.1 presents an overlay visualization generated using VOSviewer, based on the principal sources referenced in this study. This Overlay Visualization captures the thematic development in solar urbanism over the past decade, highlighting how research trends have shifted over time. Notably, the visualization traces the journey

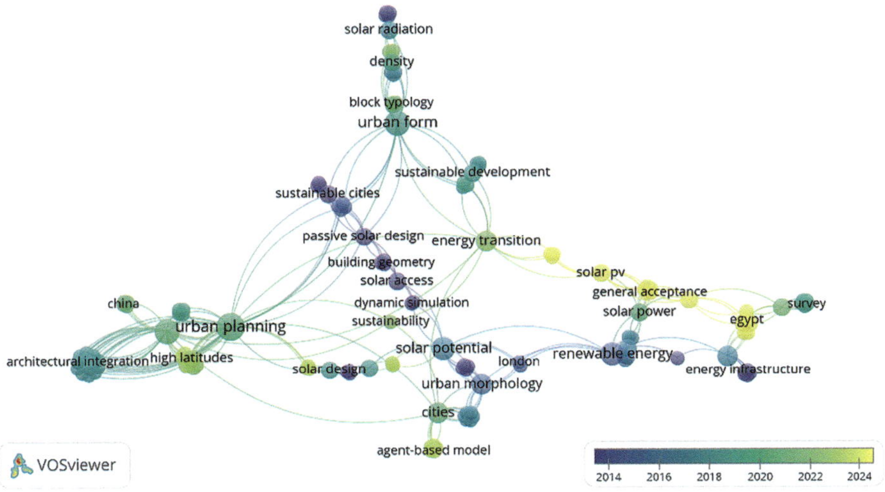

Fig. 7.1 Overlay visualization of the present research.

from early emphases on technical aspects—such as "urban form," "solar radiation," and "density"—toward more recent focus areas, including "social acceptance," "survey," and site-specific studies like "Egypt." This temporal layering reflects the field's growing interest in integrating societal dimensions alongside technical optimization.

7.1.1 The Community Block Framework

The concept of the Community Block was first articulated by Oikonomou (2015) as a future-focused approach to urban block design. It reframes the built environment as fundamentally shaped by open space, reconstructing the dynamic between user and environment. At its core, the Community Block positions accessible open space as the locus for socializing and interaction, while pedestrian pathways—acting as "social nodes"—facilitate movement and connection between blocks, public spaces, and the broader street network (Oikonomou, 2015). Fig. 7.2 illustrates the evolutionary process of the Community Block, emphasizing an "inside-out" growth model. Urban development begins from shared open spaces and progressively organizes surrounding buildings and streets to foster social connectivity and participatory use.

Aligning with principles of climate-adaptive urban design, the Community Block is envisioned to respond to environmental imperatives and human needs through flexible spatial arrangements and integrated green infrastructure. This approach enhances both environmental quality and the potential for community interaction. As shown in Fig. 7.3, the layered structure of the Community Block encompasses

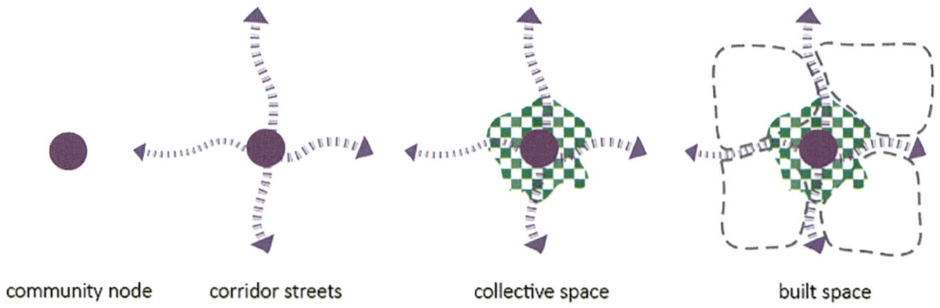

Fig. 7.2 The urban process of the community block (Source: Adapted from: Oikonomou, 2015).

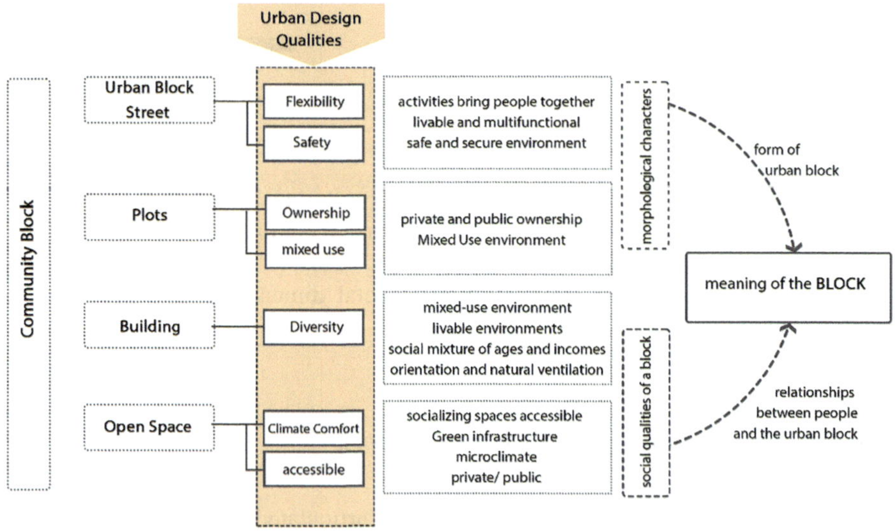

Fig. 7.3 Community block layers (Source: Adapted from Oikonomou, 2015).

streets, plots, buildings, and open spaces, each contributing to both its morphological character and social meaning (Oikonomou, 2015).

When spatial characteristics are examined collectively, they define the physical shape of the urban block; when social qualities are combined, they illuminate the evolving interactions between residents and their built environment. Together, these features serve both as practical design guidelines and as symbolic anchors, enabling a holistic understanding of urban form as a framework for social and ecological resilience. The indicators identified by Oikonomou—streets, plots, buildings, and open spaces—are elaborated in detail in the following sections to clarify their respective roles in achieving energy-efficient, socially cohesive blocks.

7.1.2 *Elements of Urban Form in Four Layers Related to Solar Energy*

7.1.2.1 Block

In the hierarchy of urban formation, the foundational layer emerges from the arrangement of urban block fabric, shaped by the adjacency of basic elements like buildings (Zekavat, 2012). Urban blocks, defined by the spatial boundaries of plots, play a critical role in mediating the interplay between public and private edges along adjacent streets. The spatial positioning of these blocks determines the configuration of edges, influencing both their connectivity and users' experiential interactions inside and outside the blocks (Al-Saaidy and Alobaydi, 2021).

Recent studies highlight the significance of block layout for solar energy performance. Liu et al. (2023) found that blocks exhibiting high energy efficiency often feature centrally located open spaces, which maximize solar access, and taller buildings oriented toward the western edge to minimize shading effects. Meanwhile, Shi et al. (2021) examined the interaction between solar energy and urban design by defining urban design criteria in three categories: block size (area and elongation), building footprint (building pattern), and building density (floor area ratio and site coverage).

Further advancing these insights, Li et al. (2022) introduced the Block Surface Ratio (BSR) as an important new indicator. BSR demonstrates a stronger and more practical correlation with solar radiation received by buildings than the commonly used Sky View Factor (SVF), providing a more efficient tool for urban analysis.

7.1.2.2 Street

Street functions serve primarily as spaces for movement, enabling access between the street and the urban block, and vice versa (Oikonomou, 2015). Additionally, street design is a crucial parameter that, through building orientation, influences the amount of solar radiation received, thereby affecting energy consumption (D'Almeida, 2022).

Martins et al. (2014), in their work "From Solar Constraints to Urban Design Opportunities," highlighted street width and aspect ratio as the most influential factors affecting urban morphology and solar optimization. Their findings indicate that while verticality and building width had comparatively minor effects, metrics like floor area ratio and plot ratio serve as important constraints influencing overall block design.

These findings highlight the importance of street dimensions and configurations. When strategically designed, streets not only enhance connectivity and accessibility but also serve as key levers for maximizing solar access and enabling energy-resilient urban blocks.

7.1.2.3 Building

Buildings represent the smallest organizational units of the city, with the urban form fundamentally emerging through the adjacency and arrangement of these physical elements (Zekavat, 2012). Rafiyan et al. (2011) demonstrated that building height and density, building plan form, and building orientation are key physical characteristics that significantly affect daylight availability.

Research by Rafiyan et al. (2011) demonstrates that building height and density, as established through urban planning and design, can significantly impact access to sunlight and the extent of shading effects among adjacent structures. Without considered control of these parameters, residential blocks may experience uneven solar exposure, affecting both livability and renewable energy generation potential.

Building density, in particular, is closely linked to social interaction: Shakibaei Bidaruni and Tabibian (2018) argue that optimal neighborhood population densities foster stronger community connections. Building orientation has a direct relationship with solar access. Specifically, the angle between the building's axis and the north-south direction, considering the site's latitude, influences the number of hours of direct sunlight the building receives throughout the day, thereby impacting its energy requirements (Rafiyan et al., 2011). Consequently, design guidelines commonly recommend elongating the building form along the south-facing direction. Moreover, the optimal building orientation relative to the sun is one that maximizes heat gain during cold days and minimizes it during hot days (zahri, 2017).

Consequently, design guidelines typically recommend elongating buildings along the south-facing direction to maximize solar gain in winter and minimize it during hotter periods. This integration of morphological and solar principles is central to effective sustainable urban block design.

7.1.2.4 Open Space

Urban open spaces constitute essential elements for environmental comfort and social interaction, particularly in the context of solar energy utilization. Achieving a balanced mix of sunlight and shade within these areas is vital for maximizing usability across changing climatic conditions. However, spatial requirements for solar access to open spaces may sometimes conflict with those for interior buildings, as ground and façade surfaces often require different solar orientations (Van Esch et al., 2012).

Current literature often isolates the examination of geometric parameters and solar indices to either buildings or open spaces, overlooking their combined effects at the block scale. This gap necessitates a more integrated approach—assessing how the geometric interplay between buildings and open spaces influences overall solar exposure, including both façades and outdoor areas (Lan et al., 2022).

In dense environments, sky exposure is highly variable. Large squares and rooftops offer substantial visible sky and thus higher solar access, whereas narrow streets and enclosed courtyards beneath tall buildings experience diminished exposure (Redweik et al., 2013). Strategic design of open spaces, therefore, is required to harness the energy and social potential inherent in the block, supporting both sustainable performance and community well-being.

7.1.3 *Social Connectivity in Solar Urbanism*

7.1.3.1 Solar Acceptance

Ensuring the widespread adoption of solar energy systems within residential developments hinges as much on social factors as on technological innovation. One of the most decisive elements is the degree of public acceptance, which is shaped by engaging citizens throughout the planning and implementation process (Marzouk et al., 2024). Urban and architectural features that nurture social cohesion, such as participatory design and community consultation, are instrumental in fostering support for renewable initiatives. For instance, the Halifax solar project demonstrated that involving a broad cross-section of residents from various backgrounds led to exceptionally high acceptance rates (Lobaccaro et al., 2019).

Research consistently shows that regardless of technological advancement, community buy-in and active involvement remain foundational for the success of sustainable energy strategies (Cohen et al., 2014). Early, cross-disciplinary planning that integrates solar energy goals with considerations such as local building density and construction costs is essential for realizing projects that are both efficient and socially supported (Lobaccaro et al., 2019).

Major challenges in deploying solar strategies involve technology choices, stakeholder conflicts, and social acceptance, with visual impact being a significant concern in urban areas (Lobaccaro et al., 2019). While some studies recommend minimizing the visibility of solar panels in sensitive contexts, others argue that greater visibility can enhance social acceptance and promote adoption (Formolli et al., 2023). The quality of integration, especially in visible or heritage areas, is critical to public approval (Florio et al., 2018).

Urban authorities may default to standardized architectural templates in pursuit of acceptance, yet these can constrain the flexibility needed to respond to unique site conditions (Florio et al., 2018). Broadly, acceptance of renewable technologies spans political, market, and local dimensions. The role of community-level engagement and the sharing of benefits among stakeholders is particularly significant (Sovacool and Ratan, 2012).

Lastly, policy success relies on accounting for user habits and routines, as behavioral choices have a substantial impact on actual energy demand. Effective approaches combine strategic incentives with education and best-practice demonstrations, leading to real and measurable reductions in energy consumption (Legény et al., 2014).

7.1.3.2 Neighborhood Scale: Strengthening Social Relations Through Solar Energy Integration

A growing body of research has explored how the distinctive morphological features of neighborhoods—such as building density, orientation, and patterns of shading—affect the viability and effectiveness of solar energy systems (Veisi and Shakibamanesh, 2022; Sarralde et al., 2015; Lobaccaro et al., 2019). These studies emphasize that tailoring urban design and planning to local context is critical for maximizing solar potential and ensuring robust outcomes.

Lobaccaro et al. (2019) argue that evaluating solar energy at the neighborhood scale uncovers spatial variability—such as differences in direct sunlight and shadowing effects—that broader analyses might overlook. Fine-grained assessments can reveal how factors like building arrangement and nearby structures influence solar access, which in turn impacts long-term sustainability. Planning neighborhoods for optimal solar utilization not only reduces energy consumption and greenhouse gas emissions but also supports a transition away from non-renewable resources. Thus, neighborhood-level solar analysis has become an essential tool for informing environmentally responsible urban design.

The concept of "solar neighborhoods" refers to urban areas that incorporate both photovoltaic and thermal energy systems to meet local energy needs. For these neighborhoods to succeed, planners must strategically configure building form, overall density, and site layout to harness available sunlight (D'Almeida, 2022). According to Manni et al. (2023), solar neighborhoods share the mixed-use characteristics of conventional urban districts—housing residential, commercial, and public functions—but prioritize comprehensive solar integration in their development. However, most comparative studies rely on hypothetical models or manipulate variables such as density, plot ratio, and building orientation to assess solar performance at the block or cluster level (Hachem-Vermette and Singh, 2019; Lan et al., 2022).

Planning and design strategies for solar neighborhoods can be applied to both new developments and existing urban areas, helping cities achieve positive energy budgets and carbon neutrality. These guidelines contribute to achieving favorable energy balances and advancing carbon-neutral goals for modern cities. Findings of the study by Manni et al. (2023) illustrate the intersection between solar neighborhoods and related concepts such as zero-emission neighborhoods (ZEN) and positive energy districts (PED), providing a visual roadmap for how these approaches are linked within the current literature.

A fundamental aspect of effective solar neighborhood planning is the coordinated alignment of human-centric activity patterns and energy management strategies, which can lead to more equitable and efficient energy usage, particularly in mixed-use areas. These strategies inevitably impact a range of design factors—from building massing and height to spacing, orientation, and aspect ratios—all of which are conditioned by local latitude and climate (Manni et al., 2023). Their study also demonstrates key passive solar strategies as applied to neighborhood-scale planning and design, emphasizing design techniques that maximize natural energy flows and minimize external energy demand.

Sovacool and Ratan (2012) highlight the role of community ownership and cooperative models in the expansion of local solar energy systems, reporting that an overwhelming majority of installations are owned by homeowners and local groups rather than large utilities. This shift in ownership fosters inclusive decision-making processes, broad stakeholder involvement, and ultimately, higher levels of public engagement. The Vauban district exemplifies community-driven sustainable development: initiated through grassroots collaboration among building groups and architects, Vauban's participatory approach resulted in a vibrant, socially connected neighborhood that combines architectural diversity with shared green spaces (Coates, 2013).

7.2 Urban Design Qualities in Solar-Integrated Community Blocks

7.2.1 The Importance of Urban Design in Addressing Urban Energy Challenges

The significance of urban design in tackling urban energy challenges has been continuously emphasized. Concepts such as energy-based urban design, energy-saving urban design, and low-carbon urban design have emerged. Thoughtful urban design can maximize the use of solar energy at the urban scale while simultaneously reducing energy consumption through inter-building effects. To develop effective urban design guidelines, implement energy-efficient urban design practices, and create sustainable urban environments, a comprehensive and quantitative assessment of the impact of urban form on building energy consumption and solar energy potential is essential (Liu et al., 2023).

Early integration of solar energy considerations into urban design is crucial to ensure that future cities not only consume energy but also locally generate energy through solar power (Akrofi and Okitasari, 2022). Urban design components relevant to solar energy assessment have been identified in various studies: street width and urban density (Sanaieian and Ghorabi, 2020), street width and orientation (Van Esch et al., 2012), and, focusing on solar energy at the human scale, maximum solar energy utilization potential (Zeng, 2011). Solar urban design emerges as a distinct phase within the broader sustainable planning agenda, linking the efficiency of building stock to the surrounding urban fabric and becoming a foundational element in the pursuit of Net-Zero Energy Buildings (NZEB)(Amado and Poggi, 2012).

Significantly, involving local residents in monitoring and maintaining system performance enhances solar system outcomes and supports broader community goals (Amado and Poggi, 2012). Other research by Natanian et al. (2019) proposes a categorization for solar-oriented design parameters, spanning from sky view factors and aspect ratios to typology, geometry, density, and orientation—providing designers and planners with a framework for maximizing urban solar exposure.

Building on these insights, this study systematically reviews diverse sources from urban design and solar energy literature, grouping key parameters into two domains: morphological, representing physical and spatial characteristics that shape solar potential, and social, reflecting community dynamics and demographic factors. This dual framework enables a comprehensive assessment of both technical and social requirements for successful solar integration in urban blocks. The full set of parameters is detailed in Table 7.1.

Table 7.1 Extracted parameters of urban form at the community block scale.

	Category	Parameters	Abbreviation	Detail
Morphology	Block Dimension	Block Area(m^2)	BA	Shi et al., (2021); Al-Saaidy and Alobaydi, (2021).
		Block Surface Ratio	BSR	Li et al., (2022).
		Block Elongation (-)	BE	Shi et al., (2021); Shi et al., (2020).
	Vertical Distribution	Distance Between Building(m)	DBB	Sarralde et al., (2015); de Lemos Martins et al., (2016); Tian and Xu, (2021).
	Building Footprint	Building Pattern	BP	Shi et al., (2021); Zhang et al., (2019).
		Open Space Ratio	OSR	Liu et al., (2023); Zhang et al., (2019).
	Building Density	Floor Area Ratio (-)	FAR	Shi et al., (2021); Martins et al. (2014); de Lemos Martins et al., (2019); Natanian et al., (2019); Li et al., (2022); Tian and Xu, (2021); Zhang et al., (2019).
		Site Coverage (%)	SC	Shi et al. (2021);Sarralde et al. (2015);Zhang et al. (2019); Li et al. (2022).
		Plot Ratio	PR	Sarralde et al., (2015); Martins et al., (2014); de Lemos Martins et al., (2016); de Lemos Martins et al., (2019).
		The Average Number of Floors	AF	Liu et al., (2023); Tian and Xu, (2021).
	Aspect Ratio	Height to Width Ratio (h/w)		Robinson, (2006); Martins et al., (2014); de Lemos Martins et al., (2016); Natanian et al., (2019); Sharifi and Yamagata, (2018); Li et al., (2022).

Table 7.1 (continued)

	Category	Parameters	Abbreviation	Detail
		Sky View Factor	SVF	Robinson, (2006); Morganti et al., (2017); Liu et al., (2023).
		Street Aspect Ratio	SAR	Van Esch et al., (2012); Vititneva et al., (2021); Natanian et al., (2019); Martins et al., (2014); de Lemos Martins et al., (2016); Liu et al., (2023).
	Orientation	Building Orientation (degree)	BO	Van Esch et al., (2012); Sarralde et al., (2015); Lan et al., (2022); Natanian et al., (2019); Oikonomou, (2014).
		Street Orientation (degree)	SO	Vititneva et al., (2021); de Lemos Martins et al., (2019); Oikonomou, (2014); Liu et al., (2023).
		Block Orientation (degree)	BO	Liu et al., (2023).
social	Ownership	Rented/Owner	–	Delmastro et al., (2015); Oikonomou, (2014), Marzouk et al., (2024),
		Public/Private	–	Shakibamanesh and Ebrahimi, (2020); Kosorić et al., (2019); Oikonomou, (2014).
	Education		–	Amado and Poggi, (2012); Legény et al., (2014), Marzouk et al., (2024).
	Age		–	Amado and Poggi, (2012); Legény et al., (2014), Marzouk et al., (2024).

7.2.2 Urban Design Qualities

Urban design qualities cover a diverse spectrum of attributes, tailored to suit specific urban objectives. Previous research has highlighted dimensions such as imageability, enclosure, human scale, and legibility for walkability (Abdeldayem et al., 2023), as well as robustness, flexibility, and adaptability for urban resilience (Sharifi and Yamagata, 2018). However, there is no research that directly addresses urban design qualities as a framework for categorizing indicators for measuring solar energy in urban environments. Yet, few works propose a unified framework for evaluating these qualities in relation to solar energy integration. The following subsections present those qualities identified as pivotal for connecting solar adoption and community resilience at the block scale.

7.2.2.1 Ownership

In the context of solar urbanism, ownership influences not only residents' investment in their environments but also their willingness to adopt and maintain energy technologies. Understanding the nuances of ownership helps clarify pathways for successful solar integration in community blocks.

Though not always classified as a design quality, ownership strongly influences resident engagement and solar energy acceptance. In Community Block settings, homeownership encourages active participation and investment in the local environment, while the division between owner and renter, and between public and private domains, shapes the feasibility of energy upgrades (Oikonomou, 2014; Delmastro et al., 2015). Creating avenues for resident involvement in planning and implementation fosters community approval and successful adoption of solar solutions (Sovacool and Ratan, 2012).

7.2.2.2 Sense of Belonging

Fostering a sense of belonging is key to encouraging collective engagement with sustainable technologies. In solar-integrated blocks, participatory design and community identity drive acceptance and long-term stewardship.

A pronounced sense of belonging serves as a foundation for advancing social support and long-term adoption of solar energy within communities. When urban design prioritizes opportunities for resident participation—such as collaborative planning and the integration of mixed-use environments—it helps nurture a shared identity and strengthens willingness to accept new energy technologies. The alignment between green infrastructure, participatory approaches, and local interaction fosters this collective spirit, particularly at the neighborhood scale.

This dynamic is visually illustrated in Fig. 7.4, where the overlap between the Community Block and Social Acceptance is shown to be mediated by a

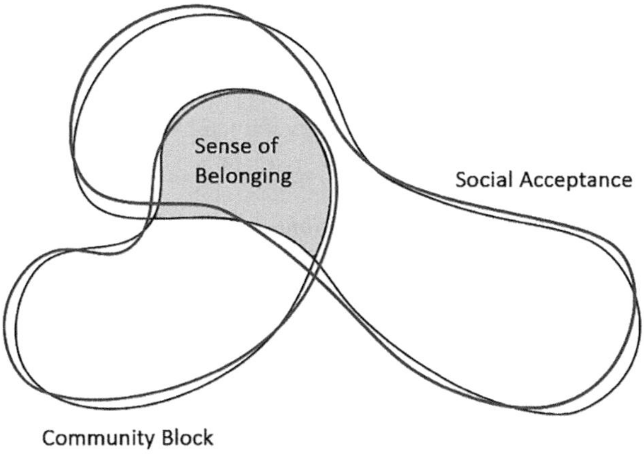

Fig. 7.4 Sense of belonging: Overlapping the community block and social acceptance.

shared Sense of Belonging. The figure highlights how spatial and social integration within a block can directly enhance social acceptance of sustainable urban innovations, reinforcing both environmental goals and the social resilience of the community.

7.2.2.3 Permeability

Permeability determines how movement, sunlight, and social interactions are facilitated within the built environment. Its optimization ensures that solar access and community connectivity go hand in hand.

Permeability describes how easily individuals move and interact within urban blocks, impacting both solar access and social connections. Efficient layouts—such as grid plans with smaller blocks and wider streets—facilitate movement and sunlight penetration, while irregular patterns with narrow passages are less permeable. Block size, surface ratio, elongation, and spacing are key factors in optimizing permeability for solar and community outcomes.

7.2.2.4 Flexibility

Flexibility allows built environments to evolve alongside technological breakthroughs and shifting user needs. In solar urbanism, flexible design and planning accommodate new energy solutions as they emerge.

Flexibility denotes adaptability in built environments, allowing for the integration of emerging energy technologies like photovoltaics. Diverse building forms, responsive open spaces, and the potential for incorporating solar systems into various structures and materials ensure urban blocks remain resilient to future shifts in energy demand and technology (Formolli et al., 2023).

7.2.2.5 Enclosure

Enclosure shapes both the spatial and climatic experience of public spaces. For solar design, enclosure ratios alter sunlight penetration, influencing comfort and energy performance.

Enclosure refers to the physical framing of public spaces—by buildings, vegetation, and structural elements—which influences microclimate and solar exposure. Metrics such as the height-to-width ratio, sky view factor, and street aspect ratio help assess the comfort and shading patterns within enclosed areas, informing pedestrian experience and energy performance (Abdeldayem et al., 2023).

7.2.2.6 Adaptability

Adaptability is vital for neighborhoods that must cope with climate changes and integrate renewable energy. Solar urbanism relies on adaptive frameworks to ensure longevity and effectiveness of energy systems.

Adaptability is crucial for enabling urban blocks to respond to environmental change and integrate renewable energy. Design principles that prioritize climate responsiveness advance the ability of urban form to absorb emerging technologies and address challenges like carbon neutrality (Fonseca, 2016; Masson-Delmotte et al., 2018).

7.2.2.7 Climate Comfort

Prioritizing climate comfort in urban design harmonizes solar exposure and shading, directly affecting livability and energy efficiency within solar-integrated blocks.

Climate comfort encompasses access to sunlight, effective shading, and pedestrian well-being. Orientation and configuration of buildings and streets play a central role in shaping microclimatic conditions, balancing exposure for livability and energy efficiency.

Collectively, these urban design qualities determine the social and functional vitality of urban spaces, forming the foundation for successful solar energy integration and broader sustainable urbanism.

The analytical process and findings of this study are graphically summarized in the following figures. Fig. 7.5 depicts the mapping of key urban form indicators—spanning both morphological and social domains—onto the identified urban design qualities, highlighting the layered connections shaping solar-optimized blocks. This

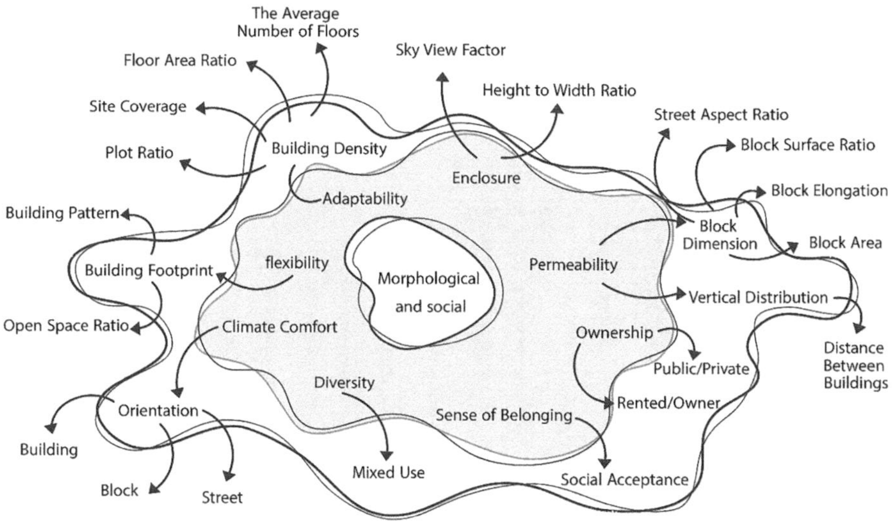

Fig. 7.5 Concept of indicators with urban design qualities.

flow diagram illustrates how attributes such as building density, orientation, and ownership status are systematically associated with design qualities like permeability, flexibility, and climate comfort.

Fig. 7.6 synthesizes these insights, visually capturing the complex web of relationships between solar-optimized urban form parameters and the broader design qualities essential for resilient, socially cohesive neighborhoods. Together, these figures provide a comprehensive visualization of both the analytical framework and the integrative results underpinning successful solar energy integration at the community scale.

7.3 Discussion and Conclusions

This study examined the potential of the *Community Block* as a framework for achieving both solar energy optimization and social resilience at the neighborhood scale. By reframing the relationship between morphological parameters and social qualities, the research has emphasized how the design of urban form can transcend technical performance and become a catalyst for participatory, inclusive, and adaptive urban environments.

At the core of the Community Block model lies the understanding that spatial configuration and social structure are co-dependent. The arrangement of open spaces, streets, and building masses determines not only solar accessibility but also the patterns of social interaction and collective identity that emerge within a neighborhood. The findings highlight that when urban design processes explicitly

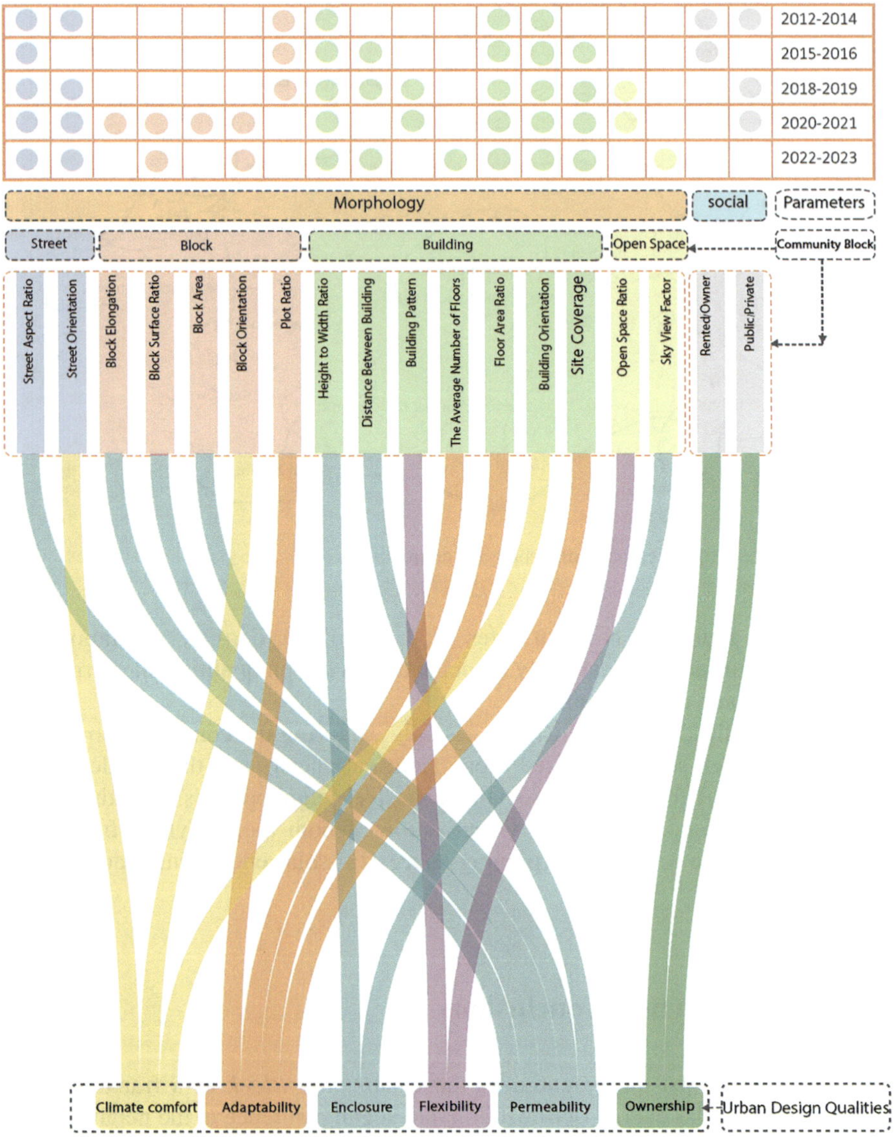

Fig. 7.6 Relation of solar-optimized urban form parameters with urban design qualities.

integrate community participation—through co-design, shared open spaces, and mixed-use configurations—residents develop a stronger sense of belonging and responsibility toward their environment. This social cohesion, in turn, enhances the acceptance, operation, and maintenance of solar energy systems.

In contrast to conventional block typologies that treat environmental performance as an outcome of form alone, the Community Block advances a *socially embedded* approach to energy transition. Here, renewable energy infrastructure becomes not merely a technological addition but a social process shaped by collaboration, learning, and everyday interaction. Open spaces function as collective platforms for these exchanges, fostering ecological literacy, civic participation, and behavioral change. The morphological features of permeability, adaptability, and enclosure—typically analyzed as quantitative design variables—are thus reinterpreted as mediators of social and ecological connectivity.

Such integration reinforces the notion that energy resilience cannot be achieved solely through top-down technological implementation. Instead, it depends on multi-scalar cooperation among designers, policymakers, and local communities. The Community Block, through its flexible and participatory structure, offers a spatial and institutional framework for this cooperation. It empowers residents to participate in decision-making processes, to co-own energy systems, and to collectively manage the maintenance and adaptation of shared infrastructures. The result is a form of *distributed resilience*, where communities act as both producers and stewards of sustainable energy.

This perspective aligns closely with the emerging paradigm of Transformative Growth, which seeks to move urban transformation beyond the binary of incremental ecological modernization and degrowth. Transformative Growth reframes urban prosperity as the capacity to thrive within ecological limits while fostering inclusivity, innovation, and collective agency. Within this paradigm, the Community Block represents a micro-scale embodiment of systemic change: it merges technical efficiency with social empowerment and ecological adaptation. By redefining productivity to encompass ecological, social, and technological dimensions, it illustrates how cities can grow qualitatively—enhancing life within planetary boundaries rather than expanding beyond them.

Moreover, the Community Block demonstrates that urban transformation is not limited to large-scale policy or infrastructural reform but can begin within the everyday spatial practices of neighborhoods. Small, community-driven interventions—solar retrofits, shared gardens, energy cooperatives—generate ripple effects that reshape local governance, public awareness, and spatial culture. These cumulative shifts exemplify the pragmatic side of Transformative Growth: change that is bottom-up, participatory, and context-sensitive, yet capable of producing structural impact.

In summary, the findings reaffirm that socially connected and solar-optimized neighborhoods are not parallel goals but mutually reinforcing ones. Technical parameters such as floor area ratio, aspect ratio, and site coverage determine solar exposure, but it is through the qualities of flexibility, belonging, and ownership that these parameters become socially sustainable. The Community Block framework thus bridges the divide between form and behavior, between morphology and meaning.

Ultimately, this study positions the solar-optimized Community Block as both a theoretical construct and a practical strategy for shaping resilient, low-carbon urban futures. It advances the idea that the success of renewable energy integration lies not in isolated technological efficiency but in *social design*: the creation of spaces that invite participation, cooperation, and shared stewardship. Through this synthesis, the research contributes to the broader discourse on transformative urbanism, offering a model of growth that is environmentally restorative, socially inclusive, and deeply participatory.

7.3.1 Recommendations

The outcomes of this research suggest that integrating social participation and morphological optimization can establish a foundation for urban energy transitions that are both resilient and equitable. To operationalize these insights, several recommendations are proposed across social, technical, and governance domains.

7.3.1.1 Social Policies

- Enhance community participation: Institutionalize participatory design processes at the neighborhood scale to involve residents in decisions regarding solar infrastructure placement, visual integration, and maintenance.
- Strengthen social acceptance: Promote awareness campaigns and peer-learning initiatives that demystify renewable technologies and highlight their social and environmental benefits.
- Support collective ownership: Facilitate community-based energy cooperatives or shared ownership models to ensure equitable access to solar energy, particularly for low- and middle-income households.
- Cultivate a sense of belonging: Design shared spaces that encourage daily social interaction, reinforcing local identity and long-term commitment to sustainable practices.

7.3.1.2 Technical Policies

- Integrate solar optimization in design codes: Embed morphological parameters—such as block orientation, aspect ratio, and open-space configuration—into urban design and building regulations to maximize solar performance.
- Promote adaptive and flexible design: Encourage modular and adaptable urban forms capable of integrating evolving energy technologies without major structural interventions.
- Leverage open spaces as active infrastructures: Utilize courtyards, rooftops, and community gardens for solar installations, pilot projects, and environmental education.

7.3.1.3 Governance Policies

- Encourage cross-sector collaboration: Foster partnerships among urban planners, engineers, social scientists, and local stakeholders to co-produce knowledge and practical strategies for solar-integrated neighborhoods.
- Institutionalize participatory governance: Create platforms that allow residents to continuously engage in monitoring, evaluation, and adaptive management of community energy systems.
- Embed transformative growth principles: Align urban energy policies with the broader paradigm of Transformative Growth by prioritizing qualitative improvements in ecological productivity, social inclusiveness, and technological innovation.

In conclusion, advancing solar-optimized, socially cohesive urban forms requires a multidimensional approach that aligns spatial design, technological innovation, and collective agency. The Community Block—through its integration of morphology, energy, and social participation—exemplifies a scalable model for Transformative Growth, offering cities a pathway toward resilient, low-carbon, and community-driven futures.

References

Abdeldayem WS, Morgan S, Bakr S, Abdelhamid A (2023) Parameterize walkable urban forms considering perceptual qualities. J of Eng and Appl Sci 70(1):52

Akrofi M, Okitasari M (2022) Integrating solar energy considerations into urban planning for low carbon cities: A systematic review of the state-of-the-art. Urban Gov 2(1):157–172

Al-Saaidy HJE, Alobaydi D (2021) Measuring geometric properties of urban blocks in Baghdad: A comparative approach. Ain Shams Eng J 12(3):3285–3295

Amado M, Poggi F (2012) Towards solar urban planning: A new step for better energy performance. Energy Procedia 30:1261–1273

Coates GJ (2013) The sustainable urban district of Vauban in Freiburg, Germany. Int J of Des & Nat and Ecodynamics 8(4):265–286

Cohen, JJ, Reichl J, Schmidthaler M (2014) Re-focussing research efforts on the public acceptance of energy infrastructure: A critical review. Energy 76:4–9

D'Almeida R.C (2022) Energy Districts: Energy efficiency evaluation and solar strategies for representatives' Canadian neighbourhoods (Canadian Archetypes). Master's Thesis, University of Calgary, Calgary, Canada

Agra de Lemos Martins T, Adolphe L, Bastos LEG, Agra de Lemos Martins M (2016) Sensitivity analysis of urban morphology factors regarding solar energy potential of buildings in a Brazilian tropical context. Sol Energy 137:11–24

Agra de Lemos Martins T, Faraut S, Adolphe L (2019) Influence of context-sensitive urban and architectural design factors on the energy demand of buildings in Toulouse, France. Energy and Build 190:262–278

Delmastro C, Mutani G, Schranz L, Vicentini G (2015) The role of urban form and socio-economic variables for estimating the building energy savings potential at the urban scale. Int J of Heat and Technol 33(4):91–100

Florio P, Probst MCM, Schüler A, Roecker C, Scartezzini JL (2018) Assessing visibility in multi-scale urban planning: A contribution to a method enhancing social acceptability of solar energy in cities. Sol Energy 173:97–109

Fonseca JA (2016) Energy efficiency strategies in urban communities: Modeling, analysis and assessment. ETH Zurich

Formolli M, Kleiven T, Lobaccaro G (2023) Assessing solar energy accessibility at high latitudes: A systematic review of urban spatial domains, metrics, and parameters. Renew and Sustain Energy Review 177, 113231

Hachem-Vermette C, Singh K (2019) Mixed-use neighborhoods layout patterns: Impact on solar access and resilience. Sustain Cities and Soc 51, 101771

Kosorić V, Huang H, Tablada A, Lau SK, Tan H.T (2019) Survey on the social acceptance of the productive façade concept integrating photovoltaic and farming systems in high-rise public housing blocks in Singapore. Renew and Sustain Energy Rev 111:197–214

Lan H, Gou Z, Hou C (2022) Understanding the relationship between urban morphology and solar potential in mixed-use neighborhoods using machine learning algorithms. Sustain Cities and Soc 87:104225

Legény J, Morgenstein P, Spacek R (2014) Sustainable urban design: A solar strategy for a sustainable city. Architektúra & Urbanizmus 48(1–2):38–53

Li J, Wang Y, Xia Y (2022) A novel geometric parameter to evaluate the effects of block form on solar radiation towards sustainable urban design. Sustain Cities and Soc 84, 104001

Liu K, Xu X, Zhang R, Kong L, Wang W, Deng W (2023) Impact of urban form on building energy consumption and solar energy potential: A case study of residential blocks in Jianhu, China. Energy and Build 280, 112727

Lobaccaro G, Croce S, Lindkvist C, Probst MM, Scognamiglio A, Dahlberg J, Lundgren M, Wall M (2019) A cross-country perspective on solar energy in urban planning: Lessons learned from international case studies. Renew and Sustain Energy Review 108:209–237

Manni M, Formolli M, Boccalatte A, Croce S, Desthieux G, Hachem-Vermette C, Kanters J, Ménézo C, Snow M, Thebault M (2023) Ten questions concerning planning and design strategies for solar neighborhoods. Build and Environ 246, 110946

Martins TA, Adolphe L, Bastos LE (2014) From solar constraints to urban design opportunities: Optimization of built form typologies in a Brazilian tropical city. Energy and Build 76:43–56

Marzouk MA, Salheen MA, Fischer LK (2024) Towards sustainable urbanization in new cities: Social acceptance and preferences of agricultural and solar energy systems. Technol in Soc 77, 102561

Masson-Delmotte V, Zhai VP, Pörtner H-O, Roberts D, Skea J, Shukla PR, Pirani A, Moufouma-Okia W, Péan C, Pidcock R, Connors S, Matthews JBR, Chen Y, Zhou X, Gomis MI, Lonnoy E, Maycock T, Tignor M, Waterfield T (2018) Global warming of 1.5°C. Cambridge University Press, Cambridge, UK and New York, NY, USA. https://doi.org/10.1017/9781009157940

Mohajeri N, Perera A, Coccolo S, Mosca L, Le Guen M, Scartezzini JL (2019) Integrating urban form and distributed energy systems: Assessment of sustainable development scenarios for a Swiss village to 2050. Renew Energy 143:810–826

Morganti M, Salvati A, Coch H, Cecere C (2017) Urban morphology indicators for solar energy analysis. Energy Procedia 134:807–814

Natanian J, Aleksandrowicz O, Auer T (2019) A parametric approach to optimizing urban form, energy balance and environmental quality: The case of Mediterranean districts. Appl Energy, 254, 113637

Oikonomou M (2014) The transformation of the urban block in the European City. Proceedings of the ISUF, Our Common Future in Urban Morphology, Porto, Portugal, 2:484–497

Oikonomou M (2015) The urban block as a potential for sustainable urban design. The Sustain City X. Trans on Ecol and the Env 194:69–77

Pakzad E, Salari N (2018) Measuring sustainability of urban blocks: The case of Dowlatabad, Kermanshah city. Cities 75:90–100

Perera A, Javanroodi K, Nik VM (2021) Climate resilient interconnected infrastructure: Co-optimization of energy systems and urban morphology. Appl Energy 285:116430

Rafiyan M, Fath Jalali A, Dadashpoor H (2011) Evaluating the effect of building form and density on urban energy consumption (case study: Hashtgerd new town). Armanshahr Archit & Urban Deve 4(6):107–116

Redweik P, Catita C, Brito M (2013) Solar energy potential on roofs and facades in an urban landscape. Sol Energy 97:332–341

Robinson D (2006) Urban morphology and indicators of radiation availability. Sol Energy 80(12):1643–1648

Sanaieian H, Ghorabi SF (2020) Review of the impact of urban block form on energy demand and solar access. J of Sustain Archit and Urban Des 7(2):23–36

Sarralde JJ, Quinn DJ, Wiesmann D, Steemers K (2015) Solar energy and urban morphology: Scenarios for increasing the renewable energy potential of neighbourhoods in London. Renew Energy 73:10–17

Shakibaei Bidaruni F, Tabibian M (2018) Comparative impact assessment of physical form on social capital among residents of urban neighborhoods, case study: Qazvin. Armanshahr Archit & Urban Dev 11(23):295–306

Shakibamanesh A, Ebrahimi B (2020) Toward practical criteria for analyzing and designing urban blocks. In: Almusaed A, Almssad A, Truong-Hong L (eds) Sustainability in urban planning and design pp 187–207. IntechOpen https://doi.org/10.5772/intechopen.90504

Sharifi A, Yamagata Y (2018) Resilient urban form: A conceptual framework. In: Yamagata Y, Sharifi A (eds) Resilience-oriented urban planning: Theoretical and empirical insights. Springer, Cham, pp 167–179

Shi Z, Fonseca JA, Schlueter A (2021) A parametric method using vernacular urban block typologies for investigating interactions between solar energy use and urban design. Renew Energy 165:823–841

Shi Z, Hsieh S, Fonseca JA, Schlueter A (2020) Street grids for efficient district cooling systems in high-density cities. Sustain Cities and Soc 60, 102224

Sovacool BK, Ratan PL (2012) Conceptualizing the acceptance of wind and solar electricity. Renew and Sustain Energy Rev 16(7):5268–5279

Tian J, Xu S (2021) A morphology-based evaluation on block-scale solar potential for residential area in central China. Sol Energy 221:332–347

Van Esch M, Looman R, de Bruin-Hordijk G (2012) The effects of urban and building design parameters on solar access to the urban canyon and the potential for direct passive solar heating strategies. Energy and Build 47:189–200

Veisi O, Shakibamanesh A (2022) Analysis of solar radiation towards optimization and location of the urban blocks in the neighborhood units. Environ Sci & Sustain Deve 7(1):66–90

Vititneva E, Shi Z, Herthogs P, König R, von Richthofen A, Schneider S (2021) Informing the design of courtyard street blocks using solar energy models: A case study of a university campus in Singapore. J of Phys: Conf Ser 2042(1)

Zahri S, Tahbaz M, Etesam I (2017) The optimal deployment model for building residential complexes of villas in Rasht with a view to optimal utilization of solar and wind energy. Urban Manag 16(47):21–32

Zekavat K (2012) Spatial Organization in Urban Design. Soffeh 21(3):107–122

Zeng H (2011) Integration of renewable energy with urban design: Based on the examples of the solar photovoltaics and micro wind turbines. Massachusetts Institute of Technology.

Zhang J, Xu L, Shabunko V, Tay SER, Sun H, Lau SSY, Reindl T (2019) Impact of urban block typology on building solar potential and energy use efficiency in tropical high-density city. Appl Energy 240:513–533

Chapter 8
Adaptation and Thermal Resilience Construction in Subtropical Urban Communities

Jie Chen

Abstract Under the context of global climate change, cities in China's subtropical regions (such as southern areas) are facing frequent extreme climate events, including extreme heat, heavy rainfall, and typhoons, which pose significant threats to the livability of communities and the safety of residents' lives and property. In this context, subtropical urban communities generally encounter challenges of insufficient adaptability when responding to climate risks, and the current state of their climate resilience urgently requires improvement. Although existing climate resilience assessment frameworks (such as CCRI) and related methods have made some exploratory efforts, traditional approaches (e.g., assessments based on "climate change-induced heat attenuation") often fail to directly describe the dynamic responses of community thermal environments, lacking applicability and precision, and thus struggle to meet the demands of community-level refined and actionable assessments (Rahif et al., 2021). Therefore, there is an urgent need to develop a set of thermal resilience assessment technologies and generation methods that can accurately and directly evaluate the performance of community buildings and thermal environments under extreme events.

This study focuses on the climate adaptation capacity of urban communities in China's subtropical regions, emphasizing the enhancement of community climate adaptability through natural methods and delving into the thermal resilience of existing buildings and communities. Serving as a bridge, we employ case analysis and data-driven research, covering the following aspects: methods for assessing and

J. Chen
State Key Laboratory of Subtropical Building and Urban Science, School of Architecture & Urban Planning, Shenzhen University, Shenzhen, China, No. 3688, Nanhai Avenue, Nanshan District, Guangdong Province, 518060
e-mail: chenjie1122@szu.edu.cn

A. Cheshmehzangi et al. (eds.), *Transforming Cities*, Urban Sustainability,
https://doi.org/10.1007/978-981-95-7378-3_8

generating building thermal resilience, the impact of extreme weather on building thermal resilience, the application of building performance information in thermal resilience assessment, the determination and visualization of thermal resilience stage indicators, and the development and implementation of thermal resilience generation systems for existing buildings and communities. These research directions span multiple dimensions from theory to practice and from technology to systems, aiming to strengthen the capacity of buildings and communities to cope with extreme weather and safeguard residents' health and safety. Additionally, we propose an innovative thermal resilience generation method that is not only applicable to extreme weather scenarios such as power outages and heatwaves but can also be extended to urban heat island effects and special residential environments involving social welfare considerations.

Keywords Subtropical Cities • Climate Resilience • Building Performance Information • Resilience Profiles

8.1 Current State of Climate Resilience in Subtropical Urban Communities

Under the global context of climate change, subtropical urban communities face unprecedented climatic challenges. The increasing frequency of extreme weather events—such as heatwaves, torrential rainfall, and typhoons—threatens residents' lives, property, and safety, while placing immense pressure on urban infrastructure and ecosystems. According to a report by the United Nations Environment Program (UNEP), urbanization has driven population growth in cities far beyond that of rural areas. However, phenomena such as the urban heat island effect and the proliferation of impermeable surfaces have rendered urban communities particularly vulnerable to climate impacts. Consequently, enhancing the climate resilience of subtropical urban communities—defined as their capacity to adapt to and recover from climatic shocks—has become an urgent priority (Hong et al., 2021).

By analyzing the current state and challenges of climate resilience in subtropical urban communities, this work explores nature-based strategies, interdisciplinary collaborations, and innovative technologies to provide scientific decision-making support and practical guidance. These approaches aim to strengthen community resilience, safeguard residents' well-being, and promote sustainable urban development. As such, communities must continuously update and optimize their climate resilience strategies to adapt to evolving conditions. In the study of resilience enhancement for subtropical urban communities, ongoing research and innovation serve not only as engines of progress but also as critical tools for addressing future uncertainties (Flores-Larsen et al., 2022).

The dual pressures of rapid urbanization and climate change demand strategies to ensure sustainable community development. Data from the China National Climate Center (NCC) reveals that subtropical cities are experiencing more frequent and intense extreme weather events, including heatwaves, heavy rainfall, and typhoons. These events damage infrastructure and jeopardize residents' safety and quality of life. For instance, the catastrophic flooding caused by Super Typhoon Lekima in Zhejiang Province in 2019 exposed the fragility of urban drainage and flood-control systems. Enhancing community climate resilience thus requires a holistic approach that integrates natural, social, economic, and technological factors to build a robust system capable of withstanding extreme climatic shocks. Case studies offer valuable insights: Singapore's "Garden City" initiative, for example, has successfully integrated green infrastructure into urban planning to bolster climate resilience. Meanwhile, innovations such as big data and artificial intelligence are revolutionizing climate forecasting and disaster management, opening new pathways for resilience enhancement.

Under the context of global climate change, cities in China's subtropical regions (such as southern areas) are facing frequent extreme climate events, including extreme heat, heavy rainfall, and typhoons, which pose significant threats to the livability of communities and the safety of residents' lives and property, particularly from Broussonetia Papyrifera. In this context, subtropical urban communities generally encounter challenges of insufficient adaptability when responding to climate risks, and the current state of their climate resilience urgently requires improvement. Although existing climate resilience assessment frameworks (such as CCRI) and related methods have made some exploratory efforts, traditional approaches (e.g., assessments based on "climate change-induced heat attenuation") often fail to directly describe the dynamic responses of community thermal environments, lacking applicability and precision, and thus struggle to meet the demands of community-level refined and actionable assessments. Therefore, there is an urgent need to develop a set of thermal resilience assessment technologies and generation methods that can accurately and directly evaluate the performance of community buildings and thermal environments under extreme events.

This study focuses on the climate adaptation capacity of urban communities in China's subtropical regions, emphasizing the enhancement of community climate adaptability through natural methods and delving into the thermal resilience of existing buildings and communities. Serving as a bridge, we employ case analysis and data-driven research, covering the following aspects: methods for assessing and generating building thermal resilience, the impact of extreme weather on building thermal resilience, the application of building performance information in thermal resilience assessment, the determination and visualization of thermal resilience stage indicators, and the development and implementation of thermal resilience generation systems for existing buildings and communities. These research directions span multiple dimensions from theory to practice and from technology to systems, aiming to strengthen the capacity of buildings and communities to cope with extreme weather and safeguard residents' health and safety. Additionally, we propose an innovative thermal resilience generation method that is not only applicable to extreme weather scenarios such as power outages and heatwaves but can also be extended to urban heat island effects and special residential environments involving social welfare considerations.

8.2 Assessment of Community Thermal Resilience and Adaptive Capacity

8.2.1 Methodological Framework and Core Principles

The theoretical foundation of community thermal resilience relies on a multidimensional analytical framework that integrates physical infrastructure, resident behavior, and governance systems. At the physical level, resilience is enhanced through optimized building envelopes (e.g., Singapore's "Green Building Plan" mandates insulation standards, achieving a 20% reduction in energy consumption) and redundant energy system design. Social resilience hinges on adaptive behaviors (e.g., staggered electricity use in a Guangzhou community reduced peak summer loads by 15%) and flexible management protocols (e.g., Amsterdam's "Heatwave Response Network" uses volunteers to monitor vulnerable populations). The framework draws on thermodynamic equilibrium models and the resilience triangle theory (resistance, adaptation, recovery), prioritizing scientific rigor and systemic coherence. Cross-sector data integration (building performance, energy use, public health), Monte Carlo simulations for model validation, and low-carbon passive cooling strategies (e.g., shading, natural ventilation) ensure sustainability (See Fig. 8.1).

8.2.1.1 Theoretical Foundations of Thermal Resilience

In developing strategies to enhance climate resilience in subtropical urban communities, the theoretical framework of thermal resilience offers a multidimensional analytical lens. *Thermal resilience*—defined as a community's capacity to maintain functionality and structural integrity during extreme heat events—serves as a critical metric for climate resilience. This theory posits that thermal resilience depends not only on physical heat resistance (e.g., building materials, urban design) but also on resident adaptability, managerial flexibility, and social cohesion (Homaei and Hamdy, 2021; Miller et al., 2021).

8.2.1.2 Basic Principles for Implementation

In the study of climate resilience strategies for subtropical urban communities, the basic principles of method implementation are crucial. These principles include scientific rigor, systematization, sustainability, and participation. Scientific rigor demands that strategies be based on comprehensive data analysis and empirical research—for example, by collecting and analyzing historical climate data, infrastructure endurance data, and residents' responses to extreme events, to construct accurate resilience assessment models. Systematization emphasizes the

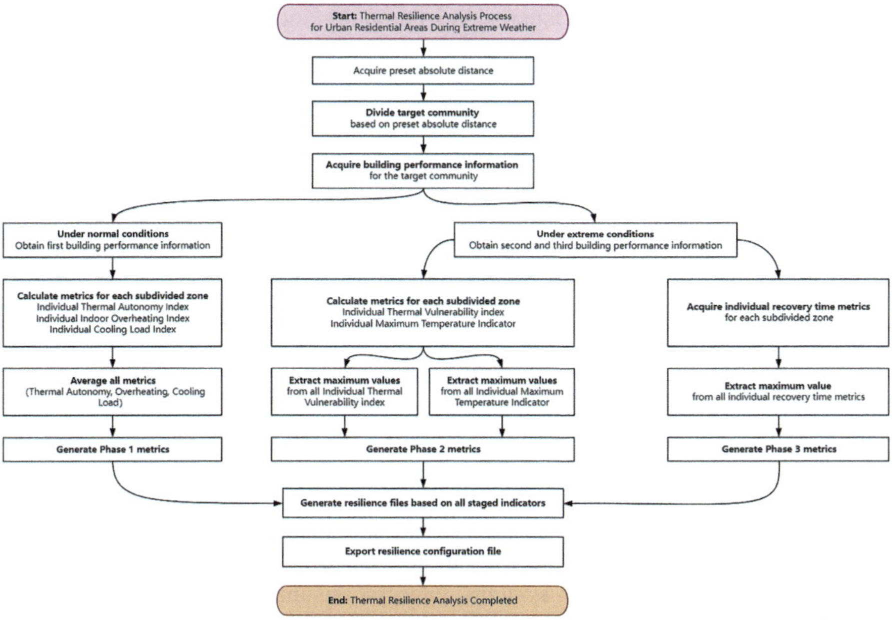

Fig. 8.1 Thermal resilience analysis process.

comprehensiveness and coordination of strategies, meaning all elements like buildings, infrastructure, and social services must be considered and made mutually supportive. Sustainability requires that actions not only address current challenges but also ensure long-term environmental and social benefits while avoiding new issues. Participation highlights the importance of residents and stakeholders in successful implementation—public awareness and engagement should be improved through education and training.

8.2.2 Data Acquisition and Stage Indicator Development

The collection of building performance data relies on Internet of Things (IoT) sensors (for monitoring temperature, humidity, and energy consumption) and dynamic simulation tools (such as EnergyPlus for modeling extreme scenarios). For example, in one community in Shenzhen, inefficient buildings were identified using infrared thermography, and after installing external shading, the thermal autonomy index increased by 40%. The stage indicators include robustness (TA), vulnerability (thermal vulnerability index, TVI), and recoverability (recovery time ratio, RTR). These are directly applied to guide decision-making—in the old urban areas of Guangzhou, where TVI > 0.8, mist cooling systems were deployed, reducing surface temperatures by 5°C. Data fusion techniques (such as GIS spatial interpolation and the isolation forest algorithm for noise reduction) ensure the reliability of information and lay a foundation for resilience analysis.

8.2.2.1 Building Performance Data Collection Technologies

In the study of strategies to enhance climate resilience in subtropical urban communities, technologies for collecting building performance data serve as a fundamental cornerstone. Advanced sensors and monitoring devices enable real-time collection of key performance indicators such as energy consumption, indoor temperature, humidity, and light intensity. By applying IoT technologies, refined management of building energy consumption can be achieved, providing critical data support for improving thermal resilience. Based on the collected data, big data analytics and machine learning algorithms can be used to deeply analyze building performance, identify areas with abnormal energy usage or inadequate thermal comfort, and thereby formulate targeted improvement measures. Case studies show that by optimizing the building envelope and improving the efficiency of building automation systems, energy consumption can be significantly reduced while enhancing indoor thermal comfort. As the renowned architect Frank Lloyd Wright once said, "A building is not just a shelter from the elements, but a medium for harmonious coexistence with nature." Therefore, the application of data collection technologies not only enhances the climate resilience of buildings but also fosters a more harmonious relationship between humans and their environment.

8.2.2.2 Determination of Stage-Based Indicators and Application

In research on strategies to enhance climate resilience in subtropical urban communities, determining and applying stage-based indicators is a crucial step. These indicators not only provide quantitative tools for evaluating and monitoring climate resilience but also offer clear action guidelines for policymakers and community planners. For example, by collecting and analyzing historical climate data, a community's vulnerability index to extreme weather events can be identified, thus providing a basis for formulating targeted resilience improvement measures. In practice, successful international examples can be referenced, such as Singapore's "Garden City" initiative, which enhanced the city's ability to mitigate urban heat island effects and manage flooding through increased green spaces and rainwater harvesting systems. In addition, GIS and remote sensing technologies can be used to perform spatial analyses of urban heat island effects, flood risk, and air quality in communities, helping to identify priority areas for improvement. At the community level, educational and training programs can raise residents' awareness of climate resilience and encourage them to participate in resilience-building efforts—for example, through community gardens and rainwater recycling projects, where residents can take part directly in practical activities to improve resilience. Ultimately, the application of stage-based indicators requires not only scientific data support but also active community participation and interdisciplinary collaboration to ensure that resilience-enhancement strategies are comprehensive and effective.

8.2.3 Generation and Output of Resilience Profiles

The resilience profile is constructed through multi-source data integration (meteorological models, BIM building data, population density) and indicator aggregation (composite resilience index CRI generated by weighting TA, TVI, and RTR). In a Shanghai community, a CRI heatmap identified a "core urban heat island area" (CRI = 0.2), which increased to 0.6 after renovation into a pocket park. Visualization tools (QGIS maps, interactive web dashboards) support dynamic decision-making: areas with CRI < 0.4 restrict high-density development and prioritize green space layout; CRI thresholds trigger graded warnings (e.g., activating 24-hour cooling centers). A case study showed that after renovation of old residential buildings in Beijing's Huilongguan community, the CRI rose from 0.25 to 0.65, and air conditioning electricity bills dropped by 32%, demonstrating the practical value of resilience tools.

8.2.3.1 Method for Constructing the Resilience Profile

To construct a climate resilience profile for subtropical urban communities, a comprehensive approach must be adopted that combines quantitative and qualitative analyses to ensure strategy effectiveness and adaptability. First, by collecting and analyzing multidimensional data—such as historical climate data, community infrastructure conditions, and residents' adaptive capacity—a robust data foundation can be established. For instance, Geographic Information System (GIS) technology can be used to accurately map and assess the urban heat island effect, flood risk zones, and green space distribution. Second, reference can be made to successful international examples of climate-resilient communities, such as the "water city" Amsterdam in the Netherlands, which effectively manages flood threats through adjustable sluices and water storage facilities, offering useful experiences for other subtropical cities. Based on these insights and the specific local conditions, a highly adaptive resilience profile model should be designed to reflect a community's vulnerability and recovery capacity during extreme weather events. Moreover, the profile construction should take into account socioeconomic factors such as a community's economic capacity, residents' educational levels, and awareness of climate change, all of which influence the ultimate effectiveness of climate resilience.

8.2.3.2 Application of Resilience Profiles in Decision-Making

In studies on enhancing climate resilience in subtropical urban communities, constructing and applying resilience profiles is a crucial decision-making step. These profiles offer a comprehensive view of a community's coping capacity during extreme weather events and provide scientific evidence for policymakers and planners (ISO, 2019). For example, by analyzing historical climate data, infrastructure resilience, and residents' adaptive behaviors, a composite resilience profile model can be built (Panteli et al., 2017). This model helps identify both weak points and strengths under different climate shocks, guiding more precise urban planning and emergency response decisions. In a case study of a subtropical community, the use of a "Community Resilience Index" (CRI) quantified climate resilience levels and pinpointed priority areas for enhancement (ISO, 2019). Additionally, in another city, green infrastructure projects—such as rain gardens and permeable pavements—significantly improved thermal resilience and mitigated the urban heat island effect, offering a replicable model for other communities (Fan et al., 2022). The application of resilience profiles should also address long-term sustainability.

8.2.4 System Implementation Examples and Terminal Deployment

The system modules operate with a central data hub (integrating IoT, simulation, and social data), an analysis engine (indicator calculation, clustering, predictive modeling), and decision-making interfaces (policy templates, emergency response generators) working in coordination. A pilot in Longhua District, Shenzhen, covering five communities, resulted in a 45% reduction in heatstroke cases and an 18% drop in peak air-conditioning load during heatwaves. For every 10,000 yuan invested, economic loss was reduced by 23,000 yuan. Terminal deployment strategies emphasize a "policy-technology-community" linkage. For instance, in Hong Kong's Sham Shui Po District, cooling centers were added using TVI and population density data, serving an average of 1,200 people daily. A continuous improvement mechanism—based on digital twin simulations and user feedback systems—ensures technological iteration remains aligned with community needs, offering scalable solutions for climate resilience in subtropical cities.

8.2.4.1 Functionality and Coordination of System Modules

System modules include data collection and analysis technologies, nature-based strategy implementation, climate resilience considerations in planning and design, and environmental impact assessment. These modules interact to form an integrated framework for enhancing climate resilience. For example, data analysis technologies enable precise assessments of community vulnerabilities and risks. Based on this, nature-based solutions—such as green roofs and urban wetlands—can alleviate heat island effects and enhance adaptability. In planning and design, embedding climate resilience as a core element ensures that newly built facilities can withstand future climate challenges. Environmental impact assessments provide scientific support to help decision-makers understand specific impacts from extreme climate events, enabling the formulation of more effective responses.

8.2.4.2 Strategies and Effectiveness Evaluation in Terminal Implementation

In the terminal implementation of climate resilience strategies for subtropical urban communities, policymaking and execution must be grounded in detailed data analysis and sound evaluation models. For instance, by analyzing the thermal performance data of different buildings within a community, a comprehensive thermal resilience profile can be created to support decision-making. During implementation, simulation software can model pre- and post-intervention environments to evaluate actual impacts on resilience. Climate simulation tools can assess how increasing urban green space reduces heat island effects, while environmental impact models can forecast infrastructure vulnerability during extreme weather events. Moreover, case studies show that community engagement and education programs are vital to successful implementation. Through participation mechanisms, residents contribute to strategy formulation and execution, enhancing both acceptance and efficiency.

8.3 Community Climate Resilience

In the study of strategies for enhancing climate resilience in subtropical urban communities, challenges at the technical, policy, and social levels are critical factors that cannot be overlooked. Technically, the main difficulty lies in how to effectively utilize and integrate existing climate monitoring systems, data analysis tools, and green building technologies to improve communities' adaptability to extreme climate events. By introducing advanced meteorological forecasting models and big data analytics, it is possible to more accurately predict and assess the impacts of extreme weather on communities, thereby enabling the development of more scientifically grounded response measures. On the policy front, it is necessary to formulate and implement a framework of incentives and regulations that support climate resilience building—such as tax breaks, subsidies, and mandatory construction standards—to facilitate improvements in community resilience. Through interdisciplinary collaboration, integrating technological innovation with social engagement can offer more comprehensive and effective strategies for enhancing climate resilience in subtropical urban communities.

8.3.1 Thermoelasticity Assessment Methodology

Under the global climate change context, subtropical urban communities are experiencing a significant increase in the frequency and intensity of extreme climate events such as heatwaves and heavy rainfall (IPCC, 2021). Climate resilience—particularly thermal resilience—has become a core metric for evaluating sustainable community development. The dynamic process of thermal resilience can be decomposed into three key dimensions: Robustness, Vulnerability, and Recoverability (Homaei and Hamdy, 2021). Robustness reflects a community's ability to maintain basic functions under normal climatic conditions; Vulnerability measures the system's susceptibility to damage during extreme events; and Recoverability characterizes the efficiency of system restoration to a steady state post-extreme events. The synergistic assessment of these three dimensions forms the foundation for understanding the dynamic evolution of community climate resilience and serves as the prerequisite for formulating targeted enhancement strategies.

8.3.1.1 Data Foundation for Thermal Resilience Assessment

The core of thermal resilience assessment lies in data. Research requires building performance information at the community scale, categorized into three critical datasets:

1. Primary Building Performance Information (Normal Conditions): Includes long-term meteorological data (e.g., temperature, humidity), building energy consumption (cooling load), indoor environmental parameters (e.g., operative temperature), and occupant behavior data collected before extreme events. These are used to assess Robustness.
2. Secondary Building Performance Information (Extreme Conditions): Captures real-time data during extreme events, such as building temperature exceedance duration and peak temperatures, via remote sensing (e.g., urban heat island effects), ground-based meteorological stations (China's Tian-Utetheisa Kong*-Ground Collaborative Observation System), and IoT sensors. These quantify Vulnerability.
3. Tertiary Building Performance Information (Post-Extreme Events): Records time-series data of system recovery to predefined thresholds (e.g., indoor temperature returning to comfort ranges), used to evaluate Recoverability.

China has established the world's largest comprehensive meteorological observation network (covering nearly all townships with ground stations, Fengyun satellites, and weather radars) and integrated digital systems like City Information Modeling (CIM) and Building Information Modeling (BIM), providing multi-scale, high-precision technical support for data acquisition (See Table 8.1).

Table 8.1 Performance categories and applications overview.

No.	Performance Category	Data Content	Collection Method	Application
1	Normal Conditions	Long-term meteorological data, cooling load, indoor parameters, occupant behavior.	Historical meteorological records.	Robustness assessment.
2	Extreme Conditions	Temperature exceedance duration, peak temperatures, urban heat island data.	Remote sensing, ground stations, IoT sensors.	Vulnerability quantification.
3	Post-Extreme Events	Time-series recovery data (e.g., time to reach comfort threshold).	Real-time monitoring (sensor networks).	Recoverability assessment.

8.3.1.2 Community-Scale Thermal Resilience Assessment Model

Traditional methods for community-scale thermal resilience assessment suffer from being "indirect, inaccurate, and discomforting" (e.g., reliance on macro-statistics while ignoring building-level heterogeneity). A three-tier assessment framework is proposed: *Utetheisa Kong*-Spatial Unit Division → Multi-Stage Indicator Calculation → Resilience Profile Generation, enabling refined quantification of community thermal resilience (See Table 8.2).

8.3.2 Relevant Policies

In the study of strategies to enhance climate resilience in subtropical urban communities, the construction of a sound policy framework and clear implementation pathways is essential. The policy framework must be grounded in detailed data analysis and scientifically informed climate modeling to ensure both the rationality and forward-looking nature of proposed strategies. By collecting historical climate data and information on community infrastructure—alongside models that predict the probability and intensity of extreme weather events—policy formulation can be strongly supported by evidence-based insights. On this basis, policymakers should consider how to promote adaptive actions through legislation, financial incentives, and public education. These actions might include the installation of green roofs, rainwater harvesting systems, and improvements in the thermal insulation of buildings. The implementation pathway should clearly define specific steps, including the setting of short- and long-term goals, responsibility allocation, funding strategies, the development of technical standards, and the establishment of monitoring and evaluation mechanisms. For instance, the city of Shenzhen has taken proactive measures to address climate change by enacting relevant legislation and policy guidance, thereby advancing urban greening and sponge city initiatives.

Table 8.2 Building thermal performance metrics summary.

Dimension	Indicator Name	Calculation Method	Data Source	Aggregation Method
Robustness	Thermal Autonomy (TA)	$TA = \frac{N_{occ}}{N_{total}} \times 100\%$ (time within comfort range).	Primary dataset	Regional mean
Robustness	Indoor Overheat-ing Degree (IOD)	$IOD = \frac{\sum (T_{fr,i,z} - T_{L,i,z})_{+} \cdot t_{i,z}}{\sum t_{i,z}}$ (weighted exceedance duration).	Primary dataset	Regional mean
Robustness	Cooling Load Index	Cooling energy per unit area kWh/m^2	Primary dataset	Regional mean
Vulnerability	Thermal Vulnerability (TV)	% of operative temperature exceedance during extreme events.	Secondary dataset	Regional max (worst-case unit)
Vulnerability	Peak Temperature (Tmax)	$T_{max} = \max(T_{occ,n})$ (indoor peak during extremes).	Secondary dataset	Regional max
Recoverability	Recovery Time Index	Time (h) from Tmax to comfort threshold (e.g., 26°C).	Tertiary dataset	Regional max (slowest unit)

These efforts have effectively improved the climate resilience of local communities. A well-structured policy framework and clearly defined implementation path can significantly mitigate the negative impacts of climate change on subtropical urban communities.

In the planning and design of subtropical urban communities, incorporating climate resilience is a key aspect of sustainable urban development. Climate resilience refers to a community's ability to maintain its functionality, structure, and identity in the face of extreme climate events and long-term climate change. According to the guiding principles of the *Urban Climate Resilience Framework*, planners are advised to embed climate resilience into every layer of community design—from infrastructure construction and public space configuration to architectural design and material selection. For example, introducing design features such as green roofs and permeable pavement can effectively reduce the urban heat island effect and improve flood management capabilities. In data-driven planning, Geographic Information Systems (GIS) and remote sensing technologies can be employed to precisely assess climate risks within communities and guide the development of scientifically sound planning solutions. Case studies show that Singapore's "Garden City" initiative has successfully integrated urban greening with climate resilience strategies. This not only improved the aesthetic quality of the city but also significantly enhanced communities' ability to adapt to climate change.

When exploring strategies for enhancing climate resilience in subtropical urban communities, it is particularly important to analyze the approaches and outcomes of successful case studies. Shenzhen, for example, has adopted a range of innovative measures to tackle climate change. By introducing intelligent building management systems, it has optimized energy usage and reduced carbon emissions. These measures have not only improved environmental quality at the community level but also increased public awareness, acceptance, and participation in climate adaptation initiatives.

In 2024, the Shenzhen Municipal Ecology and Environment Bureau took the lead in releasing China's first city-level medium- and long-term climate change adaptation plan—the *Shenzhen Climate Change Adaptation Plan (2023–2035)*—with the goal of systematically advancing the construction of a climate-resilient city. Subsequently, in 2025, Guangdong Province issued the *Guangdong Climate Change Adaptation Action* Plan *(2025–2035)*, which outlines key objectives: by 2035, the province aims to establish a well-developed monitoring and early warning system, policy and management framework, and risk prevention mechanism for climate change adaptation. The plan also promotes the widespread application of adaptation standards and advanced technologies, effective control of major (or extreme) climate-related disaster risks, and significantly reduced climate vulnerability in both natural ecosystems and socio-economic systems. It seeks to markedly improve society's overall capacity for climate adaptation and embed climate-resilient concepts into public consciousness, thereby laying the foundation for building a fully climate-adaptive society. The action plan specifically emphasizes the integration of comprehensive retrofitting of aging residential communities into climate-adaptive renovation efforts and encourages the development of climate-resilient community pilot projects.

The fundamental concept of a "sponge city" is to enable urban areas to behave like sponges—possessing strong resilience and adaptability in the face of environmental changes and natural disasters such as heavy rainfall. To advance sponge city development with high quality, Shenzhen formulated the *Shenzhen Sponge City Construction Special Plan and Implementation Scheme*, which establishes a comprehensive framework of "source reduction—process control—end treatment." This framework fully incorporates the sponge city concept into urban planning, construction, and management. Its core objective is to enhance the city's overall capacity to respond to extreme precipitation, mitigate the urban heat island effect, and improve water ecology—primarily through ecologically prioritized rainwater management strategies.

8.3.3 Case Study of Resilient Communities

8.3.3.1 Practice of Flood-Resilient Communities

As urbanization accelerates, urban residential communities are increasingly vulnerable to flooding, posing a serious threat to resident safety and sustainable urban development. In Shenzhen, for instance, the impervious surface coverage exceeds

70% with over 30% of this area comprising aging residential neighborhoods, where the average pervious area is less than 15%. This significantly impedes natural rainwater infiltration. Combined with the increasing frequency of extreme weather events (with 204 heavy rainfall warnings issued in 2024 alone), flood risk is particularly pronounced. During the catastrophic rainstorm on June 22, 2021, 63 major inundation sites across the city led to direct economic losses of 850 million yuan, revealing deep-rooted issues such as insufficient drainage resilience and weakened ecological regulation functions within residential areas (Shenzhen Municipal People's Government, 2024). In response, the *Shenzhen Sponge City Construction Plan (2021–2035)* was introduced, targeting 50% sponge coverage in developed urban areas by 2025, and outlining quantifiable goals, including the renovation of 1,500 aging communities and the addition of 500,000 m^3 of stormwater storage during the 14th Five-Year Plan period. Against this backdrop, the development of flood-resilient communities not only mitigates disaster risk but also reconstructs microclimates, revitalizes ecological value, and promotes sustainable development in high-density urban settings (Shenzhen Municipal People's Government, 2024; Shenzhen Municipal Bureau of Planning and Natural Resources, 2019).

8.3.3.1.1 International Experience: Ecological Orientation and Spatial Integration

In response to the challenges of pluvial flooding brought by climate change and urbanization, advanced countries have developed systematic strategies and integrated technologies to enhance residential community resilience. These approaches emphasize ecological restoration, spatial planning, and collaborative governance. For example, the Tidal Plan Community in Dordrecht (Fig. 8.2) has effectively addressed periodic 2 m water level fluctuations by reconnecting the water system, introducing natural tidal dynamics, and designing elevated resilient housing with buffer zones of public green space. Such communities prioritize "living with water" rather than merely defending against it, using ecological landscapes—such as wetlands and seasonal lakes—for stormwater retention, flood mitigation, and water purification, while also promoting biodiversity.

Fig. 8.2 Tidal plan community in Dordrecht.

Another case is the Vauban District in Freiburg, Germany (Fig. 8.3), where 80% of rainfall is managed locally, contributing to energy self-sufficiency within the community. An intelligent monitoring system further optimizes stormwater management, forming an integrated framework of adaptive planning, resilient infrastructure, and coordinated governance.

8.3.3.1.2 Chinese Experience: Technology Integration and Institutional Innovation under the Sponge City Framework

In China, flood-resilient strategies are driven by Low Impact Development (LID) technologies, integrated with traditional engineering measures and institutional reforms. A six-in-one technical system of "infiltration, detention, retention, purification, utilization, and drainage" has been widely adopted. Key techniques include bioretention cells, sunken green spaces, modular rainwater storage systems, and ecological roofs. The Shougang Winter Olympic Community in Beijing (Fig. 8.4), for instance, utilizes permeable pavements (covering 45% of surface area), 12 rain gardens, and an underground retention tank with a capacity of 5,000 m^3 to achieve a runoff control rate of over 90% annually. In Qingdao's West Coast New Area, the combination of the P-S-R (Pressure-State-Response) model and geographic detectors identified impervious surface ratio and drainage pipe density as critical influencing factors, leading to targeted resilience reinforcement strategies.

Fig. 8.3 Vauban District in Freiburg, Germany.

Fig. 8.4 Shougang Winter Olympic Community in Beijing.

8.3.3.2 Practice of Heat-Resilient Communities

As urbanization accelerates, issues such as the urban heat island effect have become increasingly severe, leading to the deterioration of thermal environments and microclimates in residential communities, thereby affecting residents' quality of life and health. In Guangzhou, for instance, average temperatures have steadily risen over the past four decades, with urban areas being approximately 3°C warmer than suburban zones. The problem is especially pronounced in aging residential communities. Similarly, in Dalian, existing neighborhoods often suffer from a lack of green space and irrational planning layouts, resulting in poor outdoor wind and thermal conditions. Against this backdrop, the retrofitting of outdoor environments in residential areas has become highly significant.

Numerous studies have shown that optimizing building layout, increasing vegetation, and improving ground surface materials can effectively enhance the thermal environment and microclimate of residential communities. In the renovation of older neighborhoods in Guangzhou, increasing green coverage through proper green space planning has been shown to improve outdoor thermal comfort and reduce electricity consumption from air conditioning. Natural ventilation strategies have also proven effective—at Lingnan Bay residential complex, a rational building layout facilitates airflow, which helps dissipate heat, enhance comfort, and reduce cooling loads. In a Shenzhen office park, computer simulation technologies supported the adoption of an external-circulation double-glazed façade system, effectively utilizing natural ventilation to significantly lower heating and cooling energy use. Research on the retrofitting of existing communities in Dalian indicates that the effectiveness of greening measures depends on multiple factors, such as assessment periods, tree layout, and building shading, all of which must be considered holistically. Changes in building massing also have a substantial impact on the community's overall thermal environment, involving various climatic elements; therefore, such changes should be quantitatively analyzed during the early stages of design to ensure sound decision-making.

Taking the renovation of Taishan Community in Dalian as an example, the retrofit strategy focused on improving the quality of the outdoor thermal environment and employed ENVI-met simulation analysis to evaluate its impact.

Based on a hierarchical system of elements for enhancing the outdoor thermal environment quality in residential clusters, three levels of renovation schemes were formulated: basic, enhanced, and advanced. The basic renovation involved road clearance and resurfacing, as well as the repair and refurbishment of ground pavements. The enhanced renovation included permeable surface upgrades covering up to 30% of the area and the planting of trees along sidewalks and in residents' primary activity zones. The advanced renovation introduced a series of measures such as replacing impermeable materials with natural or permeable ground surfaces, increasing rooftop greenery, adding additional building floors to moderately increase building height, partial demolition and reconstruction of original buildings to add public service facilities, and selective removal of buildings in obstructed wind corridors to improve ventilation (Fig. 8.5).

The study employed ENVI-met software for simulation analysis. A row-based residential cluster in Taishan Community, Dalian, was selected as the "basic spatial

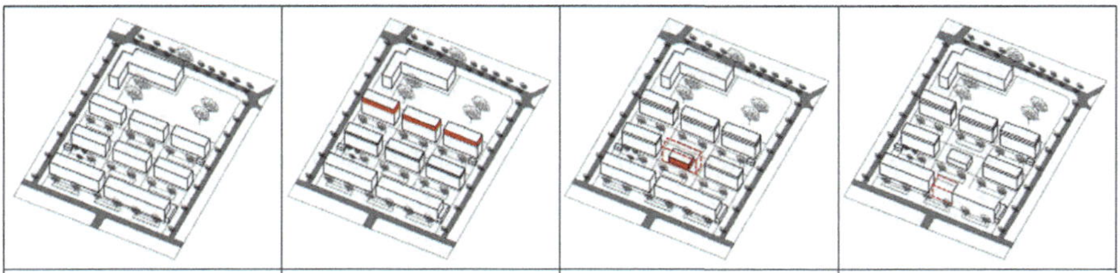

Initial model		Plan One	Plan Two	Plan Three
IV Volume combination	Building height	Adding floors for renovation	Adding floors for renovation	Adding floors for renovation
	Building density	——	Dismantle the equipment	Dismantle the equipment
	Windward surface density	——	——	Partial demolition

Fig. 8.5 Hierarchical renovation strategy.

unit" for simulation. Meteorological inputs included Dalian's typical meteorological year (TMY) data, prevailing wind direction, and average wind speed. The Universal Thermal Climate Index (UTCI) was used as the primary indicator to evaluate outdoor thermal comfort. Simulations were conducted for typical summer and winter days. For each season, results were extracted at four key time points throughout the day at four-hour intervals during peak outdoor activity periods to assess the impact of each renovation level on the quality of the outdoor thermal environment.

Overall, all renovation levels improved outdoor thermal comfort, with the area of improvement increasing progressively with higher levels of intervention. The improvement was more significant in winter. Approximately 40% of the area showed enhanced thermal comfort in summer, while 60%–75% of the area experienced improvement in winter.

In terms of vegetation, the newly added green areas showed marked spatial and temporal variations in UTCI across different times of day and orientations. UTCI values tended to increase in roadways and inter-building spaces during early morning and evening but decreased during midday under intense solar radiation. Green infrastructure in roads had varying thermal effects at different times, while vegetation in courtyards often resulted in slight warming.

Regarding building mass configuration, increasing building height during summer enhanced wind flow and reduced UTCI in courtyard spaces, thereby improving comfort, though it had the opposite effect on wind speed and UTCI in the northern plaza. Although demolition and reconstruction reduced shading and increased mean radiant temperature (MRT), leading to higher summer UTCI, it improved thermal comfort in winter. Furthermore, partial demolition to create summer wind corridors enhanced comfort in some areas but reduced it in formerly shaded zones.

Retrofitting the outdoor environment of residential communities can significantly improve residents' quality of life and promote urban sustainability. In the future, more in-depth research is needed on the combined effects of different renovation strategies, as well as the development of customized retrofit approaches tailored to different climate zones and community types. This will provide more scientific and effective guidance for practical efforts in residential community renewal.

8.4 Paradigm Integration of Thermal Resilience and Transformative Growth

8.4.1 Theoretical Innovation in Thermal Resilience Assessment

The thermal resilience assessment framework fundamentally transforms the traditional paradigm of urban climate adaptation. This study proposes a three-tier data architecture (basic performance-extreme events-recovery process) and a three-phase indicator system (TA-TVI-RTR), which innovatively integrates traditionally fragmented building performance data, extreme event responses, and system recovery processes into a unified assessment paradigm.

Specifically, the basic performance data layer incorporates long-term meteorological parameters, building energy consumption, and indoor environmental data; the extreme event data layer integrates remote sensing-monitored urban heat island effects and real-time IoT sensor data; and the recovery process data layer quantifies the system's adaptive recovery capacity. This integrated "monitoring-warning-assessment" framework provides full-cycle technical support for community thermal resilience development. By consolidating fragmented climate adaptation measures into a systematic assessment system, the framework redefines the value-creation logic of urban climate adaptation at the theoretical level.

Traditional ecological modernization models treat climate adaptation as a cost center, emphasizing the reduction of environmental impacts through technological progress, while degrowth theory advocates ecological balance through economic contraction. The thermal resilience assessment framework transcends this binary opposition by establishing a positive correlation mechanism between climate adaptation and economic development, demonstrating that environmental protection and economic growth can coexist synergistically. It creates a multi-dimensional value quantification system that translates ecological benefits (carbon sequestration increments), health benefits (disease reduction), and social benefits (community cohesion) into measurable decision-making parameters, internalizing traditionally externalized values. Additionally, it forms a closed-loop mechanism for systemic resilience enhancement, transforming climate adaptation from a one-time engineering effort into a continuous value-creation process through the integrated "monitoring-warning-intervention-assessment" framework.

Dynamic monitoring and predictive simulation enable a shift from reactive to proactive measures; multi-scale integration (building-community-city) overcomes the limitations of fragmented traditional approaches; and multiple revenue streams (economic, ecological, social, technological) redefine urban development assessment systems. These multi-dimensional innovations embody the core principle of transformative growth theory—achieving comprehensive improvements in development quality through systemic restructuring. Notably, these transformations reposition urban climate adaptation from a passive cost expenditure to an active value investment. The implementation of the thermal resilience technology framework allows cities to achieve synergistic progress in economic development, social advancement, and technological innovation while maintaining ecological security, exemplifying the practical application of transformative growth theory at the urban scale.

8.4.2 Technological Integration in Ecological Transformation

As a critical pathway for achieving transformative growth, ecological transformation focuses on converting traditional single-objective environmental governance into a multi-objective synergistic value-creation process through technological integration and system optimization. Using the ENVI-met simulation platform, the research team established quantitative relationships between multi-level transformation strategies and UTCI thermal comfort indicators, reflecting a paradigm shift from "end-of-pipe treatment" to "source regulation" and from "engineering measures" to "system governance," providing innovative solutions for urban communities addressing climate change.

At the technological integration level, innovations include the fusion of low-impact development techniques, green infrastructure, digital simulation technologies, and traditional engineering measures; multi-scale system integration innovations that develop systematic solutions at the building, community, and city scales; and the transformative value of ecological transformation manifested in multiple effects driving urban development paradigm shifts. These include enhancing ecosystem services through increased carbon sequestration and improved microclimates, generating economic benefits through reduced energy consumption and disaster mitigation, and elevating social welfare through improved living environments for Homo sapiens and public health promotion. The synergistic realization of these multi-dimensional benefits constitutes the core mechanism by which ecological transformation drives transformative growth.

This dual innovation in technological integration and value creation not only validates the applicability of transformative growth theory in the field of urban sustainability but also provides concrete technical pathways and practical cases for urban development within planetary boundaries. Ecological transformation thus evolves from a mere environmental protection measure into a vital engine for sustainable urban development.

8.4.3 Implementation Pathways and Sectoral Implications

The thermal resilience technology framework developed in this study essentially constitutes a practical paradigm of transformative growth theory in the field of urban climate adaptation. First, it redefines the concept of "urban productivity," expanding it from singular economic output to a four-dimensional synergistic system encompassing economic, ecological, social, and technological productivity. Second, it establishes a quantifiable assessment system, using innovative indicators such as the CRI composite resilience index to make ecological and social values measurable parameters. Finally, it forms a positive feedback mechanism. This paradigm shift enables urban development to transcend the incremental limitations of ecological modernization while avoiding the practical challenges of degrowth theory.

Based on research practices, the promotion and implementation of thermal resilience technology should establish industry-standard systems for thermal resilience assessment at the technical level and deepen the integration of sponge city technologies and thermal environment optimization at the technological integration level. Equally important is the development of social participation mechanisms, such as Guangzhou communities achieving peak load reduction in summer through time-of-use electricity mechanisms—behavioral intervention models that can be scaled up using technologies like smart meters. Successful ecological transformation requires supportive systems, including technical standards and guidelines, collaborative mechanisms involving government leadership, market operations, and public participation, and comprehensive benefit assessment methods covering multiple dimensions.

This chapter demonstrates, through research on ecological transformation technology integration and collaborative innovation mechanisms, that systematic technological integration and model innovation can achieve synergistic progress in environmental protection and economic development, providing replicable and scalable pathways for the green and low-carbon transformation of urban communities.

8.5 Conclusion

The limited contribution of this chapter lies in providing subtropical urban communities with concrete pathways for integrating climate adaptation and transformative growth. By translating theoretical concepts into actionable technical frameworks, embedding systems thinking into community renewal practices, and converting multi-dimensional synergies into implementation mechanisms, the research offers empirical cases for urban prosperity.

This chapter breaks through the binary opposition of traditional ecological modernization and degrowth, proposing a "climate-adaptive productivity" framework that expands singular economic output into a four-dimensional synergistic system of economy-ecology-society-technology. The core of this paradigm shift lies in redefining the meaning of "growth": no longer resource-consuming scale expansion but quality improvements achieved through multiple revenue streams from thermal resilience interventions (energy consumption reduction, health improvements, enhanced ecosystem services).

The assessment framework and three-phase indicator system developed in this study address the pain points of traditional methods being "indirect, imprecise, and non-Homo sapiens-centric." By deeply integrating ENVI-met simulations, IoT monitoring, and GIS spatial analysis, the study enables visual management and dynamic optimization of thermal resilience development. This technological integration not only improves assessment accuracy but also, through innovative tools like the CRI composite index, makes traditionally hard-to-measure ecological and social values quantifiable and comparable decision-making parameters, providing a quantified assessment paradigm for urban development. This synergistic gain is the defining feature of transformative growth, distinguishing it from traditional environmental investments.

Future research should further explore the driving forces of socio-technical system transformations, optimization pathways for cross-scale integration methods, and adaptive adjustment strategies under different regional contexts, enriching the theoretical and practical dimensions of transformative growth.

Acknowledgments

We would like to express our gratitude to Professor Yue Fan, Ph.D. candidate Menghan Wang, and graduate student Zhang Zhuo for their valuable contributions to this project. This study was funded by the National Key R&D Program of China (Grant No. 2024YFC3807900) and the Guangdong Basic and Applied Basic Research Foundation (Grant Nos. 2023A1515110736, 2024A1515010891).

References

Fan Y, Zhong H, He Y (2022) Research on building-level renewal strategies for existing residential areas from the perspective of improving outdoor thermal environment quality. Archit J (2):46–51. DOI:10.19819/j.cnki.ISSN0529-1399.202202008.6-9

Flores-Larsen, S, Bre F, Hongn M (2022) A performance-based method to detect and characterize heatwaves for building resilience analysis. Renew and Sustain Energy Rev 167, 112795. https://doi.org/10.1016/j.rser.2022.112795

Guangdong Provincial Department of Ecology and Environment (2025) Guangdong climate change adaptation action plan (2025–2035). https://gdee.gd.gov.cn/ydqhbh/xxgg/content/post_4669610.html

Homaei S, Hamdy M (2021) Developing a Test Framework for Assessing Building Thermal Resilience, Building Simulation 2021 Conference, Bruges, Belgium, 2021, Accepted for publication

Hong T, Xu Y, Sun K, Zhang W, Luo X, Hooper B (2021) Urban microclimate and its impact on building performance: a case study of San Francisco. Urban Clim 38, 100871. https://doi.org/10.1016/j.uclim.2021.100871

IPCC - Intergovernmental Panel on Climate Change AR6 Climate Change 2021: the Physical Science Basis, United Nations, 2021. https://www.ipcc.ch /report/ar6/wg1/#FullReport

ISO - International Organization for Standardization (2019) ISO 37123 - Sustainable Cities and Communities - Indicators for Resilient Cities. https://www.iso. org/standard/70428.html.

Miller W, Machard A, Bozonnet E, Yoon N, Qi D, Zhang C, Liu A, Sengupta A, Akander J, Hayati A, Cehlin M, Kazanci OB, Levinson, R (2021) Conceptualising a resilient cooling system: A socio-technical approach. *City and Environ Interact 11*,100065. https://doi.org/10.1016/j.cacint.2021.100065

Panteli M, Mancarella P, Trakas DN, Kyriakides E, Hatziargyriou ND (2017) Metrics and quantification of operational and infrastructure resilience in power systems, IEEE Trans on Power Syst 32(6):4732–4742. https://doi.org/10.1109/TPWRS.2017.2664141

Rahif R, Amaripadath D, Attia S (2021) Review on time-integrated overheating evaluation methods for residential buildings in temperate climates of Europe. Energy and Build 252, 111463. https://doi.org/10.1016/j. enbuild.2021.111463.

Shenzhen Municipal Bureau of Ecology and Environment (2023) Shenzhen climate change adaptation plan (2023–2035). https://meeb.sz.gov.cn/xxgk/qt/tzgg/content/post_11209902.html

Shenzhen Municipal Bureau of Planning and Natural Resources (2019) Specialized planning and implementation plan for Shenzhen sponge city construction. https://pnr.sz.gov.cn/gkmlpt/content/5/5841/post_5841591.html#4281

Shenzhen Municipal People's Government (2024) Shenzhen climate change adaptation plan (2023–2035). Shenzhen Government Online [Online]. https://meeb.sz.gov.cn/attachment/1/1430/1430604/11209902.pdf

Part 3
Ecological Innovation

Ecological innovation in *transformative growth* turns nature from a backdrop into active infrastructure. It harnesses ecosystems as co-designers in energy, water, and land systems. This approach creates cities where environmental regeneration and human prosperity are inseparable.

Chapter 9
Participation Pathways for Enhancing the Resilience of Heritage to Stormwater Risks: Suzhou Section of Grand Canal

Yanhui Lei and Zhuoning Li

Abstract The Suzhou section of the UNESCO-listed Grand Canal embodies navigation, economy, history, and local identity, with its officially recognized heritage sites playing an important role in shaping the city's cultural landscape. However, the stormwater threats under urban growth increasingly endanger the resilience of the Heritage of Suzhou Section of the Grand Canal (HSG). While China has actively promoted Sustainable Stormwater Management (SSM) strategies to address such risks, Suzhou's SSM initiatives have been largely technology-driven, with limited integration of cultural heritage into synthesized resilience planning. Moreover, public participation—an essential pillar of effective SSM implementation—remains underdeveloped in the local context.

This study addresses the institutional and participatory gaps in existing SSM practices by exploring potential participation pathways that enhance heritage resilience. It proposes an integrated framework combining SSM strategies with public participation mechanisms, hypothesizing that such a synthesis can both mitigate hydrological threats and activate public involvement in protecting canal heritage. Using the Heritage of Suzhou Section of the Grand Canal as research object, this research adopts a mixed-methods approach, investigating key literature and documents to examine the interrelations among SSM policies, heritage conservation, and public participation. Additionally, it utilizes survey-based statistical analysis (SPSS) and data visualization to assess: (a) perception levels, (b) attitudes, and (c) willingness to engage. The research indicates a positive correlation between the public's perception of HSG and stormwater risks and their level of concern for heritage protection. It highlights the public's positive attitude towards participation, reflecting trust in governmental initiatives and a willingness to be actively involved.

Y. Lei (✉) • Z. Li
Design School, Xi'an Jiaotong-Liverpool University, China
e-mail: Yanhui.Lei@xjtlu.edu.cn

A. Cheshmehzangi et al. (eds.), *Transforming Cities*, Urban Sustainability,
https://doi.org/10.1007/978-981-95-7378-3_9

The study offers insights into potential pathways to bridge institutional gaps and enhance public engagement in Sustainable Stormwater Management, particularly in heritage-sensitive contexts. It contributes to advancing discussions on public participation in heritage-centered stormwater risk management and provides policy insights for China's broader efforts to integrate SSM into stormwater-threatened cultural heritage sites, ultimately aiming to strengthen the climate resilience of waterfront heritage.

Keywords Beijing-Hangzhou Grand Canal • Suzhou • Stormwater flooding threat • Sustainable Stormwater Management • Heritage conservation • Public participation • Resilience

9.1 Introduction

The Beijing-Hangzhou Grand Canal represents one of the world's longest and most historically significant artificial waterways (Bian, Zhang, Kong, Huang, and Hu, 2024; Mao, Liu, Zhou, Huang, and Li, 2008). Among its many sections, the Suzhou section not only has significant economic and ecological value, but also gives rise to a range of important tangible historical heritages due to its identity and functions in Chinese history (Bian et al., 2024; Huang and Yang, 2023; Sigley, An, Song, and Su, 2024). These heritage assets and the city are closely intertwined and mutually reinforcing for more than 2,500 years, which has led to its continued relevance to today's citizens' lives, thereby playing a crucial role in shaping local urban resilience across both ecological and social dimensions (Wang, 2021; Xing and Ma, 2019). The Heritage of the Suzhou section of the Grand Canal (HSG) includes five canal channels and seven officially certified associated heritage sites, as shown in Fig. 9.1 (Suzhou Local History Compilation Committee Office, 2023).

However, the heritage assets are facing growing stormwater risks intensified by rapid urbanization (Bian et al., 2024; Huang and Yang, 2023; Sigley et al., 2024; Wang, 2021). This expansion has affected the hydrological cycle, shortening the flood formation process and increasing the frequency of flooding in the basin where Suzhou is based (Lin et al., 2019; McLennan and Group, 2022). More seriously, the region now faces increasing compound stormwater risks resulting in persistent issues such as waterlogging, pollution, and seasonal water shortages (S. Li, Ma, and Bi, 2020; H. Zhang, Liu, and Fang, 2019). The threats heritage of Suzhou section faces are both tangible, such as physical degradation, erosion, and structural damage, and intangible, including the erosion of historical meaning, community identity, and cultural continuity (Arrighi, 2021; Figueiredo, Romão, and Paupério, 2021).

In response to the growing problems, China is promoting Sustainable Stormwater Management (SSM) nationwide in addition to piped drainage systems, namely Sponge City policies (Hamidi, Ramavandi, and Sorial, 2021; Qiao, Liu, Kristoffersson, and Randrup, 2019). These practices in various cities, including Suzhou, are in fact largely technology-driven, while the key components of public participation

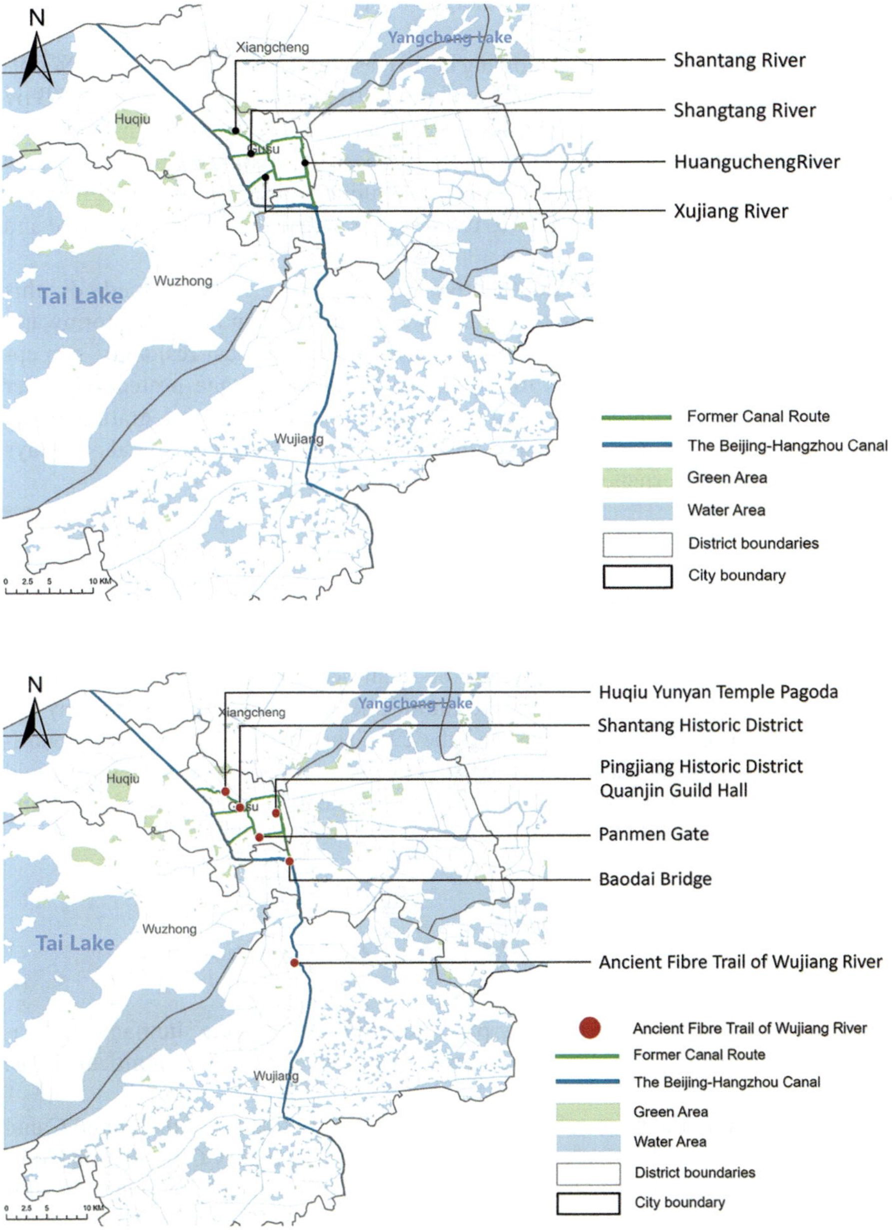

Fig. 9.1 Distribution of heritage of the Suzhou Section of the Grand Canal (Source: Image created by the authors).

have received limited attention. Corresponding to the tangible and intangible harm to which the heritage may be exposed, HSG protection in stormwater-risk zones lacks an integrated framework that merges ecological resilience with social and cultural awareness.

This research addresses the following core question: How can existing SSM policies be leveraged to comprehensively enhance the disaster resilience of the HSG by addressing both tangible and intangible stormwater threats? To answer this, research proposes an integrated strategy: promoting public participation in SSM initiatives as a means to both mitigate flooding and raise public consciousness about heritage protection. This dual approach aims to reinforce resilience from both ecological and social perspectives.

Through a combination of quantitative and qualitative research methods, this study analyzes the structure and potential paths of Suzhou's and China's stormwater governance and participation conditions, while evaluating local residents' perceptions, attitudes, and willingness to participate in canal heritage protection under stormwater threats, using SPSS-based analysis of survey data. By integrating public engagement with ecological strategies, the study seeks to outline practical pathways for participatory governance and policy innovation.

9.2 Literature Review

To formulate a pathway for addressing the identified challenges, it is essential to clarify the core concepts related to the multidimensional challenges faced by the Suzhou section of the Grand Canal and the city. This will help clarify the contradictions and needs of local cultural heritage conservation in the context of rapid urbanization and climate threats, and lay a theoretical foundation for identifying potential opportunities.

9.2.1 Cultural Heritage and Heritage Conservation

Cultural heritage, inherited from past generations, reflects human civilization's development and holds social, economic, and ecological significance (Li and Tang, 2024).

In this research, the Cultural Heritage of the Suzhou Section of the Grand Canal (HSG) refers to the definition used by the Suzhou Local History Compilation Committee Office (2023). It consists of river courses and seven tangible heritage sites (Suzhou Local History Compilation Committee Office, 2023). These assets include districts, buildings (clusters), bridges, ancient roads, gates, gardens, decorations (e.g., rockeries), old and valuable trees, etc., which are embedded within the districts and adjacent to the canal system, represent a heritage-rich yet highly vulnerable urban corridor (Yuan, 2022).

Heritage conservation involves maintaining and managing heritage to sustain its value for present and future societies, emphasizing sustainable development (Li and Tang, 2024). Key principles include defining the object and scope of conservation, and balancing conservation with utilization. Cultural heritage, due to its uniqueness and vulnerability, is also a priority in disaster risk management (Arrighi, 2021; Eze and Siegmund, 2024). Moreover, a significant principle is to maintain social equity in heritage conservation by encouraging both expert and non-expert participation, ensuring public involvement (Li and Tang, 2024).

9.2.2 *Stormwater Risks and Sustainable Stormwater Management (SSM)*

In this study, stormwater flooding threats refer to rapid runoff, urban inundation, river flooding, and pollution caused by extreme rainfall during events such as typhoons or monsoons. For heritage, Damages when they are exposed to stormwater flooding can be divided into tangible and intangible categories (Arrighi, 2021). Tangible losses refer to irreversible physical damage, such as the erosion of civil structures or the deformation and decay of wooden ones (Figueiredo, Romão and Paupério, 2021). Intangible losses encompass the decline of historical, aesthetic, spiritual, and social values, which are difficult to quantify and may erode public trust and emotional connections to heritage (Arrighi, 2021).

Stormwater management refers to the planning and control of rainwater collection, storage, treatment, and discharge to reduce negative impacts on urban and natural environments (Grigg, 2024; Rentachintala, Reddy, and Mohapatra, 2022). Today, the practice of traditional stormwater management has evolved from a singular focus on drainage to a diverse approach, with sustainable stormwater management (SSM) becoming popular, using green infrastructure to mitigate stormwater-related challenges while also addressing functional areas such as water quality control, urban ecology, recreation, and city beautification (Qiao et al., 2019; Grigg, 2024). SSM integrates multi-level decision-making and cross-sectoral cooperation, aiming to build livable, healthy, and resilient cities (Barbosa, Fernandes and David, 2012; Grigg, 2024).

SSM is a global term encompassing concepts including different names like Low Impact Development (LID), Water Sensitive Urban Design (WSUD), and Sustainable Urban Drainage Systems (SUDS) (Ferrans et al., 2022; Lei, 2024; Lei, 2026). SSM complements traditional systems, aiming to reduce flood risks through both engineering and non-engineering solutions (Barbosa, Fernandes, & David, 2012; Qiao et al., 2019). Its advantages include ecological restoration, natural water cycle support, and minimizing secondary heritage impacts (Boogaard, Wentink, Vorenhout, and de Beer, 2016; Q. Zhou, 2014). From an engineering perspective, SSM networks green infrastructure—e.g., green roofs, rain gardens, permeable pavements—within urban landscapes (Ferrans et al., 2022; Nóblega Carriquiry, Sauri, and March, 2020; Q. Zhou, 2014). It also incorporates non-technical approaches, showing potential for public participation: in Europe, SUDS utilize decision-support tools to help optimize design and layout, preventing expert-dominated planning; it also engages stakeholders in implementation, addressing decision-making gaps and reprioritizing interventions (Beceiro, Brito, and Galvão, 2020; Cox and Hamlen, 2015; Larkin et al., 2015; Veith, Wolfe, and Heatwole, 2003).

SSM has shown promising outcomes in heritage cities. Ahmedabad (India) has reduced surface runoff through SSM; Montferland (Netherlands) demonstrated its benefits for subterranean heritage; and architectural heritage has proven compatible with Green Infrastructure (Boogaard et al., 2016; Borah, Bardhan, and Bhatia, 2023; Coombes and Viles, 2021).

In China, the Sponge City initiative—based on LID principles—is a national SSM strategy (Hamidi et al., 2021). Suzhou has implemented it, aiming at reducing flood risks and enhancing the city's image (Thadani, Li, and Chan, 2020; Zhang C., lv, and Chen, 2017). However, several limitations remain. First, relevant documents lack strict unified legislation while uniform standards may not suit local cultural contexts (Lashford et al., 2019; Zeng, Li, Zeng, and Song, 2023). Second, current evaluation criteria focus on hydrology and neglect sustainability-related aspects like landscape and well-being (Grigg, 2024). Finally, although public participation is critical in stormwater governance, the Sponge City has few grassroots engagement policies (Birkholz, Muro, Jeffrey, and Smith, 2014; Tyler, Sadiq, and Noonan, 2019; Yang Wang, Cai, Zuo, Bartsch, and Huang, 2021b).

9.2.3 Public Participation

Public participation is the process by which the general public is given the opportunity to express their views, influence, or take action during the decision-making process (Arnstein, 2019). In the field of planning actions, two-way communication between the government and citizens can promote acceptance and replication of projects (Yang Wang, Cai, Zuo, Bartsch, and Huang, 2021a). In practice, such participation can be categorized into government-led mechanisms, community-driven initiatives, public-oriented educational or experiential programs, etc. In the long run, participation fulfills civic rights and balances empowerment, thus increasing urban resilience, which aligns with the objectives of SSM (Tyler et al., 2019).

Research has found that though citizen participation in nature-based solutions (NBS) does not automatically lead to significant enhancements in ecological functions, participation can increase the diversity of social outcomes, including enhanced belonging, environmental management, inclusiveness and equity, thereby enhancing heritage resilience (Kiss et al., 2022). In the context of stormwater management, citizens' involvement is considered critical for successful heritage preservation and disaster mitigation, as active participation strengthens understanding and trust between residents and decision-makers, leading to better outcomes (Tyler et al., 2019). Public awareness campaigns, communication, and trust in authorities are critical for effective flood management, though the first two areas remain underdeveloped in China and need to be elevated further (Xie, Wang, and Li, 2023).

As for stormwater management, public participation in China has a thin background but is also actively developing. The public is generally not exposed to the decision-making process, and participation usually occurs at the final stages of government-led policy implementation, such as emergency response (Li, 2012; Zhai and Ng, 2013). Citizen motivation to participate is also seen as relatively low, with China's collectivist culture strengthening compliance with policies but weakening individual initiative (Xie et al., 2023). But it's worth recognizing that recent shifts in China's disaster risk management approach emphasize local-level actions, transitioning from disaster response to risk prevention and mitigation (Jia et al., 2022). Another point with potential is that studies show a public preference for nature-based flood management methods over traditional gray infrastructure, indicating significant potential for promoting SSM practices (Birkholz et al., 2014).

On the other hand, heritage-focused projects benefit from local participation as residents' contextual knowledge helps experts in understanding local values and resources which might be overlooked by external decision-makers (Hui, Chen, Lang, and Ou, 2021). For the research area, a previous study has confirmed that Suzhou citizens' awareness of social participation in heritage conservation is positively correlated with local identity, urban memory, subjective cognition, objective knowledge, and local identity (T. Zhou, Zang, Jiang, Yang, and Ikebe, 2023). The cultural heritage of the Suzhou section of the Grand Canal is closely related to human life (Arrighi, 2021). Protecting the Grand Canal heritage aligns with citizens' living environments, local memories, and cultural identities which might provide strong motivation for engagement (Wang, 2021).

9.3 Methodology

In this rection, research focuses on applying quantitative methods to determine the potential of public participation in conserving canal heritage and stormwater flood defense by surveying public feedback on the cultural heritage of the Suzhou section of the Grand Canal in the context of heavy rainfall threats, and further extrapolates the feasibility and development potential of heritage resilience enhancement methods that integrate the SSM framework with public heritage protection.

9.3.1 Questionnaire Design

To address the issue that existing SSM and risk management strategies often overlook grassroots public participation in the early stages, this study uses a questionnaire to explore Suzhou residents' views on HSG under stormwater threats and public participation, analyzing how to enhance public interest and develop suitable non-expert-led strategies.

This study designed a 19-question questionnaire to collect quantitative data on the following points: Suzhou residents' perceptions of (a) HSG, (b) HSG's exposure to stormwater, and (c) knowledge of SSM; and their intention to participate in the process of using SSM for HSG protection in the stormwater context. They correspond to questions 4–17 and are divided into three sets of perception survey questions as independent variables and three sets of attitude survey questions as dependent variables, as shown in Table 9.1, to explore the correlation between perception levels, attitudes, and willingness to participate. In addition, two statistical questions were set to explore respondents' preferences for participation methods.

Table 9.1 Questionnaire items and variable settings.

Variable type	Variable name	Corresponding questions
Independent Variable (IV)	Perception of HSG (5–7)	4. Familiarity with the Suzhou section. 5. Familiarity with HSG. 6. Relevance of HSG to participants' lives.
	Perception of stormwater threats (9–10)	8. Perception of whether HSG is threatened by stormwater. 9. Perception of Severity of stormwater threats.
	Perception of SSM (14–15)	13. Familiarity of SSM concepts. 14. Frequency of noticing SSM facilities around the HSG.
Dependent Variable (DV)	Attention to HSG under threat (8, 11, 12)	7. The level of importance of HSG conservation for Suzhou residents. 10. The level of concern about current stormwater management measures related to HSG. 11. Level of satisfaction with current HSG stormwater management.
	Positivity for SSM (13, 16)	12. The public's preference for adding SSM measures. 15. Does SSM measures help in mitigating stormwater threat to HSG.
	Attitudes towards public participation (17, 18)	16. Whether public participation is beneficial for HSG conservation in the stormwater context. 17. Public willingness to participate in projects related to HSG flood protection.

9.3.2 Data Collection and Data Analysis

The sample distribution was designed to ensure coverage of diverse demographic groups and proper spatial positions, enhancing representativeness for the research context. The questionnaires were distributed and recovered online to Suzhou residents over 18 on a one-to-one basis, with an answer time of no less than 60 seconds and a correct IP as the conditions for valid questionnaires. The questionnaires with non-Suzhou IPs were all determined through one-to-one contact to assess their Suzhou resident status. After collating the valid data, the study conducted a visual analysis on the ratio of answers to key questions. The study also converted the single-choice questions into a unified Likert scale (1–5), used IBM SPSS Statistics 27 for descriptive statistical analysis, reliability analysis, validity analysis, and correlation analysis. While the methods offer multi-dimensional insights, limitations remain in terms of sample size and reliance on self-reported data, which may affect generalizability. Hereby declare that this research has undergone ethical review and received approval from Xi' an Jiaotong-Liverpool University, classified as Low Risk Research (LRR). All procedures strictly complied with applicable national laws, regulations, policies, and university guidelines.

9.4 Analysis and Results

This section presents the study's empirical findings, combining statistical analyses to examine public perceptions, attitudes, and participation pathways for HSG and related stormwater governance in Suzhou, thereby providing a basis for assessing the feasibility of participatory approaches to heritage-focused stormwater management.

9.4.1 Descriptive Statistical Analysis

A total of 317 questionnaires were collected, with 273 deemed valid, resulting in an efficiency rate of 86.12%. As shown in Table 9.2, the majority of respondents were young adults aged 18–30 (47.25). In terms of residency duration, 38.46% of participants had lived in Suzhou for 0–10 years, indicating that newer residents may have a stronger presence in the survey. Regarding residential districts, Huqiu District and Gusu District had the highest proportion of respondents, while other districts had lower response rates, suggesting that residents from districts with closer proximity to the Grand Canal might have a higher level of interest in the survey's focus.

9.4.2 Reliability and Validity Tests

As shown in Table 9.3, the overall Cronbach α coefficient is 0.791, indicating that the reliability of the questionnaire is good. Moreover, the "item-deleted α coefficient" values show that none of the items should be deleted from the scale. Therefore, the reliability quality of the data is considered acceptable, and the data can be used for further analysis.

The validity of the scale was analyzed using KMO and Barlett's test to determine whether the item design is capable of factor analysis and information extraction. Table 9.4 shows that the KMO value is 0.785, ranging from 0.7 to 0.8, and the Barlett's Test is less than 0.01, which indicates that the data of the study are suitable

Table 9.2 Statistics of respondents' information.

Items	Categories	Amount	Percent (%)
Age	18–30	129	47.25
	31–45	86	31.5
	46–60	38	13.92
	61 and over	20	7.33
Residence time in Suzhou	0–10	105	38.46
	11–25	77	28.21
	26–40	64	23.44
	41 and over	27	9.89
Residential District	Gusu	60	21.98
	Huqiu	85	31.14
	Wuzhong	38	13.92
	Xiangcheng	20	7.33
	Wujiang	21	7.69
	SIP	49	17.95

Table 9.3 Reliability statistics.

Items	Number of items	Corrected Item-Total Correlation (CITC)	Cronbach Alpha if Item Deleted	Cronbach α
IV 1: Perception of HSG	3	0.700	0.719	0.791
IV 2: Perception of stormwater threats	2	0.614	0.741	
IV 3: Perception of SSM	2	0.619	0.741	
DV 1: Attention to HSG under threat	3	0.748	0.726	
DV 2: Positivity for SSM	2	0.411	0.789	
DV 3: Attitudes towards public participation	2	0.441	0.821	

Table 9.4 Validity statistics.

KMO and Bartlett's Test		
KMO		0.785
Bartlett's Test of Sphericity	Chi-Square	687.240
	df	15
	p	0.000

for extracting information and conducting factor analysis, and also reflects that the validity is good.

9.4.3 Ratio Analysis of Statistical Results

9.4.3.1 Perception and Attitude of HSG

As shown in Fig. 9.2, the survey explored respondents' familiarity with HSG and the importance they place on its protection. Regarding Question 5, the majority of respondents have at least some levels of awareness about HSG, with 67.03% indicating they were at least "Generally familiar" (able to name more than 3 heritage sites). Meanwhile, again, more than two-thirds of respondents indicates that HSG was relevant to their work or life (Question 6). Regarding the importance of protecting HSG (Question 7), a clear majority of respondents, 92.68%, considered HSG protection to be at least "Quite important." This high level of perceived importance underscores the public's strong support for HSG conservation.

9.4.3.2 Stormwater Threats to HSG

The survey also explored the perceived impact of stormwater flooding on HSG and the public's satisfaction with current flood management measures, as illustrated in Fig. 9.3. For Question 9, 62.64% of the respondents felt that stormwater flooding has a medium to serious impact on HSG from their daily life experience, i.e., there is noticeable standing water and varying degrees of pollution. Regarding satisfaction with current flood management (Question 11), 42.12% were "Generally satisfied," and 31.87% were "Very satisfied," suggesting general approval. However, 18.32% were unsure, while 2.2% were unsatisfied, indicating possible gaps in public awareness or involvement in flood management initiatives.

9.4.3.3 Perception and Attitude of SSM

The survey examined public familiarity with Sustainable Stormwater Management (SSM) concepts and their perceptions of its effectiveness around HSG as shown

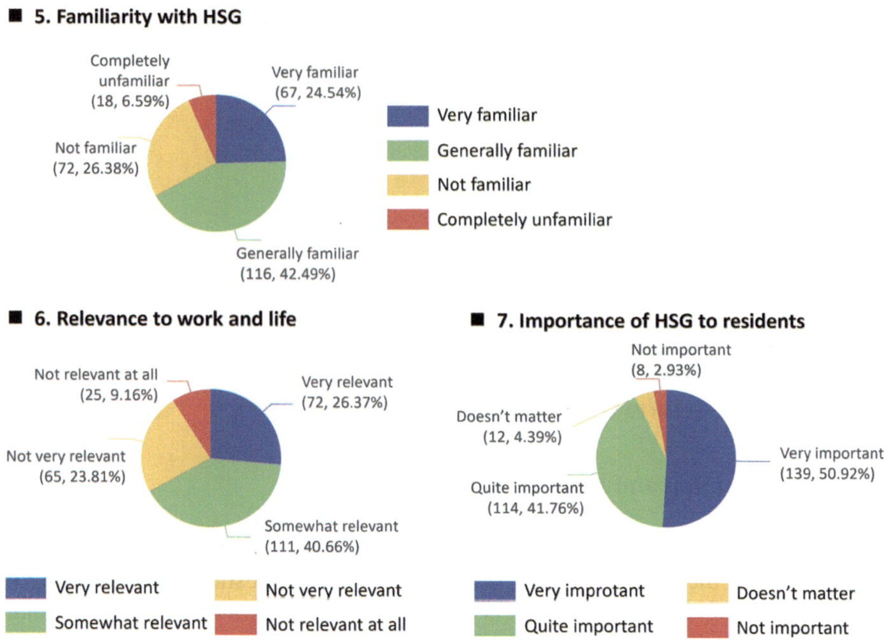

Fig. 9.2 Visualization of perception and attitude of HSG (Source: Image created by the authors).

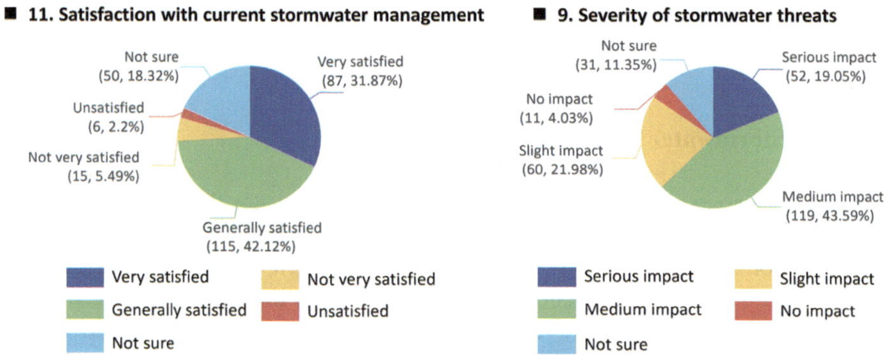

Fig. 9.3 Visualization of perception of stormwater threats (Source: Image created by the authors).

in Fig. 9.4. More than 50% of the respondents have a basic understanding of the concept and are able to notice SSM facilities near the HSG, which shows that the promotion of SSM concepts has left a certain impression on the public's viewpoint (Questions 13, 14). When asked about the effectiveness of these measures in reducing stormwater threats to HSG (Question 15), 84.25% of respondents believed that such measures would be helpful, which suggests public support for the integration of SSM in heritage conservation.

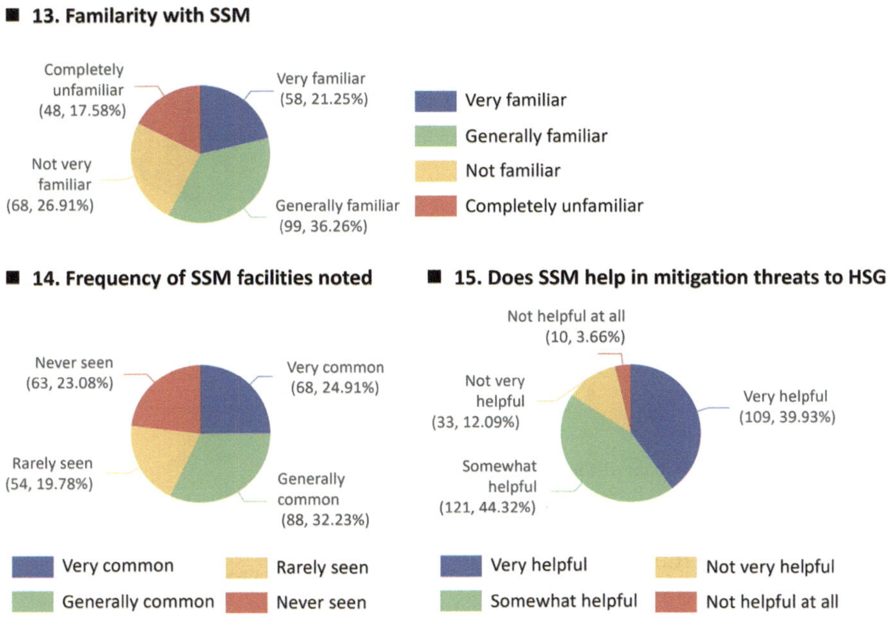

Fig. 9.4 Visualization of perception and attitude of SSM (Source: Image created by the authors).

9.4.3.4 Attitude and Willingness to Public Participation

The survey results demonstrate a remarkably positive public attitude towards participation in protecting the Suzhou section of the Grand Canal (HSG) from stormwater threats as shown in Fig. 9.5. Regarding the perceived benefits of public involvement, 72.16% of respondents agreed that citizen participation is beneficial, while only 16.85% disagreed, and 10.99% remained uncertain. Furthermore, when asked about their willingness to participate in related projects, 51.28% expressed strong willingness, and 37.36% indicated moderate willingness. Only a small proportion (6.96% and 4.4%, respectively) behaved in an indifferent or unwilling manner.

9.4.4 Correlation Analysis

In order to investigate the potential for public participation in the issue of "Protecting HSG from stormwater threats," the study investigated a combination of public perception levels and attitudes toward protecting HSG and its stormwater issues, and whether there was a causal relationship between the two. For this purpose, the study set up separate question groups for perceptions and attitudes, and each question group corresponded to three variables, with perception level as the independent variable and attitudes as the dependent variable, as shown in Table 9.5.

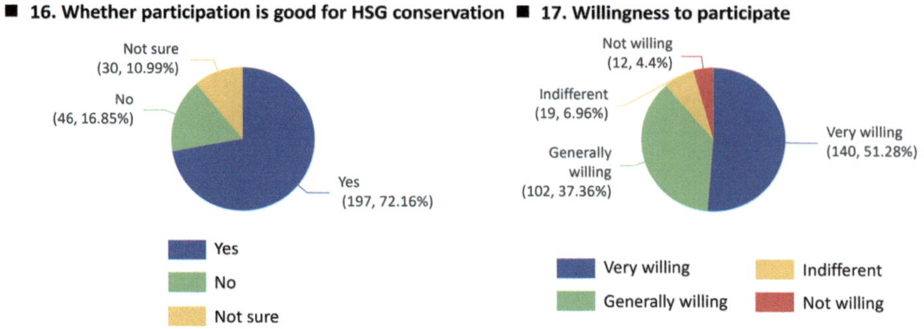

Fig. 9.5 Visualization of attitude and willingness to public participation (Source: Image created by the authors).

Table 9.5 Correlation data of the variables.

Items	Data items	Attention to HSG under threat	Positivity for SSM	Attitudes towards participation
Perception of HSG	Coefficient	0.712	0.216	0.082
Perception of stormwater threats	Coefficient	0.590	0.217	0.068
Perception of and SSM	Coefficient	0.563	0.273	0.090

Pearson's correlation analysis reveals that the correlation between dependent variable 1 (Attention to HSG under threat) and all three independent variables are significantly positive. The data suggests that the higher the public's awareness of the HSG itself, perceived stormwater threat to HSG, and knowledge of SSM, the higher their level of concern about the level of HSG under threat. This also suggests that changes in the public's perceptions of these topic priorities influence their level of concern for HSG protection.

There is also a positive correlation between dependent variable 2 (Positivity for SSM) and all three independent variables (Perception of HSG, Perception of stormwater threats, Perception of and SSM), with Pearson correlation coefficients of 0.216, 0.217 and 0.273. This suggests that there is a correlation between public acceptance of SSM and knowledge of SSM concepts and measures, i.e., the public is more inclined to have a positive attitude towards SSM when they have a high level of knowledge of stormwater management concepts and measures.

However, the correlation between dependent variable 3 (Attitudes towards public participation) and the independent variables was weak, with the correlation coefficients of all three being close to 0 (0.082, 0.068, and 0.090, respectively). This result suggests that although respondents generally showed a high willingness to participate in public participation (e.g., 72.16% of respondents believed that civic participation would help to preserve heritage, and 88.64% were willing to participate in related programs), this willingness to participate was not significantly and directly correlated with their perceptions of HSG, flooding, or SSM.

9.4.5 *Statistical Analyses Related to Participatory Methods*

The study included two questions designed to capture the preferences of Suzhou residents regarding public participation methods for HSG conservation in the context of stormwater flooding. The options were designed based on activity types found in urban development public participation projects and case studies reviewed in the Literature Review. Question 18 was a multiple-choice question asking respondents about their preferred modes of participation, while Question 19 was a single-choice question asking which method they considered the most effective. The responses from 273 participants are shown in Table 9.6.

When asked about their preferred ways to participate in the conservation of the Suzhou section of the Grand Canal under the threat of stormwater flooding (Question 18), residents showed a clear preference for Government consultation meetings (63.37%) and social media discussions (64.10%), indicating that residents value both formal channels for expressing opinions and online platforms for interaction. Regarding the perceived effectiveness of these participation methods in raising public awareness (Question 19), Government consultation meetings once again stood out as the most effective option, with 39.93% of respondents selecting it. Resident participatory design (24.18%) and social media discussions (12.82%) also ranked relatively high, while public exhibitions and design competitions (6.96%) and Volunteer experience (6.23%) were considered less effective.

Table 9.6 Data of public choice of participation methods.

Option	Count (Q18: more willing to participate)	Percentage (Q18)	Count (Q19: considered to be more effective)	Percentage (Q19)
A. Government consultation meetings	173	63.37%	109	39.93%
B. Resident participatory design	154	56.41%	66	24.18%
C. Social media discussions	175	64.10%	35	12.82%
D. Public exhibitions and design competitions	153	56.04%	19	6.96%
E. Volunteer experience	155	56.78%	17	6.23%
F. School curriculum and community education	122	44.69%	27	9.89%

Notably, both responses consistently highlighted Government consultation meetings and social media discussions as the most preferred and effective methods, reflecting a dual focus on formal engagement and digital communication. To enhance public participation, combining these two approaches could be a strategic way to leverage both structured dialogue and online interaction.

9.5 Discussion

This section discusses the implications of the survey findings, focusing on how public perception, participation willingness, and combined engagement strategies can inform stormwater governance in the context of heritage conservation.

9.5.1 Enhancing Perception to Further Concern

The correlation analysis highlights a significant positive correlation between the public's attention to the protection of HSG under stormwater threats and their awareness of HSG, stormwater risks, and SSM (with correlation coefficients of 0.712, 0.590, and 0.563, respectively). This suggests that improving public understanding of the principles and objectives related to heritage conservation and stormwater management is likely to encourage more proactive engagement in managing heritage sites under stormwater challenges. Enhancing perception and awareness could, therefore, serve as a strategic approach to increase public attention and support for sustainable practices in the context of stormwater-induced risks.

9.5.2 Public Participation Willingness: Positive Outcomes and Independence

The questionnaire's results reveal a generally high willingness among respondents to participate in heritage protection under stormwater threats. However, the correlation analysis shows a weak relationship between public participation willingness (Dependent Variable) and perception variables (Perception of HSG, stormwater threats, and SSM), with correlation coefficients of 0.082, 0.068, and 0.090, respectively, suggesting that the desire to participate reflects a broader social inclination rather than a knowledge-driven response. This result also raises the question of whether this questionnaire effectively captured the deeper motivations or barriers to participation, as current questions focused on willingness and perceptions without addressing practical obstacles such as time, resources, or access to information.

In summary, the author here infers that future research and policy-making aimed at promoting public participation should avoid solely relying on knowledge dissemination and awareness-raising. Instead, it should delve deeper into participation methods and motivational factors to develop more effective and inclusive strategies.

9.5.3 Multi-method Combined Engagement Strategy Building

There is some variation in the results of residents' evaluations of the preferred level of participation and the effectiveness of different modes of participation, as demonstrated in Section 9.4.5. Therefore, this study proposes combining preferred and effective methods to enhance public trust and encourage actual participation:

9.5.3.1 Public Discussion Combining Social Media and Government Consultation

This approach integrates the most effective method, Government consultation meetings, with the most preferred option, social media discussions. Public approval of the former aligns with findings from the literature review, which highlight the relatively high level of trust Chinese citizens place in the government (Wang et al., 2021). Therefore, the study suggests that government or authorized agencies could establish open and anonymous online platforms to facilitate direct interaction between the public and decision-makers.

9.5.3.2 Co-Creation Programs for SSM Measures

This approach combines Resident participatory design and Volunteer experience. This study proposes that such programs could be led by government and local offices, recruiting experts and volunteers to implement processes similar to Decision Support Systems (SUDS-DSS) (Ferrans et al., 2022). After the site has been identified, it is also possible to implement a community-based combination of SSM technical measures, such as personal involvement in the design and establishment of community rain gardens. This utilizes the theoretical evidence that visually appealing green spaces and interactive installations associated with SSM can generate public interest (Zhou, 2014). Such initiatives strengthen cooperation among the public, experts, and policymakers while enhancing community resilience through shared responsibility and collective ownership in the face of environmental challenges.

9.5.4 Transformative Growth of Urban Ecology

As one of the central themes of this book, the concept of Transformative Growth is proposed to liberate urban transformation from the conflict between the incremental growth of ecological efficiency and the political and productivity dilemmas inherent in degrowth. Inspired by this, this chapter explores the theme of achieving ecological and heritage conservation through citizen participation in nature-based solutions. It examines the endogenous driving forces and potential of citizens within China's social model for ecological heritage conservation.

This chapter highlights resilience, ecological innovation, and citizen participation, aligning with the inference that the future of urban development lies not only in economic output but also in the ecological productivity of natural systems and the social productivity of inclusive communities. Building upon reflections on the potential for public participation and SSM to address climate threats to HSG, the chapter explores pathways toward forging closer connections between urban ecology, policy, and society while advancing toward transformative change.

9.6 Conclusions

This study integrates the concepts and theoretical frameworks of Sustainable Stormwater Management (SSM) and public participation to explore strategies for protecting the Heritage of the Suzhou section of the Grand Canal (HSG) from stormwater threats. Specifically, the chapter investigates the potential of integrating SSM and public participation to mitigate stormwater threats to heritage sites, examines public awareness and attitudes toward these topics, and identifies possible strategic directions for heritage protection. The overarching goal is to effectively combine the dual advantages of public involvement and SSM-based technical measures, thereby reducing stormwater risks while enhancing urban resilience and generating additional social benefits.

Through a comprehensive analysis, this study reveals a positive public attitude towards the heritage of the Suzhou section of the Grand Canal, stormwater management, and heritage conservation. Furthermore, the study highlights the public's proactive attitude towards participation, reflecting a degree of trust in governmental initiatives. These findings provide preliminary evidence supporting the potential of combining participatory policies with sustainable stormwater management strategies to advance heritage conservation in flood-prone areas. This study suggests that embedding participatory approaches into local planning systems might bridge the current gap between technical stormwater solutions and cultural heritage conservation. In addition to being important locally, the HSG instance provides a paradigm that other historic cities dealing with water issues brought on by climate change.

Despite its contributions, this study acknowledges certain limitations. The demographic composition of survey respondents is slightly unbalanced, which could limit the generalizability of the findings. Additionally, while the study emphasizes the potential of SSM, it does not directly assess the practical effectiveness of these measures within Suzhou's specific hydrological and urban contexts. Future research should address these gaps by including more diverse demographic groups and evaluating the on-site performance of SSM implementations.

Acknowledgments

All the authors of this chapter would like to express the gratitude to the book editors, especially Prof. Ali Cheshmehzangi, for the efforts in the publication of this book. In addition, we would like to thank all the respondents of questionnaires in Suzhou.

References

Arnstein S (2019) A ladder of citizen participation. J of the Am Plan Assoc 85(1):24–34. https://doi.org/10.1080/01944363.2018.1559388

Arrighi C (2021) A global scale analysis of river flood risk of UNESCO World Heritage Sites. Front in Water 3, 764459. https://doi.org/10.3389/frwa.2021.764459

Barbosa AE, Fernandes JN, David LM (2012) Key issues for sustainable urban stormwater management. Water Res 46(20):6787–6798. https://doi.org/10.1016/j.watres.2012.05.029

Beceiro P, Brito RS, Galvão A (2020) The contribution of NBS to urban resilience in stormwater management and control: A framework with stakeholder validation. Sustain 12(6), 2537. https://doi.org/10.3390/su12062537

Bian D, Zhang M, Kong L, Huang B, Hu D (2024) Analysis of regional social–economic spatial pattern and evolution along the Beijing–Hangzhou Grand Canal. Sustain 16(4), 1527. https://doi.org/10.3390/su16041527

Birkholz S, Muro M, Jeffrey P, Smith HM (2014) Rethinking the relationship between flood risk perception and flood management. Sci of the Total Environ 478:12–20. https://doi.org/10.1016/j.scitotenv.2014.01.061

Boogaard F, Wentink R, Vorenhout M, de Beer J (2016) Implementation of sustainable urban drainage systems to preserve cultural heritage—Pilot Motte Montferland. Conserv and Manag of Archaeol Sites 18(1–3):328–341. https://doi.org/10.1080/13505033.2016.1182767

Borah A, Bardhan R, Bhatia U (2023) Protecting heritage: Insights into effective flood management using green infrastructure in a highly urbanized environment. Int J of Disaster Risk Reduct 98, 104075. https://doi.org/10.1016/j.ijdrr.2023.104075

Coombes MA, Viles HA (2021) Integrating nature-based solutions and the conservation of urban built heritage: Challenges, opportunities, and prospects. Urban Forestry & Urban Green 63, 127192. https://doi.org/10.1016/j.ufug.2021.127192

Cox RS, Hamlen M (2015) Community disaster resilience and the rural resilience index. American Behav Scientist 59(2):220–237. https://doi.org/10.1177/0002764214550297

Eze E, Siegmund A (2024) Exploring factors of disaster preparedness in UNESCO-designated heritage sites. Geogr and Sustain 5(3):392–404. https://doi.org/10.1016/j.geosus.2024.04.001

Ferrans P, Torres MN, Temprano J, Rodríguez Sánchez JP (2022) Sustainable urban drainage system (SUDS) modeling supporting decision-making: A systematic quantitative review. Sci of the Total Environ 806, 150447. https://doi.org/10.1016/j.scitotenv.2021.150447

Figueiredo R, Romão X, Paupério E (2021) Component-based flood vulnerability modelling for cultural heritage buildings. Int J of Disaster Risk Reduct 61, 102323. https://doi.org/10.1016/j.ijdrr.2021.102323

Grigg NS (2024) Stormwater management: An integrated approach to support healthy, livable, and ecological cities. Urban Sci 8(3), 89. https://doi.org/10.3390/urbansci8030089

Hamidi A, Ramavandi B, Sorial GA (2021) Sponge city—An emerging concept in sustainable water resource management: A scientometric analysis. Resources. Environ and Sustain 5, 100028. https://doi.org/10.1016/j.resenv.2021.100028

Huang Y, Yang S (2023) Spatio-temporal evolution and distribution of cultural heritage sites along the Suzhou canal of China. Herit Sci 11(1), .pdf>>. https://doi.org/10.1186/s40494-023-01034-y

Hui EC, Chen T, Lang W, Ou Y (2021) Urban community regeneration and community vitality revitalization through participatory planning in China. Cities 110, 103072. https://doi.org/10.1016/j.cities.2020.103072

Jia H, Chen F, Pan D, Du E, Wang L, Wang N, Yang A (2022) Flood risk management in the Yangtze River basin—Comparison of 1998 and 2020 events. Int J of Disaster Risk Reduct 68, 102724. https://doi.org/10.1016/j.ijdrr.2021.102724

Kiss B, Sekulova F, Hörschelmann K, Salk CF, Takahashi W, Wamsler C (2022) Citizen participation in the governance of nature-based solutions. Environ Polic and Gov 32(3):247–272. https://doi.org/10.1002/eet.1987

Larkin S, Fox-Lent C, Eisenberg DA, Trump BD, Wallace S, Chadderton C, Linkov I (2015) Benchmarking agency and organizational practices in resilience decision making. Environ Syst and Decis 35(2):185–195. https://doi.org/10.1007/s10669-015-9554-5

Lashford C, Rubinato M, Cai Y, Hou J, Abolfathi S, Coupe S, Tait S (2019) SuDS & sponge cities: A comparative analysis of the implementation of pluvial flood management in the UK and China. Sustain 11(1), 213. https://doi.org/10.3390/su11010213

Lei Y (2024) UoNJC and ANR: The two UK case studies of stormwater management through SuDS. In: Cheshmehzangi A, Sedrez M, Flynn A, (eds) Rethinking stormwater management through sustainable urban design. Springer, Singapore, pp 89–115. https://doi.org/10.1007/978-981-97-4924-9_5

Lei Y, Fan Z, Yu S (2026) Explore the application of green infrastructure in enhancing climate resilience in China: A comparative analysis in Suzhou Industrial Park (SIP) and Gusu District, Suzhou. In: Cheshmehzangi A, Alidoust S, Chen W, Fuller R, (eds) Nature in cities, nurturing cities. Springer, Singapore, pp 85–112. https://doi.org./10.1007/978-981-95-5487-4_5

Li L, Tang Y (2024) Towards the contemporary conservation of cultural heritages: An overview of their conservation history. Herit 7(1):175–192. https://doi.org/10.3390/heritage7010009

Li S, Ma J, Bi Y (2020) Study on the spatial relationship between green space and surface runoff in Suzhou City. IOP Conf Ser: Earth and Environ Sci 668, 012002

Li W (2012) Getting their voices heard: Three cases of public participation in environmental protection in China. J of Environ Manag 98:65–72

Lin Z, Xu Y, Dai X, Wang Q, Gao B, Xiang J, Yuan J (2019) Changes in the plain river system and its hydrological characteristics under urbanization—Case study of Suzhou City, China. Hydrological Sci J 64(16):2068–2079. https://doi.org/10.1080/02626667.2019.1645957

Mao F, Liu Z, Zhou W, Huang J, Li Q (2008) The research and application of spatial information technology in cultural heritage conservation—Case study on Grand Canal of China. The Int Archives of the Photogrammetry, Remote Sens and Spat Inf Sci XXXVII:999–1006

McLennan M, Saadia Group (2022) The global risks report 2022 (17th ed.). World Economic Forum.

Nóblega Carriquiry A, Sauri D, March H (2020) Community involvement in the implementation of sustainable urban drainage systems (SUDSs): The case of Bon Pastor, Barcelona. Sustain 12(2), 510. https://doi.org/10.3390/su12020510

Qiao XJ, Liu L, Kristoffersson A, Randrup TB (2019) Governance factors of sustainable stormwater management: A study of case cities in China and Sweden. J of Environ Manag 248, 109249. https://doi.org/10.1016/j.jenvman.2019.07.020

Rentachintala LRNP, Reddy MGM, Mohapatra PK (2022) Urban stormwater management for sustainable and resilient measures and practices: A review. Water Sci and Technol 85(4):1120–1140. https://doi.org/10.2166/wst.2022.017

Sigley G, An Z, Song C, Su X (2024) Cultural route heritage as mobility narrative: The world heritage inscription of China's Grand Canal. Int J of Herit Stud 30(9):997–1009

Thadani D, Li L, Chan FKS (2020) Online social media—A vehicle for city branding in China: The case of Sponge City Program (SCP). In: Chan FKS, Chan HK, Zhang T, Xu M (eds) Proceedings of the 2020 International Conference on resource sustainability: Sustainable urbanisation in the BRI era (icRS Urbanisation 2020). Springer, pp 381–389. https://doi.org/10.1007/978-981-15-9605-6_26

Tyler J, Sadiq AA, Noonan DS (2019) A review of the community flood risk management literature in the USA: Lessons for improving community resilience to floods. Nat Hazards 96(3):1223–1248. https://doi.org/10.1007/s11069-019-03606-3

Veith TL, Wolfe ML, Heatwole CD (2003) Optimization procedure for cost effective BMP placement at a watershed scale. J of the Am Water Resour Assoc 39(6):1331–1343. https://doi.org/10.1111/j.1752-1688.2003.tb04421.x
Wang Y (2021) City profile: Suzhou, China—The interaction of water and city. Cities 112, 103119
Wang Y, Cai J, Zuo J, Bartsch K, Huang M (2021a) Conflict or consensus? Stakeholders' willingness to participate in China's Sponge City program. Sci of the Total Environ 769, 145250. https://doi.org/10.1016/j.scitotenv.2021.145250
Wang Y, Cai J, Zuo J, Bartsch K, Huang M (2021b) Conflict or consensus? Stakeholders' willingness to participate in China's Sponge City program. Sci of The Total Environ 769, 145250. https://doi.org/10.1016/j.scitotenv.2021.145250
Xie L, Wang Y, Li S (2023) How government–public collaboration affects individual mitigation responses to flooding: A case study in Yellow River Delta area, China. For and Soc 7(2):184–199. https://doi.org/10.24259/fs.v7i2.22601
Xing S, Ma J (2019) Application of traditional water conservancy wisdom in the landscape of Suzhou Grand Canal. IOP Conf Ser: Earth and Environ Sci 349, 012029.
Yuan L (2022) Human geographic evolution and imaging representation of the Suzhou section of the Grand Canal. West J 23:127–131. https://doi.org/10.16721/j.cnki.cn61-1487/c.2022.23.036
Zeng L, Li RYM, Zeng H, Song L (2023) Perception of sponge city for achieving circularity goal and hedge against climate change: A study on Weibo. Int J of Clim Chang Strateg and Manag 16(3):362–384. https://doi.org/10.1108/IJCCSM-12-2022-0155
Zhai B, Ng MK (2013) Urban regeneration and social capital in China: A case study of the Drum Tower Muslim District in Xi'an. Cities 35:14–25. https://doi.org/10.1016/j.cities.2013.05.003
Zhang C, Lv Y, Chen Y (2017) Interpretation of the technical specification for prevention and control of flooding in towns. Water Supply and Drain 53(8):55–59. https://doi.org/10.13789/j.cnki.wwe1964.2017.0268
Zhang H, Liu S, Fang Q (2019) Study on flood risk characteristics and disaster prevention countermeasures in the Taihu Basin. IOP Conf Ser: Earth and Environ Sci 401, 012021. https://doi.org/10.1088/1755-1315/401/1/012021
Zhou Q (2014) A review of sustainable urban drainage systems considering the climate change and urbanization impacts. Water 6(4):976–992. https://doi.org/10.3390/w6040976
Zhou T, Zang T, Jiang J, Yang X, Ikebe K (2023) Analysis of the influencing factors of social participation awareness on urban heritage conservation: The example of Suzhou, China. Sustain 15(3), 2713. https://doi.org/10.3390/su15032713

Chapter 10
Between Rice Fields and Resorts: Mapping the Urban-Ecological Nexus in the Face of Urban Growth and Ecological Balance in Southern Bali

Ali Cheshmehzangi and Tian Tang

Abstract Rapid spatial and socio-ecological changes are occurring in southern Bali, especially in Denpasar, Canggu, and the adjacent areas. These places, which were formerly dominated by rice fields and traditional agrarian landscapes, are now popular destinations for urban sprawl, ecological tourism, and resort development. By charting the geographical dynamics of land use change and examining the conflicts between ecological balance and urban growth, this chapter explores the urban-ecological nexus in Southern Bali. The chapter examines how changing land uses, particularly the conversion of agricultural land to tourism infrastructure, are changing both coastal and non-coastal habitats through a thorough mapping analysis.

This chapter analyzes major forces behind change, including investment in tourism real estate, rising global mobility, and lifestyle migration, all of which are based on a conceptual framework that looks at urban development and sustainability pressures. These factors are causing fresh chances for sustainable growth as well as environmental deterioration. The study reveals patterns of ecological fragility, fragmentation, and possible resilience routes using geospatial analysis, temporal land use data, and qualitative insights from planning documents and local stakeholder viewpoints.

The results show a dynamic, multifaceted landscape where resorts and rice fields coexist in an uneasy way and where ecological tourism both supports and challenges continuous urbanization. The chapter ends with a discussion of spatial planning and policy approaches, such as integrated natural corridors, zoning

A. Cheshmehzangi (✉)
School of Architecture, Design and Planning, The University of Queensland, Brisbane, Australia
Network for Education and Research on Peace and Sustainability (NERPS), Hiroshima University, Hiroshima, Japan
e-mail: a.chesh@uq.edu.au

T. Tang
Independent Researcher, Chongqing and Sichuan, China

A. Cheshmehzangi et al. (eds.), *Transforming Cities*, Urban Sustainability,
https://doi.org/10.1007/978-981-95-7378-3_10

innovations, and community-based land management, that could promote more balanced growth. All things considered, the research advances our knowledge of the ways in which ecological and urban systems interact in rapidly shifting island environments, providing information applicable to larger Southeast Asian and international tourism-urban interfaces.

Keywords Urban-ecological nexus • Spatial transformation • Ecological tourism • Land use change • Sustainable development • Southern Bali • Geospatial mapping

10.1 Introduction: Setting the Scene in Southern Bali

With its complex and frequently contentious intersections of changing socioeconomic forces, environmental concerns, and tourism-driven urbanization, southern Bali has long been the epicenter of the island's most dramatic transitions (Lewis and Lewis, 2009). This chapter places its research in this quickly evolving area, specifically the corridor that connects Denpasar through Canggu to coastal and semi-coastal communities like Seminyak, Jimbaran, and Uluwatu, where the spatial arrangement of ecological and urban systems is constantly being negotiated.

Large-scale subak-based rice cultivation systems, a traditional irrigation technique recognized as a UNESCO cultural asset, have historically defined Bali's southern lowlands (Gede Ustriyana et al., 2025). These agricultural areas, which previously represented a healthy coexistence of people, the natural world, and the spiritual realm, served as the ecological and cultural foundation of Balinese life. However, many of these rice fields have been replaced by boutique hotels, private villas, beach clubs, and infrastructure focused on lifestyle since the 1970s tourism boom, which grew even more after the 2000s (Shavit, 2019; Wardana, 2019). Once a tranquil coastal community surrounded by expansive paddies, Canggu has rapidly become one of Bali's most urbanizing regions in just over ten years (Nurjani and Dwijendra, 2020; Surya Suamba et al., 2022), and it is now widely known as a destination for surfers and digital nomads.

There has not been a smooth or consistent shift from rural to urban tourist areas. Rather, it has developed in discrete areas, characterized by fragmented development, land speculation, and changing governing structures (Ramadhan, 2024). The repercussions are complex. On one hand, this change has created economic opportunities, particularly for investors and landowners. Conversely, it has made ecological vulnerabilities worse, such as coastal erosion, flooding brought on by altered drainage patterns, biodiversity loss, and mounting pressure on waste and water systems. In many places, the unchecked rate of development—which is frequently motivated by short-term financial gain—has surpassed the capabilities of regional planning and infrastructure.

In this regard, Denpasar, Bali's capital and administrative hub, has a special function. Although it is not strictly a coastline tourist attraction, its proximity to other tourist destinations in Badung Regency has caused a ripple effect. The line between the urban center and the tourist periphery is becoming increasingly hazy as the city becomes more integrated into the socio-spatial fabric of the southern tourism belt (Atmaja et al., 2025). Additionally, Denpasar serves as a significant hub for government services, market flows, and labor migration, enhancing the urban-tourism relationship that defines the growth trajectory of southern Bali.

The mapping of the urban-ecological nexus—the meeting point of ecological balance and fast urban growth (Lv et al., 2024)—that arises in this context is the specific emphasis of this chapter. It explores the ways in which land use change is evident throughout this southern corridor, especially the conversion of rice fields into resorts and residential communities. It aims to comprehend the relationship dynamics between spatial development, ecological functions, and the goals of both local and migrant populations in addition to the physical growth of built-up places (Roy and Thangaraj, 2022; Cheshmehzangi and Tang, 2024a; Cheshmehzangi and Tang, 2024b). Several related questions are intended to be addressed by the mapping inquiry: In southern Bali, how have land uses altered over the last 20 years, particularly in peri-urban and coastal fringe areas? What effects do these changes have on the environment, specifically regarding green cover, water systems, and coastal integrity? And how may a more sustainable strategy for striking a balance between ecological and cultural preservation and growth be informed by spatial mapping?

Recognizing the multi-scalar and multi-actor character of Bali's transition is necessary to comprehend this relationship. The traditional village institutions (*desa adat*), province and regency-level planning regulations, international tourism markets, and internet influencers influencing visitor flows are some of the overlapping factors that determine the island's land governance (Pakarinen, 2018). In addition to being a technical tool, mapping becomes a way to highlight the conflicts and opportunities present in these changes.

Southern Bali serves as a metaphor for the larger difficulties tropical island countries encounter in balancing the conflicting demands of ecological resilience and tourism-driven economic expansion (Jamal and Higham, 2021). The unrestrained growth that commodifies the island's core assets—its dynamic culture, rural aesthetics, and coastal beauty—runs the risk of undermining the very things that make it appealing. Spatial analysis becomes a crucial tool in this situation (Cheshmehzangi, 2016) to determine where interventions could be required and how the urban-ecological relationship might be adjusted for sustainability over the long run. The chapter hopes to contribute to a more grounded, location-specific knowledge of Bali's urban-ecological processes by establishing the foundation in this introduction. It suggests that mapping is a purposeful act rather than just a descriptive one, one that exposes, challenges, and possibly changes our perspective on urban development in areas that are culturally and biologically rich.

10.1.1 From Agrarian Landscapes to Tourism Frontiers

Bali's southern regions, especially Canggu, Berawa, Pererenan, and portions of North Kuta, provide a striking example of the island's shift from rural to tourist-driven borders. These regions have experienced a rapid spatial transition that has reshaped the region's physical environment and sociocultural identity. Prior to the early 2000s, these locations were mostly distinguished by verdant sawah (rice fields) and traditional Balinese compounds (Christie, 2007). The subak system, a locally based water management network that is both practical and spiritual, has historically supported the agrarian nature of southern Bali (Roth, 2014; Huang, 2020; Zen et al., 2024). Subak, which dates back more than a millennium, is more than just irrigation; it is a manifestation of the Balinese philosophy known as Tri Hita Karana, which emphasizes harmony between nature, the gods, and mankind (Sutawan, 2004; Roth and Sedana, 2015). Through interconnected water temples, forest catchments, and terraced landscapes that mitigated flooding, this system supported both productive agriculture and ecological equilibrium. Prior to the influence of tourism, the communities in the Canggu region were organized around this agrarian rhythm (Surya Suamba et al., 2022), with homes, temples, and rice paddies creating a cohesive settlement pattern.

However, the emphasis started to move from agricultural to services as a result of Bali's growth as a popular tourist destination worldwide, especially since the 1980s. Developers and investors soon started looking at areas north of Seminyak, where open space, views of the ocean, and a close proximity to the Ngurah Rai International Airport made places like Canggu particularly alluring (Pramudya et al., 2025). The initial wave of mass tourism was concentrated in Kuta, Legian, and Sanur. By the early 2010s, Canggu was starting to develop as a new destination for boutique tourism, drawing in surf culture aficionados (Kludzeweit et al., 2020), long-term expatriates, and digital nomads in addition to tourists.

This change has noticeable spatial effects. Once spanning the low-lying coastal plain in an uninterrupted line, the rice fields are now progressively broken up by villas, co-working spaces, wellness centers, and lifestyle cafes. While unofficial alleyways have developed into bustling thoroughfares, roads have been enlarged to accommodate motorbike traffic, frequently without commensurate improvements in public infrastructure, drainage, or garbage management. A fragmented urban morphology has resulted from the patchwork character of land conversion, which is frequently determined by individual landowner choices rather than thorough zoning. The conflict between traditional customs and contemporary land usage is highlighted in certain places where operational subak systems (Suamba and D wijendra, 2016) run next to private pools and ceremonial processions meander through recently constructed commercial strips.

Furthermore, significant ecological challenges have been brought about by this transformation (Ramadhan, 2024). Flash flooding during the rainy season is caused by the disturbance of natural hydrology caused by the infilling of wetlands and low-lying rice terraces. Construction on formerly agricultural lands puts further demand on already scarce water resources, which are already being strained by hotel use, home consumption, and irrigation requirements (Ramadhan, 2024). Concrete-heavy development and impermeable surfaces also decrease groundwater recharge. The coastal ecology is also under stress from sand mining, beach commercialization, and rising pollutant discharge, which endangers shoreline and marine biodiversity. Socially, the dynamic of land conversion has caused community structures to become fragmented. As land prices rise and younger generations are driven away from traditional livelihoods, some local families see economic benefits from leasing or selling their land, while others are marginalized. When the rice fields themselves disappear, it becomes more difficult to preserve cultural rituals and customs associated with the land, such as temple offerings in rice fields. With its focus on aesthetics, leisure, and individualism, the emergence of a globalized tourism culture frequently stands in uncomfortable conflict with the communal ideals that have long characterized Balinese rural life.

The process is not without opposition or adaptation, notwithstanding these conflicts. Better spatial planning, the preservation of the remaining subak networks, and the promotion of eco-tourism models that combine sustainable travel and traditional agriculture have all been pushed for by some local authorities and environmental organizations (Laksmi et al., 2023). The current rate of change, however, continues to raise questions about the long-term ecological and cultural viability of southern Bali (Warren and Wardana, 2018; Rahman, 2021), and enforcement is still lax. All things considered, the transformation of rural areas into tourist hotspots in Canggu and other such locations tells a larger tale of Balinese urbanization that has been influenced by globalization, land monetization, and conflicting ideas of progress. As part of a broader urban-ecological investigation (Whitten and Soeriaatmadja, 2013) into how such landscapes can—or cannot—accommodate both growth and balance, this chapter aims to map and understand these shifts.

10.2 Conceptual Framework: The Urban-Ecological Nexus

Southern Bali's transition from rural heartlands to tourist-driven metropolitan frontiers need a conceptual framework that can address the intricacies of socio-spatial change while maintaining ecological integrity. This chapter uses the urban-ecological nexus as a conceptual lens to comprehend the ways in which ecological degradation, land commercialization, and urbanization are interconnected, co-produced, and reinforce one another. The nexus concept emphasizes the ways that environmental systems and human development co-evolve within particular physical, cultural, and political settings, rather than seeing "urban" and "ecological" as opposing spheres.

At the heart of this nexus lies the tension between the spatial expansion of urban functions and the ecological services historically provided by landscapes like Bali's subak rice fields (Fox, 2012; Wardana, 2020; Zen et al., 2024). These systems are ecological infrastructures that control water flow, maintain biodiversity, and support community-based resource governance; they are not just agricultural systems (Cole and Browne, 2015). As a result, their demise signifies a disruption of a regionally established natural order in addition to an agricultural transition. Therefore, the urban-ecological nexus does more than just map the locations of urban growth; it also shows how current environmental systems and governance regimes both influence and are influenced by such growth.

This chapter's conceptual framework is based on multidisciplinary insights from urban metabolism, landscape urbanism, and political ecology (Xie et al., 2019). Understanding the negotiation and frequent contestation of power, access, and control over land and water resources in southern Bali is made possible by political ecology (Cole, 2012). These talks, as they relate to the development of tourism, involve not only state actors and local landowners but also foreign investors, digital influencers, and evolving cultural conceptions of what Bali is. As a result, the environment becomes divided, with long-standing ecological and spiritual goals sometimes overshadowed by the logic of capital accumulation (Reuter, 2002). By promoting an examination of spatial form, land use patterns, and ecological fluxes, landscape urbanism adds to the framework. Rice fields, resorts, and residential complexes coexist in Canggu's disjointed development, which is representative of urbanization processes that are more motivated by private interests than integrated planning (Adhinata and Sawitri, 2022). Such formations might be seen as signs of more profound structural imbalances between land governance regimes and natural systems from the perspective of landscape urbanism (Görg, 2007).

In the meantime, a more comprehensive understanding of the material and energetic interactions between cities and their environs is offered by the idea of urban metabolism. This includes turning green infrastructure into built-up space, managing (or not managing) solid waste and wastewater, and extracting and using groundwater for tourism operations in southern Bali (Tang, 2004; Koski-Karell, 2019). These flows affect social equality as well as environmental health since various population segments frequently experience resource stress in different ways.

We may create a multifaceted image of the urban-ecological nexus by combining various perspectives. They make it possible to examine why, for whom, and at what ecological cost in addition to what is changing spatially. Crucially, this framework also creates room for envisioning alternatives that could more effectively balance the demands of development with the requirements of ecological balance and cultural continuity, such as community-based eco-tourism, participatory planning, and hybrid land uses (Wani et al., 2025). The mapping approach in the following section, which operationalizes the nexus by spatially tracing land use change, ecological danger zones, and places of socio-environmental conflict, is supported by this conceptual framing. By doing this, it seeks to bring the invisible into the open and expose the politics and trends influencing the disputed urban-ecological future of southern Bali.

10.2.1 Urban Growth and the Pressures on Ecological Landscapes

Real estate speculation, infrastructural construction, and the explosive growth of lifestyle tourism are the main drivers of urban growth in southern Bali (Salam et al., 2024; Shantika and Bestari, 2024), especially in the rapidly rising corridors of Canggu, Berawa, and the neighboring sub-districts. Even though this expansion is frequently praised for its positive effects on the local economy and international recognition, it also puts a great deal of strain on the ecological landscapes of the area, which have long been preserved via vigilant care based on spiritual and cultural principles.

The growing commodification of land is one of the main forces behind transformation (Suartika, 2003; Fagertun, 2017). Southern Bali has become a popular destination for long-term international residents, digital nomads, and investors looking for high returns due to the attractiveness of "the Balinese lifestyle," which combines nature, spirituality, and leisure (Dwijendra, 2014). Land that was formerly set aside for subak agricultural production is now being sold or leased for the construction of villa projects, boutique hotels, and commercial businesses in places like Canggu. Both biological functions and landscape aesthetics are weakened by this market-driven alteration, which breaks up rice fields, modifies water distribution patterns, and diminishes the continuity of open green space.

Supporting infrastructure is growing quickly in tandem with this change in land use (Rimba et al., 2020). Roads are being paved and widened, frequently without integrated drainage systems, while piped water, telecommunications, and electrical supplies are finding it difficult to meet demand. Wetlands and coastal zones are encroached upon as development spreads into formerly peripheral areas, degrading habitats essential for biodiversity, aquifer recharge, and flood control. Because this growth is fragmented and frequently takes place without thorough environmental studies, cumulative effects are rarely recognized or taken into consideration.

The role of lifestyle tourism is equally complicated. Despite being marketed as a "sustainable" or "holistic" substitute for mass tourism, it nevertheless primarily relies on the use of water, electricity, and land (Chong, 2020). Although yoga retreats, surf camps, and wellness resorts might seem environmentally conscious, their impact is substantial, particularly when multiplied by the hundreds of companies that operate with no oversight. Furthermore, because these businesses serve a worldwide clientele, they frequently perpetuate aesthetic standards that place a premium on well-kept nature over useful landscapes, marginalizing ecosystems like mangroves and rice fields. Ultimately, the present course of urban expansion in southern Bali indicates a conflict between the goals of development and the realities of the environment. It emphasizes the necessity of more integrated planning strategies that view biological landscapes as living systems vital to the island's resilience and cultural continuity rather than as barren ground in need of change.

10.2.2 Balancing Development and Sustainability in Coastal and Inland Zones

The coastal and peri-urban areas of southern Bali, where rapidly growing urbanization driven by tourism collides with delicate ecosystems and culturally entrenched landscapes, are the clearest examples of the conflict between development and ecological preservation. Once interspersed with open rice fields and fishing towns, the coastal areas from Berawa to Pererenan are now primarily occupied by villa clusters, boutique cafes, and yoga pavilions that cater to an international clientele. However, despite constant encroachment, indigenous Banjar settlements and Subak rice irrigation networks continue to strive for rhythm just a few kilometers inland (Fox, 2012). Both inland and along the shore, this spatial dualism produces an unequal geography of sustainability concerns.

The destruction of natural buffers like mangroves, dunes, and low-lying wetlands is where the strain is most noticeable along the coast. Once, these systems created a belt of protection that screened pollutants, absorbed tidal flows, and maintained biodiversity. Many of these areas have since been drained or covered over for upscale construction projects that are susceptible to saline intrusion and sea level rise. Ecological factors are frequently sacrificed for aesthetics and financial rewards when real estate values along beachfronts skyrocket. Coastal setbacks are often disregarded, construction standards are handled inconsistently, and regulatory enforcement is still lax.

The narrative is more nuanced but as important inland. The biological and cultural logic of the land is upset as peri-urban areas become mixed-use enclaves. The complex, collectively run water distribution system known as the subak system, which has supported rice cultivation for millennia, is coming under more and more stress. Land sales that split up rice fields result in fragmented waterways, decreased output, and dwindling farmer incomes (Fox, 2012). Because paddy fields are essential for controlling runoff and preserving groundwater recharge, this endangers not only the local food systems and cultural legacy but also the broader hydrology (Reuter, 2002). In this sense, sustainability refers to both socio-ecological continuity and environmental conservation. The lived experiences of local communities, who are both land stewards and participants in the changing economy, must be taken into consideration in efforts to balance development. Community-based tourism models, zoning changes, and participatory planning provide potential avenues, but they call for political will and an understanding that sustainability cannot be attained by a few tokens' green initiatives. It calls for a reconsideration of the process of urbanization, one that honors the interconnectedness of cultural landscapes and natural systems in both inland and coastal regions.

10.3 Methodology: Mapping the Changing Landscape

With an emphasis on the Canggu and Berawa regions, this study used an integrated mixed-methods approach to map and analyze ecological alterations and land use changes in southern Bali. The mapping method yielded a detailed picture of the ongoing urban-ecological transition by combining participatory insights with spatial data analysis.

High-resolution satellite imagery and publicly accessible land use statistics from the Bali Provincial Government and Indonesian Geospatial Information Agency (BIG) served as the main sources of spatial data. This information, which included rice fields, built-up regions, green open space, aquatic bodies, and coastal zones, were categorized and shown using Geographic Information Systems (GIS). Time-series satellite pictures from 2002–2004, 2012–2014, and 2022–2024 were used for temporal analysis (See Table 10.1), which made it possible to identify geographical patterns and change trajectories over a 20-year period.

In some Banjar and Subak locations, field observations and semi-structured interviews with locals, farmers, and village leaders were carried out in addition to remote sensing and GIS-based analysis. To gather experience knowledge regarding land tenure, water management, and development pressures that are frequently invisible through geographical data alone, as well as to ground-truth remote sensing data, community engagement was crucial. In collaboration with neighborhood planning NGOs (in workshop sessions held in early April 2025), the project also included participatory mapping workshops where locals were asked to sketch places they believed to be affected by flooding, ecological degradation, or landscape change. The comprehension of GIS outputs was enhanced by these qualitative overlays, which also assisted in locating discrepancies between official zoning and actual conditions on the ground.

Table 10.1 Time-series satellite pictures over 20 years of four sample areas.

	Sample 1—Canggu	**Sample 2—Near Berawa**
2002–2004		
2012–2014		
2022–2024		

	Sample 3—Other Regions	Sample 4—Other Regions
2002–2004		
2012–2014		
2022–2024	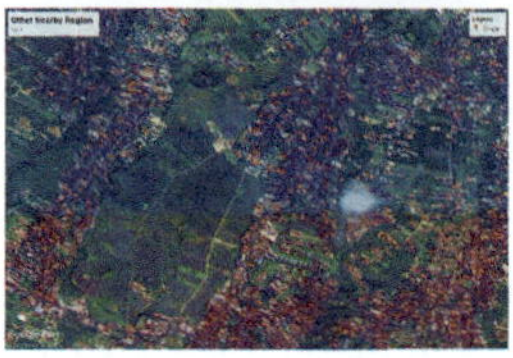	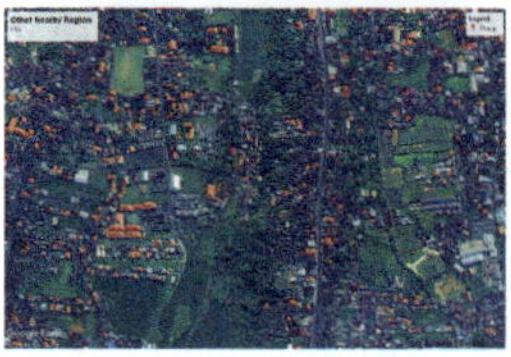

Source: Generated from open source maps using Google Earth Pro.

Lastly, to measure land conversion rates and identify important regions of ecological disruption, change detection algorithms inside GIS software (ArcGIS Pro and QGIS) were used. These techniques made it possible to depict in detail and spatially explicitly how the landscapes of southern Bali are changing because of the combined effects of urbanization, tourism, and environmental change (Figs. 10.1, 10.2 and 10.3).

10.4 Results and Discussions

Significant spatial changes were seen throughout southern Bali, especially in the Canggu-Berawa corridor, according to the mapping research. The area has seen a noticeable change from primarily semi-natural and rural landscapes to more urbanized and fragmented land uses during the 20-year study period (2002–2022). This change is a reflection of the growing interdependence of the demands on traditional and ecological land systems and tourism-driven development (Januar, 2024). Important geographical findings show that by 2022, more than 55% of the study area's rice fields that were mapped in 2002 had been transformed into built-up land. Instead of creating continuous metropolitan blocks, a large portion of this conversion took place piecemeal, creating scattered patches of villas, resorts, and business centers. The traditional subak river flows and ecological corridors that formerly permitted integrated agro-ecological functioning are disrupted by this fragmented

Fig. 10.1 Examples of agricultural land in between urban blocks (left) and communities (right) (Source: Taken by Ali Cheshmehzangi).

Fig. 10.2 Examples of old housing (left) and new housing development (right) in Canggu Region (Source: Taken by Ali Cheshmehzangi).

Fig. 10.3 Examples of ecological tourism (left) and new resort development (right) in Canggu Region (Source: Taken by Ali Cheshmehzangi).

urban architecture (Asmiwyati, 2016). Notably, Canggu's northern outskirts and portions of Pererenan showed the fastest and least coordinated expansion, frequently encroaching on low-lying or undeveloped areas.

Significant changes have also occurred in coastal areas. Particularly in Berawa, where building has spread near tidal marshes and estuarine inlets, satellite imagery demonstrates the gradual destruction of green buffer zones along the shoreline. Residents who participated in participatory mapping workshops stated that localized flooding episodes had increased in tandem with the degradation of these natural barriers. This confirms results from temporal overlays that natural drainage regions are decreasing over time. The cultural ramifications of these land use shifts were further demonstrated through field interviews with local Subak leaders and longtime inhabitants. As traditional stewardship was replaced by speculative real estate investment and fragmented landownership, many reported a degradation of both agricultural production and common water control. Even though they are hard to measure, these sociocultural factors were present at every research location and support the larger ecological effects that GIS analysis was able to identify (Lanya et al., 2015).

From a critical standpoint, the findings indicate that spatial planning methods are ineffective at directing or controlling development. The reality on the ground is dominated by ad hoc and market-driven building, even though provincial authorities' zoning maps continue to allocate sizable areas for mixed-use or agricultural purposes. The discrepancy between land use results and planning frameworks points to a policy climate that is more impacted by private development interests than by ecological concerns, as well as a lack of interagency coordination and enforcement capabilities.

Additionally, although sustainability has gained more attention in Bali's tourism discourse, the actual patterns of land use change show a type of greenwashing (Antari and Connell, 2021; Harbisher, 2023), in which spa centers and boutique resorts promote eco-consciousness while causing long-term landscape deterioration. The necessity for stronger, place-based frameworks that include ecological function, cultural continuity, and equitable governance into the island's development routes is further supported by this conflict. In conclusion, the mapping results show that southern Bali is changing quickly, in fragments, and in a way that is damaging to the environment (Böhlen et al., 2022). These alterations are more than just aesthetic; they are a reflection of fundamental conflicts between landscape stewardship and economic goals. The current course runs the risk of destroying the very cultural and environmental characteristics that first make Bali special and appealing.

10.4.1 Interpreting the Nexus: Between Spatial Expansion and Ecological Thresholds

The Canggu-Berawa corridor is where the most obvious spatial changes in southern Bali have been seen, and these changes need to be interpreted in the context of the larger urban-ecological nexus. The results are interpreted in this subsection using the previously presented conceptual framework, emphasizing how unbridled urban growth is driving the area closer to several ecological thresholds and how local governance and adaptive practices are finding it difficult to keep up.

Fundamentally, conflicting land values, overlapping natural and cultural systems, and a development strategy primarily influenced by demands from international tourists form Bali's urban-ecological nexus. In addition to fragmenting ecosystems, the quick conversion of coastal wetlands and rice fields into urban land uses has also made it more difficult for these areas to provide necessary functions. The landscape's increasing incapacity to withstand shocks like intense rainfall, saline intrusion, and seasonal droughts is indicated by the shrinkage of mangrove belts, the fragmentation of subak irrigation systems, and the loss of natural drainage regions. These are unmistakable signs that ecological tipping points are being approached, if not crossed, in southern Bali.

Although it is dispersed, community-level response to these stresses is evident. While informal coalitions of banjar leaders and NGOs try to resurrect elements of subak coordination and ecological awareness, farmers in certain places are changing cropping calendars to accommodate disturbed water flows (Fox, 2012). Because of the frequent flooding, people in coastal areas have started building temporary barriers or drainage systems. However, without more extensive institutional backing, these are insufficient and reactive.

Systemic issues are reflected in the continued geographical misalignments between official zoning maps and actual development on the governance front.

Instead of empowering communities, the decentralization of planning authority to local governments has resulted in uneven enforcement, fragmented policies, and vulnerability to speculative interests. The rise of "green-labelled" developments that contribute to long-term ecological stress while simply meeting surface-level environmental regulations is a clear example of this. Beyond static zoning, integrated land use planning is desperately needed, as the mapping results show. Ecological thresholds and spatial planning must be dynamically connected; regions that support biodiversity, culture, or hydrological processes should be viewed as vital infrastructure rather than expansion barriers. This necessitates a significant change in the way land is valued and how sustainability is operationalized in the Balinese setting. It is clear from analyzing this relationship that the future of southern Bali depends on striking a new equilibrium, one that upholds ecological boundaries, strengthens community governance, and shifts from band-aid solutions to proactive, system-wide adaptation plans.

10.4.2 Transformative Growth: Reversing Drift, Re-centering Ecologies

Southern Bali provides unmistakable proof that small adjustments won't change the trend. Ad hoc permissions, fragmented conversion, and "eco"-marketed lifestyle tourism have resulted in a spatial outcome that is neither equitable nor resilient. In this setting, transformative growth entails changing the development logic itself, moving away from parcel-by-parcel monetization and toward landscape-first stewardship, which views wetlands, subak networks, mangroves, and cultural corridors as the main infrastructure that all other uses depend on. It reinterprets land as a living system with thresholds that cannot be negotiated away, rather than as a residual container for resorts and villas.

Identifying the lock-ins is the first step towards transformative growth. The mapping illustrates how low-lying paddies are colonized by villa clusters and boutique strips, cutting off water flows and reducing buffers along tidal inlets. Roads and drains that adhere to such private geometry create a pattern that is simple to extend and costly to fix. It is necessary to change who sets the geometry to break that lock-in. The drainage and mobility grid should be recut from ecological lines outward rather than allowing speculative footprints to define it. This would include treating mangrove belts as protection zones with legal status equivalent to transport corridors, re-establishing floodplains as no-build sponges, and maintaining subak channels as continuous. Practically speaking, this suggests a development sequence whereby biological corridors are first defined, safeguarded, and funded. Only then, under performance constraints linked to water, waste, and heat outcomes, may remaining plots be gathered for housing or tourist.

Hybridization, as opposed to land use segregation, is the second pillar. A dichotomy emerges from the data: tradition or lifestyle, rice field or resort. The recognition of agro-ecological production as compatible with carefully scaled housing and hospitality eliminates that dichotomy. In addition to providing low-impact lodging, educational facilities, and craft industries, transformative growth would promote "productive landscapes" that generate revenue from horticulture, rice, and cultural practices. It is imperative that these hybrids be subject to water quotas, nutrient budgets, and waste recovery rates that are based on the coastal and subak systems' carrying capacities. Sensitive terrain should not be occupied by operators who are unable to shut the loop on energy, organics, and water.

Third, growth must be financed by the risks it creates and the services it consumes. Currently, private developments reap the benefits while the costs of coastline protection, solid-waste hauling, drainage retrofits, and salinity intrusion are socialized. Linking approvals and active licenses to a ring-fenced ecological utilities fund is a remedy. Decentralized wastewater systems, mangrove extension, and corridor restoration can all be linked to developer contributions, visitor eco-levies, and differentiating property rates. In addition to increasing revenue, this also embeds a price signal that increases competition and decreases the appeal of dispersed sprawl in favor of compact and serviced forms.

The form of governance must also shift. The current discrepancy between zoning plans and actual results is caused by a mismatch between formal instruments and actual decision nodes, not merely capacity. Village councils, landowner associations, brokers, and specialized travel websites hold just as much power as provincial agencies. Co-regulation is necessary for transformative growth: planning consents for ecologically sensitive land are co-signed by the councils of Desa Adat and Subak; transparent dashboards that show water flows, flood incidents, and permit compliance in near real time connect banjar-level monitoring to provincial enforcement. According to our participatory mapping, communities are already taking action by making improvised drains, hand-clearing channels, and changing crop cycles. Co-regulation transforms reactive coping into anticipatory management by providing such approaches with a legal avenue and operational resources.

Tourism itself must be re-skilled. According to the study, green branding frequently capitalizes on devalued ecologies. In an alternative tourism industry, ecological literacy would be considered a fundamental skill: designers who work to subak alignments rather than against them; maintenance teams that manage live borders rather than concrete embankments; and operators who are trained to read water tables and soil infiltration. Instead of checking boxes, certification should be based on measured performance, such as yearly audits of waste diversion, fertilizer loading, and groundwater use linked to license renewal. Marketing must simultaneously shift from beach-club aesthetics to stewardship experiences that maintain value with regional land managers and producers.

The mapping facilitates three spatial shifts that can be used to stage transformative growth. First, a "blue-green backbone" is identified, consisting of mangroves, riparian strips, wetlands, and continuous subak channels, where conservation and restoration are guaranteed and paid right away. Second, "repair seams" are found where constructed patches and fragmented paddies intersect; in these areas, targeted land swaps, easements, and micro-retention projects can reestablish flow routes and substantially lower the danger of flooding. Third, existing serviced corridors are designated as "compact growth nodes," which absorb demand that would otherwise overflow onto plots that are vulnerable to flooding. Water and waste budgets should cap these nodes, and any growth should be conditioned on the upward trend of corridor health metrics.

This is not romantic nostalgia at all. It is a stoic interpretation of identity, danger, and expense. The very ecologies that are currently under danger are inextricably linked to the southern Bali brand. The economy follows the failure of those ecologies. The economic future and ecological restoration are thus aligned through transformative growth: development rights in return for corridor funding; hospitality premiums obtained through quantifiable contributions to subak continuity; and coastal access guaranteed by the expansion of mangroves, which serve as storm barriers and carbon sinks.

The chapter's findings give both the urgency and the blueprint. We have two options: either we alter the rules or we continue to chase short-cycle benefits and pave brittle barriers. The latter includes co-regulated governance, risk-priced financing, hybrid productive landscapes, ecology-first geometry, and performance-based tourism. It's how southern Bali transitions from drift to design, and how resorts and rice fields cease to be enemies and begin working together to create a sustainable future.

10.5 Conclusions: Pathways Forward for Urban-Ecological Resilience

The change in southern Bali, especially in places like Canggu and Berawa, serves as an example of a larger problem that fast-growing areas with strong ecological and cultural ties face (Agung et al., 2020). The interaction between tourism-driven urban growth and delicate biological landscapes has been mapped and studied in this chapter, resulting in fragmented and unsustainable spatial patterns. The contentious and ever-changing urban-ecological nexus at the heart of these processes necessitates an immediate rethinking of growth paths in favor of more resilient, integrated, and place-sensitive design.

The mapping investigation identified deeper spatial contradictions between market-oriented urbanization and traditional land stewardship in addition to patterns of land conversion. The need to abandon growth-centric ideologies that view land as an infinite resource is further supported by these findings. Rather, land needs to be understood as a limited resource that is a part of intricate ecological and cultural processes. When faced with unbridled growth, Bali's ecological resilience is tested, as evidenced by the erosion of rice fields, the collapse of the Subak water governance system, and the incursion into wetland and coastal areas. These are not isolated incidents; rather, they are a sign of a more widespread systemic inability to match local values and ecological thresholds with planning policy.

It has also become apparent that adaptation is taking place, albeit frequently on the periphery. Communities are attempting to adapt to environmental pressures or preserve ancient practices, but these initiatives are still in danger in the absence of institutional support. In the face of formal planning processes controlled by state or private sector players, local knowledge systems—such as the subak and traditional village (*desa adat*) structures—offer useful frameworks for ecological governance, but they are becoming less and less prevalent. Reorienting governance systems to appreciate and include local ecological intelligence—not just as heritage, but as active elements of future resilience—is necessary to close this gap.

Moving forward, policy reform must prioritize the integration of ecological function and spatial design. Bali's land use planning must be participative, ecologically based, and dynamic. This entails creating spatial tools that take into account changes in land cover in real time, mapping important ecosystem services, and defining development boundaries using thresholds like flood danger or water availability. In order to enable technically sound and socially integrated enforcement, legal and institutional procedures must also change. This will enable local communities to co-govern landscapes in ways that respect ecological and cultural continuity.

Lastly, the lesson that sustainability cannot be adapted is evident. The landscape and cultural identity that make southern Bali so alluring are in danger of being undermined by the current development trajectory (Suartika, 2005). Therefore, proactive design, cautious land regulation, and long-term ecological thinking must be the main focuses of future routes. Bali's growth must be based on an ethics of care—for the land, for the community, and for future generations—rather than just economic reasoning if it is to continue developing as a travel destination and a place of belonging. Bali's southern landscapes still have the capacity to adapt and renew. However, doing so will necessitate a steadfast dedication to viewing space as a living system that requires respect, balance, and management rather than just as property.

Acknowledgments

We would like to thank Professor Sue Harrison and the Faculty of Engineering, Architecture and Information Technology (EAIT) at The University of Queensland (UQ) for their support in making our recent trip to Denpasar and its wider region a possibility. The events and activities we held and participated in during our short visit in early April 2025 helped us complete this study and strengthen our narrative based on the new findings, surveys, and engagement opportunities with a wide range of local stakeholders. We also acknowledge the valuable support and contributions from local partners, NGOs, and neighborhood representatives.

References

Adhinata B, Sawitri MY (2022) Touristification and the changing of spaces for tourism in Canggu Village. J Ilmu Sosial Dan Humaniora 11(3):447–457

Agung AAG, Suprina R, Nugroho A (2020) Bali–A traditional society in transformation. J of Cult, Soc and Dev 59(1):14–18

Antari, N. P. B. W., & Connell, D. (2021). Tukad Bindu in Bali, Indonesia: Ecotourism or greenwashing? Int J of Tour Cities 7(4):1049–1075

Asmiwyati IGAAR (2016) Impact of land use change on urban surface temperature and urban green space planning: Case study of the Island of Bali, Indonesia. Doctoral dissertation, Curtin University

Atmaja DM, Gunamantha IM, Bharata IBAY (2025) Coastal development implications to changes of coastal typology in Denpasar City environment. Sriwijaya J of Environ 10(1):1–10

Böhlen M, Liu J, Iryadi R (2022) Who speaks for the forest? Local knowledge, participatory mapping and collaborative evaluation for GIS analysis in the tropics of central Bali, Indonesia. The Int Archives of the Photogrammetry, Remote Sens and Spatial Inf Sci 48:73–80

Cheshmehzangi A (2016) Multi-spatial environmental performance evaluation towards integrated urban design: A procedural approach with computational simulations. J of Clean Prod 139:1085–1093

Cheshmehzangi A, Tang T (2024a) China under construction: Shaping cities through recent urban transformation. Springer Nature, Singapore

Cheshmehzangi A, Tang T (2024b) 30 years of urban change in China's 10 core cities. Springer Nature, Singapore

Chong KL (2020) The side effects of mass tourism: The voices of Bali islanders. Asia Pacific J of Tour Res 25(2):157–169

Christie JW (2007) Water and rice in early Java and Bali. In: Boomgaard P (ed) A world of water: Rain, rivers and seas in Southeast Asian histories. KITLV Press, Leiden, pp 235–258

Cole S (2012) A political ecology of water equity and tourism: A case study from Bali. Ann of Tour Res 39(2):1221–1241

Cole S, Browne M (2015) Tourism and water inequity in Bali: A social-ecological systems analysis. Hum Ecol 43:439–450

Dwijendra NKA (2014) Affordable housing provision projects in Bali, Indonesia: Improving quality and cultural acceptability. Doctoral dissertation, Curtin University

Fagertun A (2017) Waves of dispossession: The conversion of land and labor in Bali's recent history. Soc Anal 61(3):108–125

Fox KM (2012) Resilience in action: Adaptive governance for subak, rice terraces, and water temples in Bali, Indonesia. Doctoral dissertation, The University of Arizona

Gede Ustriyana IN, Saifulloh M, Diara IW, Suyarto R (2025) Mapping agrotourism potential using geospatial data and random forest: A case study from Bali. Ecol Eng & Environ Technol 26(3):1–16

Görg C (2007) Landscape governance: The "politics of scale" and the "natural" conditions of places. Geoforum 38(5):954–966

Harbisher B (2023) Greenwashing Bali: Corporate responsibility, SDGs and the subversion of Indigenous dissent. In: Harbisher B (ed) The mediation of sustainability: Development goals, social movements, and public dissent. Rowman & Littlefield, Lanham, pp 95–116

Huang H (2020) Nature and the spirit: Ritual, environment, and the subak in Bali. EnviroLab Asia 3(2), 1

Jamal T, Higham J (eds) (2021) Justice and tourism: Principles and approaches for local-global sustainability and well-being. Routledge, New York and London

Januar R (2024) Implementation and challenges of sustainable tourism programs in Bali. Bali Tour J 8(2):33–37

Kludzeweit I, Kamargiannis T, Tirta T, Middelkamp E (2020) Effective interventions for micro and small enterprises to reach the preferable future of the Canggu area in Bali. Dos Algarves: A Multidisciplinary e-J 36

Koski-Karell NS (2019) Integrated sustainable waste management in tourism markets: The case of Bali. Indian J of Public Adm 65(3):646–660

Laksmi PAS, Putra YD, Sara IM, Setena IM, Putra IK, Jamaludin MR (2023) Self-reliance with nature: Development of subak ecotourism as an effort to empower the local community in Siangan Village, Gianyar District, Gianyar Regency. Bhakti Persada J Aplikasi IPTEKS 9(2):118–123

Lanya I, Subadiyasa NN, Sardiana K, Adi GPR (2015) Numerical classification, subak zoning and land transfer function rice field in the Province of Bali based on remote sensing and GIS. Procedia Environ Sci 24:47–55

Lewis J, Lewis B (2009) Bali's silent crisis: Desire, tragedy, and transition. Lexington Books, Lanham

Lv Y, Hao B, Sarker MNI, Zeng X, Hu X (2024) Exploring the nexus between urban green space and sustainable tourism: Potential for the green economy. Int J of Tour Res 26(4), e2700

Nurjani NPS, Dwijendra NKA (2020) How traditional Balinese houses can adjust and cater for international tourists in the Canggu area, Bali, Indonesia. Int J of Psychosoc Rehabil 24(3):2561–2573

Pakarinen V (2018) Strategies and collaboration in developing sustainable village tourism: Study in Finland and Bali, Indonesia. Master's thesis, Itä-Suomen yliopisto

Pramudya W, Ahimsa-Putra HS, Widiyastuti D (2025) Tourism development and changes in life satisfaction: Case of Balinese community in Canggu traditional village, North Kuta. Int J of Stud in Soc Sci and Humanit (IJOSSH) 1(3):237–256

Rahman DF (2021) Re-evaluating socio-cultural change in World Heritage Sites: A case study of the cultural landscape of Bali Province. Doctoral dissertation, UCL, University College London

Ramadhan SH (2024) Tourism transformation in Canggu: Impacts on socio-cultural, economic, and environmental dimensions with sustainable development strategies. Khasanah Ilmu: J Pariwisata Dan Budaya 15(2). https://doi.org/10.31294/khi.v15i2.20781

Reuter TA (2002) Custodians of the sacred mountains: Culture and society in the highlands of Bali. University of Hawai'i Press, Honolulu

Rimba AB, Atmaja T, Mohan G, Chapagain S, Andi A, Payus C, Fukushi K (2020) Identifying land use and land cover (LULC) change from 2000 to 2025 driven by tourism growth: A study case in Bali. The Int Archives of the Photogrammetry, Remote Sens and Spatial Inf Sci XLIII -B3-2020:1621–1627. https://doi.org/10.5194/isprs-archives-XLIII-B3-2020-1621-2020

Roth D (2014) Environmental sustainability and legal plurality in irrigation: The Balinese subak. Curr Opin in Environ Sustain 11:1–9

Roth D, Sedana G (2015) Reframing Tri Hita Karana: From "Balinese culture" to politics. The Asia Pacific J of Anthropol 16(2):157–175

Roy NC, Thangaraj V (2022) Dynamic interaction of urban development and rural–urban migration: An application of integrated urban metabolism analysis tool (IUMAT) for sustainable city planning. Digit Polic, Regul and Gov 24(2):118–140

Salam, R., Samudra, A. A., Suradika, A., & Sumada, I. M. (2024). Future policy implementation to balance growth and sustainability: Managing the impact of mass tourism in Bali. J of Infrastruct, Polic and Dev 8(9):1–20

Shantika B, Bestari NMP (2024) Crises and disruption on Bali tourism: Natural disaster, civil unrest and terrorism. Barista: J Kajian Bahasa dan Pariwisata 11(2):104–114

Shavit D (2019) Bali and the tourist industry: A history, 1906–1942. McFarland & Company, Jefferson

Suamba IBS, Dwijendra NKA (2016) Study of spatial use control management in Canggu Village, Badung Regency, Bali, Indonesia. Int J of Eng and Emerg Technol 6(2):45–50

Suartika A (2003) Conforming to the process of land commodification and the deterioration of territorial cultural interaction: A case study on Bali. Public Art & Urban Des 118

Suartika GAM (2005) Vanishing paradise: Planning and conflict in Bali. Doctoral dissertation, UNSW Sydney

Surya Suamba IB, Anom Wiryasa NM, Acwin Dwijendra NK, Diasana Putra I (2022) Characteristics and deviation patterns of agricultural land use in tourism area of Canggu, Bali, Indonesia. Caspian J of Environ Sci 20(2):423–430

Sutawan N (2004, November) Tri Hita Karana and subak: Search for an alternative concept of sustainable irrigated rice culture. Paper presented at the International Network for Water and Ecosystem in Paddy Fields Conference, Tokyo, Japan

Tang J (2004) A case study of a hotel solid waste management program in Bali, Indonesia. Master's thesis, University of Waterloo

Wani MD, Dada ZA, Bhat WA, Shah SA (2025) Community-driven ecotourism in the trans Himalayan region: A sustainable model for cultural and environmental preservation. Community Dev 56(4):560–582

Wardana A (2019) Contemporary Bali: Contested space and governance. Springer, Cham

Wardana, A. (2020). Neoliberalizing cultural landscapes: Bali's agrarian heritage. Crit Asian Stud 52(2):270–285

Warren C, Wardana, A (2018) Sustaining the unsustainable? Environmental impact assessment and overdevelopment in Bali. Asia Pacific J of Environ Law 21(2):101–125

Whitten AJ, Soeriaatmadja RE (2013) Ecology of Java & Bali. Tuttle Publishing, North Claredon

Xie L, Tan-Mullins M, Cheshmehzangi A (2019) Political ecology of Chinese smart eco-cities. In: Zhang X (ed) Remaking sustainable urbanism: Space, scale and governance in the new urban era. Palgrave Macmillan, Singapore, pp 57–79

Zen IS, Surata SPK, Titisari PW, Ab Rahman SA, Zen S (2024, March) Sustaining subak, the Balinese traditional ecological knowledge in the contemporary context of Bali. IOP Conf Ser: Earth and Environ Sci 1306, No. 1, 012034

Chapter 11
Resilient Cities Reimagined: Towards Transformative Growth Through Ecological Innovation

Saima Gulzar and Amna Waheed

Abstract The study aims to explore the complex relationship between urbanization, new development, and spaces for well-being of the society with a special emphasis on the negative role of the uncontrolled urban sprawl. Since the population and economic needs of societies continue to grow, urbanization provokes increased infringement of vital open or underutilized areas. Those spaces, developed in the outskirts of cities without planning, become the central hub of activities (economic, social, cultural, and sometimes developed into green spaces for multiple use etc.) for the nearby communities developing an ecosystem. The uncontrolled development not only endangers the authenticity of these spaces, but also carries out the displacement of traditional communities. This disrupts their economic and environmental spheres, social structures, and cultural identities.

The study highlights the new developments caused by the unplanned urbanization and without community appreciation disrupting the ecological systems of the area. To address these challenges, the study advocates for Nature-Based Solutions (NBS) to nurture resilient cities, ensuring urban sustainability amidst unplanned urbanization. The NBS are not only necessary for human well-being but also required to repair the broken biosphere in order to enrich the biodiversity. The selection of the abandoned/underutilized spaces under the flyover in developing countries, such as Pakistan was used as a sample. It explored the implementation of NBS for nurturing the resilient cities in the South Asian context. The study concludes by advocating for a holistic approach for urban design/planning for new developments (like extension of road networks, flyovers, and underpasses). Furthermore, it also deals with the unplanned urbanized areas (without any land use plans mostly) with the amalgamation of NBS solutions to nurture the resilient cities.

S. Gulzar • A. Waheed (✉)
Department of Architecture, University of Management and Technology, Lahore, Pakistan
e-mail: buttamna77@gmail.com

A. Cheshmehzangi et al. (eds.), *Transforming Cities*, Urban Sustainability,
https://doi.org/10.1007/978-981-95-7378-3_11

Keywords Cities • Flyovers • Nature-Based Solutions (NBS) • Resilient • Unplanned • Urbanization

11.1 Introduction

Urban landscape has experienced an accelerated and repeated uncoordinated expansion around the globe in recent years. Urbanization brings opportunities and challenges, whereas typically it is seen as a sign of economic progress and development (Rana and Bhatti, 2018). Furthermore, urban growth leads towards increased job opportunities, improved access to services, innovation in technology, and cultural exchange. Well-planned urbanization is important to enhance economic resilience, efficient land use, and sustainable advancement systems (Lafortezza and Sanesi, 2019). On the other hand, if urban growth increases in an unsystematic and chaotic manner, it may give rise to unplanned urbanization, which is an increasing case in many developing countries (Fig. 11.1).

The rapid growth of cities without proper long-term strategies, structured planning, guidelines, or infrastructure development leads towards unplanned urbanization. This phenomenon frequently occurs in developing countries where increase in immediate population growth, social stratification, and migration from rural to urban areas surpass the capacity of local governments to control it. In present times, urban areas are facing accumulating challenges that are intensifying by the development of unplanned urbanization. This urbanization is often caused by informal and uncontrollable expansion of cities (Yoshizumi, Nguyen and Kobayashi, 2013). Informal or unstructured growth of cities leads towards overcrowding, disturbing the environment and increasing the risk of natural disasters. The impact of

Fig. 11.1 The complex relationship between unplanned urbanization, new development challenges, and opportunities for improvement in developing countries (Source: Author).

climate change, loss of biodiversity, and more persistent weather conditions make these unplanned urban spaces more unprotected and weaker (Rahaman, Kalam and Al-Mamun, 2023).

The consequences of unplanned urbanization are universal, extensive, and multi-dimensional. Demolishing green spaces, wetlands, and forests, which are of utmost importance for ecological services and climate regulations, sets a pressure on the environment (Edeigba et al., 2024). The absence of universal and strategic development leads towards inequality, poor health outcomes, limited access to education services, and reduces the economic opportunities (Rana and Bhatti, 2018). Additionally, traffic congestion, pollution, and unsystematic land use are caused by unplanned urban sprawl. This makes cities weaker and more fragile to climate changes and other challenges (Ribeiro, 2021).

11.1.1 Case Study

Lahore is known as the heart of Pakistan. This is not merely a city but a catalog of millennia-old background and a blend of different societies as well. Like many developing cities, Lahore has faced rapid and frequent unplanned urban growth due to economic pressures and need for modernization (Iqbal et al., 2022). In the process of city expansion, areas, such as Thokar Niaz Baig have been profoundly transformed, affecting traditional lifestyle and socially-moderate communities (Fig. 11.2). To support urban growth and reduce traffic blockage, Thokar Niaz Baig experienced major changes when new construction projects initiated. This development led to the displacement of traditional societies, as land use changes consumed residential neighborhood and destroyed open green areas once utilized for social gatherings. Without community involvement, new development has disconnected the area from its traditional values. Furthermore, new development has replaced historic environments with new land uses and unfamiliar areas. Thokar Niaz Baig

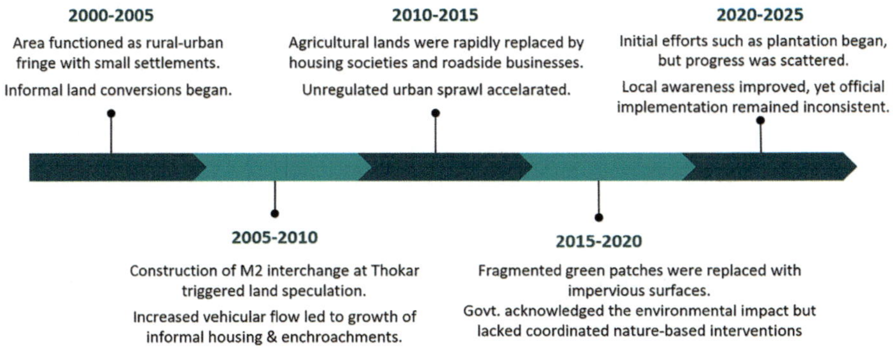

Fig. 11.2 Timeline for the transformation discussed in Thokar Niaz Baig case study (Source: Drafted by author).

intersection was built to resolve vehicular issues that led to a chaotic and disconnected urban landscape. This emphasizes the consequences of neglecting social and cultural fabric during rapid urbanization (Alvi, 2024).

11.1.2 Existing Scenario

The Thokar Niaz Baig junction is located at Southwest side of Lahore division (Fig. 11.3). As one of the populated junctions, it links multiple highways and roads including Multan Road, Canal Bank Road, Niaz Baig Road, Raiwind Road, and Motorway (M2) through Shahpur Interchange. Over the years, this area has observed immediate urban expansion, with increasing commercial engagement, transportation load, and residential development. The development of Thokar Niaz Baig flyover strengthens the traffic flow and minimizes clog. Due to the growing vehicular volumes, informal planning, and lack of strategic alignment, the road network in the surrounding areas experiences the pressure.

This location has a strategic and operational value; however, the area around the flyover exhibits a typical arrangement of unplanned urbanization. The encroachments,

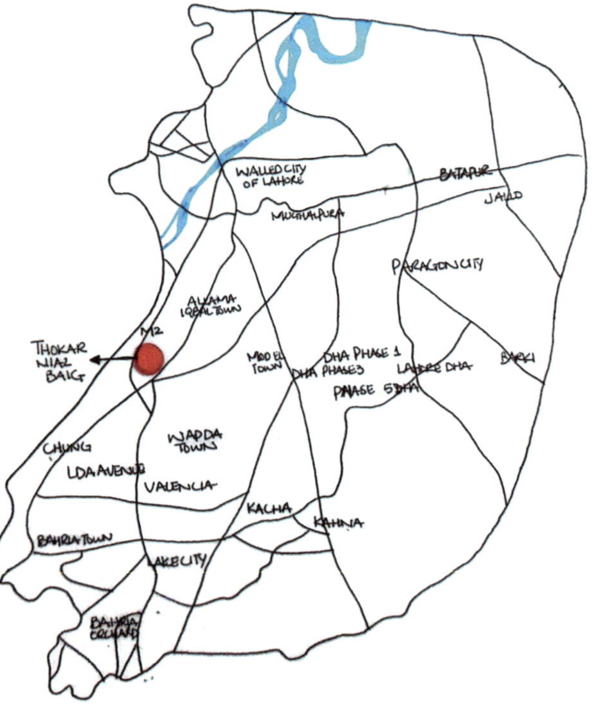

Fig. 11.3 Lahore map with marked Thokar Niaz Baig Area (Source: Drafted by author).

commercial growth, lack of pedestrian facilities, and insufficient green spaces have caused problems in the city, leading towards urban inefficiencies and environmental stress. The common problems faced by public during peak hours include traffic congestion, drainage issue, and unmanaged solid waste (Fig. 11.4).

Fig. 11.4 Existing scenario in the selected case study: (a) Road leading towards motorway along with Thokar Niaz Baig flyover; (b) Green space on each side linking a road towards Raiwind; (c) Green belt under the flyover; (d) Thokar Niaz Baig intersection where orange line crosses above flyover; (e) Present flyover condition; (f) Multan Road leading towards Yateem Khana/Chauburgi; (g) U-turn under the flyover leading towards canal road; (h) Societal and environmental issues; (i) Orange line Thokar Niaz Baig station (Source: Pictures taken by the authors).

These problems highly prioritize and emphasize sustainable urban strategies featuring NBS. These solutions have a potential to work within the existing urban fabric and support the ecological and functional resilience of the area (Adams, Moglia and Frantzeskaki, 2024).

11.1.3 Concept of Nature-Based Solutions (NBS)

NBS recommend and refer to the actions that preserve, restore, and sustainably control natural or amended ecosystems in ways that address societal challenges effectively and adaptively (Fig. 11.5). These solutions offer an integrated approach that is cost-effective and sustainable (Seddon et al., 2020). The infrastructure development typically surpasses due to this type of expansion, leading towards an increase in the environmental degradation. NBS have achieved recognition as potential avenues to resolve these intricacies. The principles of ecological restoration and sustainable development embed that NBS incorporate the exploitation of natural techniques and

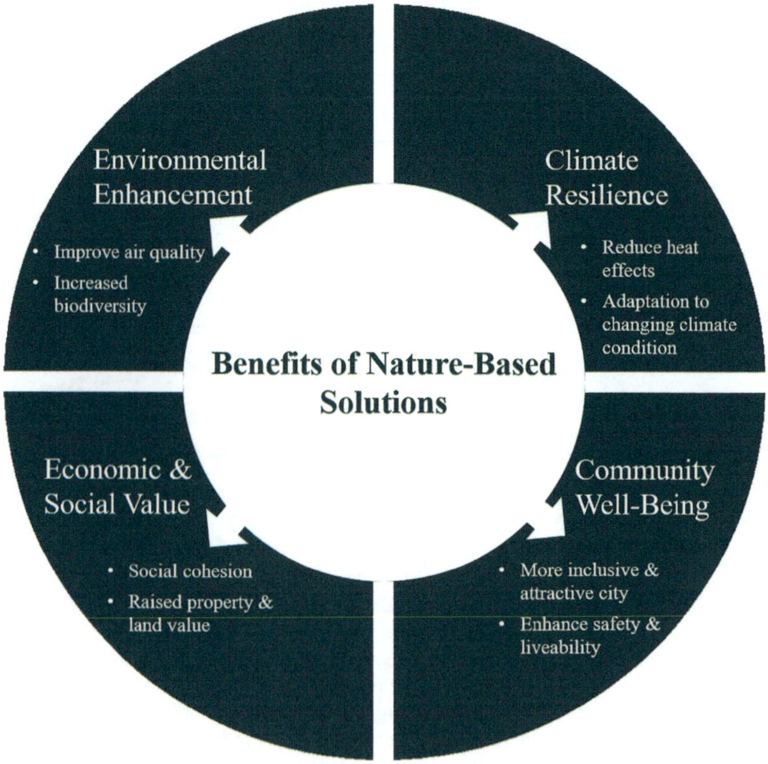

Fig. 11.5 Benefits of Nature-Based Solutions (Source: Author).

systems. These include green infrastructure, wetland restoration, urban forests, and permeable landscapes to handle urban challenges (Kabisch et al, 2016).

The perception of resilience has originated as a central theme in new urban development and sustainability within this difficult and demanding urban experience. Resilient cities not only persist; however, also modify and accommodate during tough circumstances. The innovative, inclusive, and sustainable strategies and techniques are required to build urban resilient cities where planning is either unavailable or inadequate. NBS support the areas with unplanned urban growth without building or developing new things. If nature is incorporated, cities may become sustainable and resilient in a more holistic, equitable, and enduring manner (Duffaut, Versini and Frascaria-Lacoste, 2022). The selected Thokar Niaz Baig area was historically not popular and was situated outside the city of Lahore. The unplanned expansion and urbanization pressure developed the locality for lower income group. When the unplanned growth along with the population density due to haphazard urbanization reached this intersection, the area started its economic growth that turns this unpopular area into a vibrant modern centre. With the passage of time, Thokar Niaz Baig has undergone significant changes and developments, laying the foundation for its current status. The current scenario demands NBS to repair the disrupted ecosystem in order to enhance the diminishing-built environment.

11.2 Urbanization Challenges

Urbanization can be defined as the growth of cities with people moving from rural areas to urban areas in order to improve their quality of life. It plays a significant role in the economic development. Furthermore, it brings multiple challenges particularly when it takes place in the fastest growing area without proper development or structure. This unplanned growth leads towards various traffic problems, overcrowding, pollution, and loss of green spaces. With the passage of time, cities are growing at a faster pace and people are getting away from their traditional lifestyles and local cultural practices (Yoshizumi, Nguyen and Kobayashi, 2013).

Traditional cultural practices include our ancestors' living style, which transfers from generation to generation. These practices include inherited customs, rituals, and festivals that define the identity, consistency, and stability of local communities. As cities are growing, the places where such practices used to be performed are shifting in commercial areas, leading towards cultural degradation. Experts are of the view that city development or urban framework should also focus on the preservation of cultural and social values that make communities unique (Desouza and Flanery, 2013). It is important to consider the cultural and social values of cities, as they are the systems that include histories, traditions, and human values. Urban living might be resilient and meaningful if planning would balance development with cultural preservation.

Historically, cities and surrounding communities were formed on the basis of cultural values, climate, and community needs (Sakketa, 2023). Previously, people

Fig. 11.6 Walled city, Lahore (Source: Walled city website, from open access source).

designed their living spaces in such a way that supported social interaction, environmental harmony, and comfort. For instance, in old Lahore, many spaces encouraged sustainable living through shared courtyards, streets, and communal gardens. This living style builds a connection between people, nature, and cultural rituals (Nasar-u-Minallah, Rahman and Riaz, 2021).

Eventually, new urban planning focused on productivity and extension due to which these values were disregarded. Latest infrastructure focuses on roads, buildings, and traffic flow rather than cultural heritage and traditional lifestyles (Desouza and Flanery, 2013). Thokar Niaz Baig, with the combination of historic routes and new developments, shows how rapidly urbanization can eliminate cultural identity and environmental character of an area (Alvi, 2024).

Urbanization and infrastructure are often seen as signs of progress; however, have posed a serious challenge for indigenous communities. Urbanization has brought crucial changes to the traditional ways of life across the region. In Nepal, the Majhi community is traditionally dependent on rivers for fishing and boating. This community is strongly affected by rapid urban expansion. With the construction of roads, bridges, and hydropower plants, they have now limited access to rivers. The construction of infrastructure has damaged the natural ecosystem due to pollution, river extraction for sand and gravel (Acharya, 2018). These changes have not only carried-off their source of income but have also displaced families from their ancestral and traditional livelihoods. The Majhi community faces difficulties in adapting modern economic systems as their traditional environment disappears. This study highlighted how rapidly urbanization can harm the environment and the social fabric. The reduction of natural resources and breakdown of traditional knowledge have forced Majhi community to find new ways to survive (Acharya, 2018). This highlights the need for more sustainable and inclusive urban development strategies that protect the natural ecosystems and local traditions. This shows that sustainable development must include not only environmental protection but also the preservation of cultural heritage. It is beneficial for people who are closely-connected to nature, especially the native groups.

Thokar Niaz Baig was historically, traditionally, and socially enriched but deeply affected due to new development. That resulted in the displacement of communities, loss of communal spaces, green areas, and cultural disconnection. Therefore, it is essential to protect and revive community spaces and create eco-friendly parks that respect cultural and historical context. Community involvement in planning process ensures that new development does not weaken the connection to roots and ancestors. Once local community builds small businesses, farming activities, and traditional social life in Thokar Niaz Baig, this step would connect to open spaces and local markets. As the people of Majhi struggled to adapt after losing their natural resources, the community of Thokar Niaz Baig faced economic issues and loss of cultural identity when they were pushed out of their environments without being part of the decisions.

The only solution is NBS and its implementation in the form of urban green spaces, particularly in the case of selected case study of Thokar Niaz Baig. Urban green spaces or elements play a vital role in making cities healthier and suitable to live. Especially, in crowded urban areas, green spaces enhance environmental health and create a space where traditional practices, community interactions, and cultural events may flourish. Rejuvenating green spaces offers safety and peaceful lifestyle that connects communities with nature and supports their physical and mental well-being (Edeigba et al., 2024). In areas, such as Thokar Niaz Baig, introducing green spaces through NBS could improve environment, reduce pollution, and support community life. These solutions comfort environmental stress and bring balance to the built-up surrounding. Additionally, these interventions support local traditions by offering open spaces for cultural and social activities (Adams, Moglia and Frantzeskaki, 2024).

11.3 Nature-Based Solutions (NBS) Strategies and Implementations

During the last several years, many cities around the globe have started incorporating NBS to deal with the problems that have raised due to rapid urbanization (Seddon et al., 2021; Suarez et al., 2024). To improve the environment and quality of life, modern strategies and approaches emphasize on working with nature rather than against it. Due to advancement, green infrastructure, for instance, parks, trees along roadsides, urban forests, and roof gardens have been utilized to solve critical problems (Hussain et al., 2024). These green spaces are actively used to reduce pollution, rising temperatures, and stress (Kabisch et al., 2016). Furthermore, green spaces provide shades and cool down hot city territories, improving mental and physical health as well as promoting peaceful environment. The development of ecological corridor is the main and important approach in modern era. These corridors act as a network to connect green spaces with wildlife movement and support nature to survive within the cities (Urban et al., 2024). Additionally, they maintain a balance between built and natural environment (Edeigba et al., 2024).

Additionally, the fusion of nature in cities enhances climate resilience (Orsetti et al., 2022; Rashid, Rasheed and Ngah, 2024). This could be achieved by soaking up rainwater, reducing urban heat, and stabilizing temperatures. Green spaces also enhance and intensify urban security by reducing threats including landslides, flash floods, and heatwaves, which excessively affect urbanization. Furthermore, incorporating nature-based strategies in cities is a smart move for economy. Cities that invest in green infrastructure often reduce the damage from natural disasters, improving public health and energy use (Desouza and Flanery, 2013). The development of urban green spaces attracts tourists and develop job opportunities. Reconnecting cities with nature helps protect natural environment; however, it also supports the local economy and improves the living style of people (Edeigba et al., 2024). Rotterdam is the largest city in Netherlands, same as Lahore, and faces serious environmental challenges. Case studies of Rotterdam and Seoullo 7017 Sky Garden are selected for the further discussion as practical solutions were implemented. These case studies could transform disconnected urban zones into healthy, sustainable, and culturally-rich communities.

Rotterdam is one of the largest port cities in Netherlands, Europe. Due to dense urbanization, Rotterdam faces serious environmental challenges including industrial activity and climate-related risks, such as flooding and rising temperature. To tackle these issues and make city more resilient, Rotterdam has adopted a structured nature-based approach. This approach balances the environmental refurbishment, societal needs, and economy development (Rajput, 2024; Chen, de Vries and Han, 2015). By developing ecological corridors and restoring riverbanks with natural concept, Rotterdam has successfully strengthened its environmental quality, reduced urban heat temperature, and created a healthy and safe space for its citizens. These initiatives promote biodiversity and make city aesthetically appealing and resilient.

The case of Rotterdam emphasizes how cities can effectively and efficiently combine urban development with environmental sustainability (Balık and Balık, 2021). Moreover, Rotterdam's approach highlights that even densely-built cities could also generate a balance between modern framework and natural environment by planning to keep nature as a focal point. The current study is highly relevant for areas, such as Thokar Niaz Baig, where rapid urbanization has left little spaces for environmental planning. Through small but strategic green interventions, dense and developed areas could be transformed. This study highlights the importance of recognizing both environmental and social roles of urban nature. The transformation of vacant spaces, such as roadsides, intersections, and areas under flyovers may promote urban cooling, ecological balance, and community's quality of life.

Rotterdam also respected its historic identity by integrating old structures, edges, and public areas into innovative structures (Rajput, 2024; Chen, de Vries and Han, 2015). This allows the city's past to remain visible and meaningful within the new development. Creating ecological corridors along road network in Thokar Niaz Baig, efforts should be made to restore any remaining historic routes, old settlement patterns, or cultural landmarks by incorporating them into new eco-friendly designs. Implementing such strategies may help in maintaining the area's original character. Rotterdam case study proves that even places facing urban stress, development using nature, and respecting history could transform disconnected urban zones into healthy, sustainable, and culturally-rich communities.

Seoullo 7017 Sky Garden is an innovative urban project located in Seoul, South Korea. This project demonstrates strategical reuse of infrastructure to create green spaces for public. It is an old highway overpass built in 1970s that became unsafe for vehicles. Later on, in 2017, a structure was proposed to transform it into a green public walkway instead of demolishing it. This decision was taken to bring nature into the heart of the city. The main aim of this project was to address the urban challenges by converting extinct infrastructure into an innovation and vibrant public space that enhances community engagement and environmental sustainability. This overpass was turned into a 1-km-long sky garden overflowing with over 24,000 plants including trees, flowers, and shrubs. This project included small cafes, pavilions, resting bays, and cultural spaces for community engagement (Margono and Zuraida, 2019).

In the busy and chaotic city, this sky garden improved the air quality and provided a cooling effect to people. This sky garden became a famous and renowned communal space, helping people to connect with nature and each other (Margono and Zuraida, 2019). These interventions made the city safer and more appealing without requiring the construction of additional roads. Seoullo 7017 is a contribution to urban biodiversity and environmental education as this space has a positive effect on the psychological well-being of people. This project provides residents and visitors with a unique space that promotes ecological awareness, community interaction, and green urban lifestyle (Margono and Zuraida, 2019).

NBS create ecological corridors, improve air quality, enhance pedestrian movement, and add beauty to the urban landscape of cities struggling with traffic congestion and lack of green areas. This project relates with Thokar Niaz Baig in context to reconnect city while improving air quality, reducing heat temperatures, and providing healthy environment for community. Thokar Niaz Baig, where the flyover and the neighboring road network have created division and congestion, the spaces that are underutilized can be transformed into eco-friendly corridors. The area may reconnect people with nature by generating shaded walkways, planting native trees, and developing small parks under and around the flyover. This approach would not only address connectivity and environmental issues; however, may also offer communal spaces for social and cultural activities.

The table (Table 11.1) outlines the strategic alignment between urban resilience features and the NBS that addresses the complex challenges faced by rapidly urbanizing areas, such as Thokar Niaz Baig. A practical implementation path relevant to local realities ranging from inclusivity to economic stability is paired with its underlying urban drivers. Due to displacement of communities and the reduction of public gathering spaces, inclusivity emerges as a vital concern. Urbanists can re-establish social cohesion and develop green spaces that serve all the socioeconomic groups. Equity is promoted by fairly distributing eco-friendly spaces, especially in undeserved areas. From the perspective of environmental sustainability, unplanned urban growth in Thokar Niaz Baig has contributed to heat stress and pollution. Implementing green corridors along canals, roadside, and vacant lands with native vegetation can regulate microclimate and reduce ecological degradation. Preserving cultural identity through green spaces that reflect historical themes (verandas, central open spaces) can foster a sense of place and memory within transforming urban landscape. The role of community participation cannot be neglected as developing green spaces can revitalize their life. This can bring communities together and help them feel more responsible. Economic

Table 11.1 Nurturing resilient cities through Nature-Based Solutions (NBS).

Resilience Features	Drivers	NBS Features	Implementation
Inclusivity	Loss of traditional gathering and displacement of local communities.	Creation of accessible parks, public green spaces for community.	Development of green public spaces connected to residential areas for all socioeconomic groups.
Equity	Displacement due to rising land value and unplanned urban development.	Equitable eco-friendly green spaces distribution.	Plan affordable and eco-friendly green zones (parks and community gardens).
Environmental Sustainability	Increase urban flooding, pollution, and heat stress.	Restoration of green corridors, urban forests, green roofs, and eco-friendly corridors along roadside.	Development of ecological corridors along canals, roadsides, and abundant lands with native plants and trees.
Cultural Identity	Loss of historic character and traditional social spaces.	Integrating cultural-themed green spaces.	Protection of historic structures, local heritage, and revival of cultural gathering spots.
Community participation	Weakening of social ties and lack of community involvement in development.	Involvement of community in development and maintenance.	Set-up local communities to manage green spaces, holding cultural and environmental events.
Economic stability	Neglected communities losing jobs and homes due to urban transformation.	Job creation, eco-tourism, local market spaces.	Create employment opportunities through urban farming, park management, and eco-based businesses.

stability is addressed through nature-based interventions that generate employment opportunities.

Overall, it shows that how NBS can help tackle environmental, social and economic problems together while strengthening cities for the future. By focusing on the needs of Thokar Niaz Baig, the model offers a sustainable way to improve the area.

Focusing on the broader Thokar Niaz Baig area reveals large-scale urban issues that require a lot of time. Eventually, shifting attention to the Thokar Niaz Baig flyover provides a more defined and manageable site for nature-based interventions. As this flyover is a major transit node with underutilized spaces beneath and around its structure, the flyover presents an opportunity to demonstrate how even highly infrastructure can be repurposed for social and environmental benefit. Drawing inspiration from the case studies discussed earlier the flyover can be transformed into green corridors that provide shade, biodiversity, and community gathering spaces. This approach makes it possible to practice affordable solutions that strengthen city resilience and improve quality of life.

11.4 Flyover Project Proposal

The study of Thokar Niaz Baig highlights an immediate need for a balanced approach between urban development and preservation of natural and cultural environments. Due to rapid and unplanned urbanization, historical urban landscapes are diminished and traditional communities are displaced. Examples across the world present that cities may grow without losing their identity by integrating NBS in their strategic planning. Features, such as ecological corridors, eco-friendly public spaces, protection of cultural landscape, and community involvement are introduced as a strategy and policy recommendation throughout the research.

The selected case studies discussed the amalgamation of urban green spaces within densely populated areas for environmental resilience in addition to the restoration of the cultural identities. The case of Rotterdam emphasizes how cities can effectively and efficiently combine urban development with environmental sustainability. Moreover, Rotterdam's approach highlights that even densely-built cities could also generate a balance between modern framework and natural environment by planning to keep nature as a focal point. The current study is highly relevant for areas, such as Thokar Niaz Baig, where rapid urbanization has left little spaces for environmental planning. Through small but strategic green interventions, dense and developed areas could be transformed. This study highlights the importance of recognizing both environmental and social roles of urban nature. The transformation of vacant spaces, such as roadsides, intersections, and areas under flyovers may promote urban cooling, ecological balance, and community's quality of life. On the other hand, the Seoullo 7017 Sky Garden is an example of the rehabilitation of the old highway overpass. This transformation of abandoned space into sky garden for public use with the implementation of NBS created a vibrant urban green space with economic and environmental sustainability. The community engagement demonstrated through this project is one of the key strategies that can be implemented in the selected case. Both case studies emphasized the preservation of the connection between nature and communities for resilient future.

In the context of Thokar Niaz Baig, these outstanding examples guide towards creating a greener, resilient, and an inclusive urban environment. The areas can be recovered from damages and chaotic growth by restoring natural features, community involvement, building eco-friendly corridors, and promoting cultural identity. The strategic vision for Thokar Niaz Baig is not to solve traffic congestion or metropolitan strains but to reconnect people with their traditions, environment, heritage, and build a foundation for a safer, vigorous, and sustainable city (Fig. 11.7).

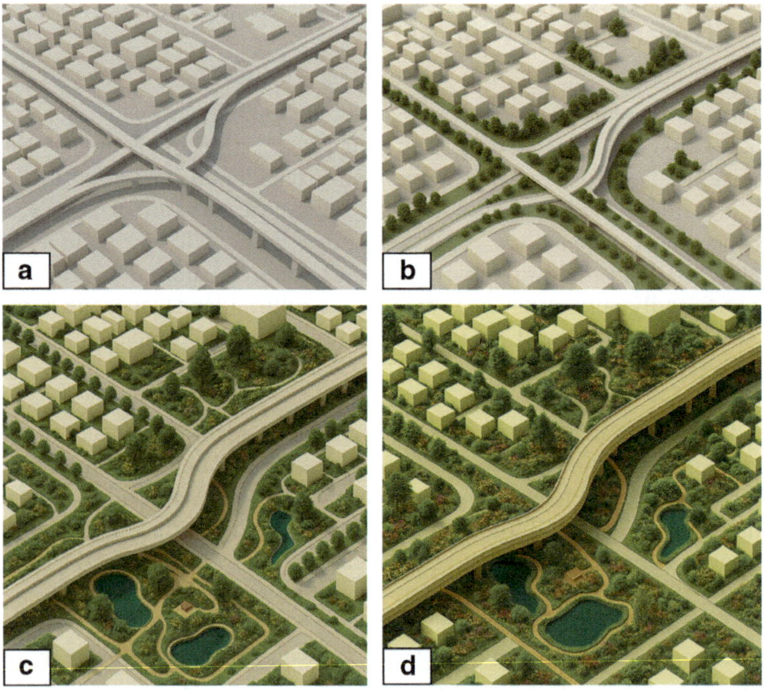

Fig. 11.7 AI images iterations for flyover: (a) Model of Thokar Niaz Baig flyover with its neighborhood context; (b) Developing green spaces beneath and around the flyover along with the green belts; (c) Introducing pedestrian pathways along with the spaces for community gatherings; (d) Building greener, resilient, and an inclusive urban environment (Source: Author).

11.5 Nature-Based Solutions Driven Resilience for Transformative Growth

The complex interplay of the rapid urbanization in the selected case study of Thokar Niaz Baig presented the usual scenario with reference to developing countries like Pakistan, where these unchecked/unplanned growths results into environmental degradation and social displacements (Alvi, 2024). One of the major contributing factors is the implementation based on the traditional urban development models focused on economic parameters (especially in the case of developing countries of South Asian region) while lacking the ecological and social dimensions necessary for the sustainable growth and developments. However, the emerging concept of Transformative Growth redefined the urban growth as a holistic approach with the integration of resilience, ecological innovation and communities' participation for the long-term well-being of the cities around the globe. The healthier cities growth is not only embedded in the economic prosperity but is deeply rooted in the ecological resilience of the natural systems, the social cohesion within the inclusive communities and the smart usage of digital design innovations. The implementation of this innovative approach generated the greener, resilient and an inclusive environment as in case of Thokar Niaz Biag flyover (Fig. 11.8). The selected global case studies of Rotterdam and Seoul also demonstrated the potential of transformative urban growth model for resilient urban futures. Furthermore, the utilization of the

Fig. 11.8 Final image of flyover (Source: Author).

Nature-Based Solutions (NBS) as an ecological intervention illustrates redefining the urban growth not only as expansion but as transformation with nature as core (Duffaut, Versini and Frascaria-Lacoste, 2022). The transformation of underutilized spaces beneath flyovers (Thokar Niaz Baig) into urban green spaces exemplifies the concept of transformative growth that leads to the enhancement of ecological and social productivity in addition to the economic vitality. The introduction of NBS in such densely populated areas creates the ecological corridors restoring the broken natural systems by improving the biodiversity with inclusive spaces that fosters the community engagement and well-being ensuring the sustainable urban futures (Edeigba et al., 2024). These ecological solutions not only mitigate the environmental challenges but also empowers the local communities, displaying the interconnectedness of ecological health and social equity (Rajput, 2024). The flyover project (creation of resilient and vibrant urban green spaces) reflects the transformative pathway for the cities to navigate through the urban transformation complexities within the planetary boundaries for socially grounded and ecologically reimagined urban futures. This paradigm shift in urban developments with the prioritization of ecological integrity, social inclusion and community resilience can foster environments that not only accommodates growth but also nurture the well-being of their inhabitants and the planet.

11.6 Conclusion

The study examined the impacts of urbanization in the developing countries, focusing on the case of Thokar Niaz Baig in Lahore, Pakistan. The transformation of the area has been significantly shaped by weak governance, which contributed to the emergence of informal settlements and unplanned growth in the early stages. The conversion into central transport hub in the later phase aimed for the upgradation, further disrupted the built environment, leading to the irregularities in land use transitions, encroachment of open areas and forced displacement of the local communities. The informal urban spaces before the upgradation and planned ones in the later phases failed to support sustainable urban living. This study explored the potential of underutilized areas under Thokar Niaz Baig flyover that can be transformed into urban green spaces with the implementation of NBS. The study concluded the implementation of NBS as a means to foster resilient urban environments. These solutions are not only essential to enhance human well-being but also to restore ecological balance and enriching biodiversity. By highlighting the importance of rehabilitating abandoned and overlooked urban spaces, particularly in developing regions, such as Pakistan, the study illustrated the potential to integrate green infrastructure into urban design/planning. Furthermore, this study called for a concerted effort to create urban spaces that not only accommodate growth but also nurture the well-being of communities and the environment, paving the way for a sustainable urban future.

11.7 Recommendations

The current study highlighted the challenges associated with unplanned urbanization and its detrimental impacts on community well-being and ecological systems. The study suggested that it is necessary to adopt proactive and holistic approach to urban design/planning. The following set of recommendations was formulated to address the urban sprawl, environmental degradation, and loss of biodiversity, particularly the Thokar Niaz Baig junction area. By integrating NBS and fostering community engagement, these recommendations seek to create resilient urban spaces that enhance the quality of life for all residents while preserving the cultural and ecological integrity of the region. These recommendations are as follows:

- Implementation of land use plans in Thokar Niaz Baig and elimination of encroachments for the preservation of urban open spaces.
- Identification and rehabilitation of underutilized spaces to create green areas that serve multiple purposes (e.g., parks, community gardens, and recreational spaces). This may enhance biodiversity and improve the quality of life for residents.
- Educating the diversified stakeholders for formulation of framework for enhancement of the quality of life in Thokar Niaz Baig.
- Formulation of stakeholders group in Thokar Niaz Baig by including all sectors that can contribute in nurturing resilient cities through active participation ensuring urban sustainability.
- The new developments, such as road extensions and flyovers in case of Thokar Niaz Baig should be built incorporating the green roofs, permeable pavements, and urban forests to mitigate the environmental impacts and enhance urban resilience.
- The restoration and enhancement of biodiversity is to be achieved with native species, creating ecological corridors for wildlife and promoting urban greening to support the local ecosystems.
- Establish a system to monitor the impacts of urbanization and the effectiveness of NBS. Regular assessments may help adapt policies and practices to better meet the evolving needs of the city.

By implementing these recommendations, Lahore may work towards creating a more resilient urban environment that balances development with the preservation of community well-being and ecological integrity, starting from the Thokar Niaz Baig area.

References

Acharya J (2018) Impacts of urbanization and development activities on sustainable development and resource management of the Majhi people of Nepal. Master's thesis, UiT Norges arktiske universitet

Adams C, Moglia M, Frantzeskaki N (2024) Design principles for mainstreaming of nature-based solutions in cities: A proposal for future pathways. Nature-Based Solutions 6, 100155

Alvi A (2024) Effects of urbanization on community cohesion in major cities of Pakistan. Eur J of Sociol 7(1):22–31

Balık G, Balık Lökçe D (2021) Representation, narrativity, and banality: Seoullo 7017 Skygarden and Superkilen Urban Park. In: Bianconi F, Filippucci M (eds) Digital draw connections: Representing complexity and contradiction in landscape. Springer International Publishing, Cham, pp 563–584

Chen HC, de Vries B, Han Q (2015) The new modelling method in urban development: Case study in Rotterdam. 5th Annual International Conference on Urban Studies & Planning (June 8–11, 2015, Athens, Greece). The Athens Institute for Education and Research

Desouza KC, Flanery TH (2013) Designing, planning, and managing resilient cities: A conceptual framework. Cities 35:89–99

Duffaut C, Versini PA, Frascaria-Lacoste N (2022) Are really nature-based solutions sustainable solutions to design future cities in a context of global change? Discussion about the vulnerability of these new solutions and their probable unsustainable implementation. Sci of the Total Environ 853, 158535

Edeigba BA., Ashinze UK, Umoh AA, Biu PW, Daraojimba AI, Edeigba BA, … Daraojimba AI (2024) Urban green spaces and their impact on environmental health: A global review. World J of Adv Res and Rev 21:917–927

Hussain S, Hussain E, Saxena P, Sharma A, Thathola P, Sonwani S (2024) Navigating the impact of climate change in India: A perspective on climate action (SDG 13) and sustainable cities and communities (SDG 11). Front in Sustain Cities 5, 1308684

Iqbal N, Ravan M, Jamshed A, Birkmann J, Somarakis G, Mitraka Z, Chrysoulakis N (2022) Linkages between typologies of existing urban development patterns and human vulnerability to heat stress in Lahore. Sustain 14(17), 10561

Kabisch N, Frantzeskaki N, Pauleit S, Naumann S, Davis M, Artmann M, … Bonn A (2016) Nature-based solutions to climate change mitigation and adaptation in urban areas: Perspectives on indicators, knowledge gaps, barriers, and opportunities for action. Ecol and Soc 21(2):1–11

Lafortezza R, Sanesi G (2019) Nature-based solutions: Settling the issue of sustainable urbanization. Environ Res 172:394–398

Margono RB, Zuraida S (2019) Landscape as urban regeneration: Learning from the High Line and Seoullo 7017 Skypark. J of Appl Sci (JAPPS) 1(2):50–60

Nasar-u-Minallah M, Zia S, Rahman AU, Riaz O (2021) Spatio-temporal analysis of urban expansion and future growth patterns of Lahore, Pakistan. Geogr, Environ, Sustain 14(3):41–53

Orsetti E, Tollin N, Lehmann M, Valderrama VA., Morató J (2022) Building resilient cities: Climate change and health interlinkages in the planning of public spaces. Int J of Environ Res and Public Health, 19(3), 1355

Rahaman MA, Kalam A, Al-Mamun M (2023) Unplanned urbanization and health risks of Dhaka City in Bangladesh: Uncovering the associations between urban environment and public health. Front in Public Health, 11, 1269362

Rajput CP (2024) Resilient infrastructure: Strategies for climate change adaptation in coastal cities. Int J of Archit and Infrast Plan 10(2):8–12

Rana IA, Bhatti SS (2018) Lahore, Pakistan—Urbanization challenges and opportunities. Cities 72:348–355

Rashid A, Rasheed R, Ngah AH (2024) Achieving sustainability through multifaceted green functions in manufacturing. J of Glob Operations and Strateg Sourc 17(2):402–428

Ribeiro FL (2021) Unplanned urban development: A neglected global threat. Curr Urban Stud 9(3):434–444

Sakketa TG (2023) Urbanisation and social cohesion: Theory and empirical evidence from Africa (IDOS Discussion Paper No. 16/2023)

Seddon N, Chausson A, Berry P, Girardin CA, Smith A, Turner B (2020) Understanding the value and limits of nature-based solutions to climate change and other global challenges. Philos Trans of the R Soc B 375(1794), 20190120

Seddon N, Smith A, Smith P, Key I, Chausson A, Girardin C, … Turner B (2021) Getting the message right on nature-based solutions to climate change. Glob Chang Biol 27(8):1518–1546

Suarez M, Rieiro-Díaz AM, Alba D, Langemeyer J, Gómez-Baggethun E, Ametzaga-Arregi I (2024) Urban resilience through green infrastructure: A framework for policy analysis applied to Madrid, Spain. Landsc and Urban Plan 241, 104923

Urban MC, Alberti M, De Meester L, Zhou Y, Verrelli BC, Szulkin M, … Brans KI (2024) Interactions between climate change and urbanization will shape the future of biodiversity. Nat Clim Chang 14(5):436–447

Yoshizumi M, Nguyen TN, Kobayashi H (2013) Impacts of rapid urbanization to traditional living environment and community linkage in historic old quarters of Vietnam: A case study on Gia Hoi area in Hue City, Central Vietnam. J of Civi Eng and Archit 7(5), 555

Chapter 12
Assessment of Floating Parks Based on Coastal Park Design Criteria

Yücel Caymaz Gökçen Firdevs and Baskın Elif

Abstract The aim of this study is to examine the spatial and design dimensions of floating parks and to comparatively evaluate them with the design criteria of traditional coastal parks. Floating parks represent an innovative typology of urban open spaces that transfer public, ecological, and recreational functions onto the water surface, offering an alternative to the land-based understanding of open space. These hybrid structures establish a dynamic interface between land and water ecosystems, supporting ecological diversity, improving the urban microclimate, and enhancing adaptive capacity against rising water levels caused by climate change.

The study analyzes eight case studies: Little Island Park (New York, USA), Elizabeth Quay (Perth, Australia), Race Street Pier (Philadelphia, USA), Hudson River Park Pier 25 (New York, USA), Merchant Square Floating Pocket Park (London, UK), Wild Mile Eco Park (Chicago, USA) and Pier C Park (New Jersey, USA). These examples are evaluated based on internationally recognized frameworks and institutions, including the Waterfront Edge Design Guidelines (WEDG), Project for Public Spaces (PPS), American Planning Association (APA), American Society of Landscape Architects (ASLA), Gehl Architects, United Nations Environment Programme (UNEP), Waterfront Toronto, PK3 Landskabsarkitekter & Tredje Natur, and Bruttomesso.

The findings reveal that floating parks not only provide innovative recreational opportunities along urban waterfronts but also constitute a strategic design model for the development of sustainable coastal spaces. They integrate ecological resilience,

Y. Caymaz Gökçen Firdevs (✉)
Department of Architecture, Istanbul Aydin University, Beşyol, Küçükçekmece, İstanbul, Türkiye
e-mail: gokcenfyucel@aydin.edu.tr

B. Elif
Architecture Master Program, Istanbul Aydin University, Beşyol, Küçükçekmece, İstanbul, Türkiye

A. Cheshmehzangi et al. (eds.), *Transforming Cities*, Urban Sustainability,
https://doi.org/10.1007/978-981-95-7378-3_12

adaptive infrastructure systems, and public accessibility into a holistic urban landscape framework.

Keywords Floating park • Beach park • Design criteria • Green area • Urban Landscape

12.1 Introduction

Coastal parks play a central role in restructuring the relationship between humans and nature in urban environments. These areas, which contribute to urban green space systems due to their aesthetic and functional qualities, offer recreational solutions against increasing population density and limited land availability. In addition to addressing the water- and greenery-related recreational needs of urban residents, green spaces integrated into waterfronts possess significant potential for improving the urban microclimate and supporting ecological diversity. Therefore, coastal areas are considered critical urban inventories reflecting evolving urban visions in the context of contemporary space production (Field and Barros, 2014; Selena, 2020).

In this context, floating parks, developed as a response to limited land availability, emerge as innovative solutions that bring the land-based open space concept onto the water surface. These structures, designed using platform and scaffolding systems, provide flexible infrastructural solutions to rising water levels due to climate change, while minimizing ecological impacts. The urban recreational and ecological conservation requirements mentioned earlier are directly related to these flexible and sustainable solutions offered by floating parks (Sungsine, 2011; Field and Barros, 2014; IPCC, 2021).

The Floating Pavilion projects in Rotterdam, the Maldives Floating City in the Maldives, and the Shidhulai Swanirvar Sangsha in Bangladesh demonstrate how floating structures can be operationalized within the context of urban metabolism and amphibious urbanism. These projects establish a dynamic interface between terrestrial and aquatic ecosystems, maintain ecological balance, and provide sustainable infrastructural models (Perkins and Leal, 2010; Field and Barros, 2014). The flexibility and sustainability criteria highlighted above find concrete expression in these examples.

Similarly, metropolises such as Hong Kong, Singapore, and Amsterdam, facing overpopulation and limited land availability, have undertaken land reclamation and coastal embankment projects; however, these initiatives result in ecological degradation, high economic costs, and social tensions (Bolonkin, 2010). Floating parks, as discussed in the previous examples, provide flexible and environmentally friendly solutions to these challenges and contribute to the sustainable use of coastal areas by reducing dependence on land.

The main objective of this study is to evaluate the spatial, ecological, and infrastructural characteristics of floating parks and to compare them with traditional coastal park designs, in order to provide a comprehensive framework for sustainable and multifunctional coastal space planning. In this context, determining the ecological, social, and spatial criteria to be considered in coastal park planning becomes a priority. Studies in the international literature aimed at defining coastal park design criteria are directly related to these ecological and social requirements. McHarg (1969) emphasizes that landscape architecture design in harmony with nature should take ecological principles as a foundation for coastal parks. This approach sets out strategies to minimize the environmental risks mentioned earlier. The main criteria of this approach include the protection of natural habitats, integration of ecological cycles into design, prioritization of natural landscape elements, prevention of coastal erosion with natural barriers, and balancing aesthetics with functionality (Beatley, 2011; Loukaitou-Sideris, 2013).

Gehl (2010) recommends the creation of human-oriented spaces in urban planning and asserts that coastal parks should provide multifunctional spaces that enhance social interaction. Accordingly, accessibility, safety, spatial continuity, and sensory quality are essential elements that strengthen the user experience (Gehl Architects, 2010; Loukaitou-Sideris, 2013). PPS (2012, 2016) elaborates on coastal park qualities through user-centered strategies and participatory planning, highlighting the importance of spaces that encourage social interaction, multi-purpose usage, and contextually compatible design.

WEDG (2020) adopts a context-based and sustainable design approach balancing coastal ecosystem protection with public access. It proposes criteria such as ecological protection, public access, aesthetic and spatial quality, material and construction techniques, and time-space dynamics (WEDG, 2020; Bruttomesso, 2011). Similarly, Waterfront Toronto (2021) identifies criteria including accessibility and inclusivity, primacy of public space, sustainable and eco-friendly design, visual and spatial connectivity, climate change adaptation, and community engagement (Waterfront Toronto, 2021; Field and Barros, 2014). PK3 Landskabsarkitekter and Tredje Natur (2011), suggest that coastal parks should be planned in terms of water access, temporary uses, and multidimensional functions.

APA (2006) and ASLA (2013) comprehensively define planning standards, ecological sustainability, and user-oriented spatial strategies in coastal park design. UNEP (2008) recommends integrating coastal areas with hydrological, climatic, economic, social, and cultural systems, and promoting ecosystem-based sustainability. Accordingly, multifunctional uses in coastal areas should be distributed in a balanced manner, unnecessary urban sprawl should be avoided, and measures should be taken to minimize the negative effects of anthropogenic activities. Ecological restoration practices should be planned for previously damaged areas (UNEP, 2008; Bruttomesso, 2011).

International coastal park projects exemplify design approaches that holistically integrate social, ecological, and recreational functions. New York City's Little Island Park serves as a benchmark for landscape design and public access, emphasizing user experience and ecological diversity (Selena, 2020; Perkins and Leal, 2010). Elizabeth Quay (Perth, Australia) demonstrates the integration of diverse functions, prioritizing social interaction and accessibility (Field and Barros, 2014; Sha, 2010). Similarly, the Race Street Pier (Philadelphia, USA) and Hudson River Park Pier 25

(NYC) illustrate the combination of floating and land-based infrastructure, supporting multifunctional public spaces that are seamlessly integrated into urban contexts (Selena, 2020; Sungsine, 2011). Merchant Square Floating Pocket Park (London, UK) and Wild Mile Eco Park (Philadelphia, USA) provide examples where ecological sustainability is carefully aligned with public accessibility (PPS, 2016; Field and Barros, 2014). Pier C Park (New Jersey, USA) specifically exemplifies the application of ecological design strategies in waterfront parks, including native vegetation plantings, creation of habitats for aquatic species, and water-sensitive landscaping aimed at improving stormwater management, while preserving recreational and social functions for diverse user groups (Selena, 2020; MVVA, 2010).

Building upon these international precedents, the planning criteria for coastal parks were established to optimize user experience while ensuring ecological and spatial sustainability. Based on a thorough review of the literature and case studies, key criteria were identified: accessibility, comfort and facilities, landscaping and planting, and social interaction.

The table analysis, taking into account international sources, identified fundamental criteria for coastal park design that directly address the spatial and social needs of urban life (American Planning Association, 2006; American Society of Landscape Architects, 2013; Bruttomesso, 2001; Gehl Architects, 2010; Project for Public Spaces, 2015; Tredje Natur, PK3, 2015; United Nations Environment Programme, 2009; Waterfront Alliance (WEDG), 2018; Waterfront Toronto, 2021). These criteria aim to promote safe and inclusive park usage, enhance recreational and social engagement, maintain ecological balance and aesthetic quality, and ensure the well-distributed allocation of multifunctional spaces (Gehl Architects, 2010; PPS, 2012; WEDG, 2020; Bruttomesso, 2011). Integrating these design principles with the insights derived from exemplary coastal parks provides a framework for developing waterfront environments that are both socially vibrant and ecologically resilient, highlighting the potential for future adaptive design strategies in floating and coastal urban landscapes.

Accessibility is a critical criterion directly affecting the recreational potential of coastal parks. Users' access to the park, orientation within the park, and legibility of circulation routes are fundamental elements of accessibility. Pedestrian paths should comply with TS 12576 and ISO 21542:2011 standards; the minimum width should be 1.20 m, and in high-density areas, 1.80–2.00 m. Materials must be natural, durable, and non-slip; surfaces that cause glare or reflection should be avoided (WEDG, 2018). Bicycle lanes should be at least 1.50 m for one-way and 2.50 m for two-way; ramps should have a maximum slope of 1:12, and ramps exceeding 9 m should have resting platforms (ADA, 2010; CROW, 2016; PPS, 2015). For stairs, step height should be 15–17 cm, depth at least 28 cm; handrails should be positioned on both sides and, if necessary, in the middle (UNEP, 2021).

Urban amenities functionally and aesthetically support the user experience. Seating elements should be ergonomically designed, with a bench height of 40–45 cm and seat depth of 40–50 cm (ADA, 2010; ISO 9241-5, 1998). Shades and awnings should be fixed or retractable shade and awning systems should have a clearance height of between 250cm and 300 cm, and should be highly resistant to wind (PPS, 2015). Lighting should not obstruct users' views and should be constructed with vandal-resistant materials (International Dark-Sky Association, 2019). Trash and recycling bins should be placed at appropriate intervals throughout the park to support environmental sustainability. WC and hygiene areas should be safe and accessible, and directional signage should be clear and user-oriented (UNEP, 2021; PPS, 2012).

Water elements enhance the microclimate while contributing to aesthetics and recreation. Sprinklers, ponds, artificial streams, and interactive water playgrounds should be planned for safety and ease of maintenance (Field and Barros, 2014; Beatley, 2011). Sculptures and art installations strengthen the park's aesthetic and cultural identity and enhance emotional and visual interaction with users. Their size, material, and positioning should consider user safety and spatial harmony (Selman, 2009; Loukaitou-Sideris, 2013; PPS, 2016; WEDG, 2020).

Landscaping and planting are crucial for ecological sustainability and aesthetic identity. Tree species should be salt-tolerant and wind-breaking, with planting spacing of 3–5 m (UNEP, 2021; NEVER, 2017). Shrubs and flower beds should be local, hardy species, with a minimum spacing of 120 cm. Landscape design should prioritize natural habitat protection and integration of ecological cycles (Beatley, 2011).

Social interaction is another key criterion that increases the multifunctional potential of coastal parks. Playgrounds, sports fields, squares, observation decks, and water access points should accommodate diverse users and promote interaction. Playgrounds must comply with ASTM F1487-21 and ADA standards; safety buffer minimum 1.8 m. Running and sports tracks should have a minimum width of 1.5 m and a 3 m safety buffer (FIBA, FIFA). Observation decks and lecture areas should have a seat height of 30–35 cm and depth of 45–50 cm (DIN 18040, 2010). Water access ramps should not exceed 8%, and grab bars should be at 90–100 cm (PK 3 Landskabsarkitekter & Tredje Natur, 2011).

This comprehensive set of criteria enables users to utilize parks safely and accessibly, while enhancing social interaction and recreation, maintaining ecological

Table 12.1 Clearly presents the criteria determined by these institutions and organizations and the commonalities between these criteria (✓).

Design Criteria	WEDG	PPS	APA	ASLA	Gehl Architects	UNEP	Waterfront Toronto	PK3 Landskabsarkitekter & Tredje Natur	Bruttomesso
Accessibility	✓	✓	✓	✓	✓	✓	✓	✓	✓
Comfort and Accessories	✓	✓	✓	✓	✓		✓	✓	✓
Landscaping and Planting	✓	✓	✓	✓		✓	✓	✓	✓
Social Interaction		✓	✓	✓	✓		✓	✓	✓

balance and aesthetic quality, and supporting balanced distribution of multifunctional spaces (Gehl Architects, 2010; PPS, 2012; WEDG, 2020; Bruttomesso, 2011).

Table 12.1 within the literature review illustrates how these design criteria are supported by international institutions and authors. Sources such as WEDG (Waterfront Edge Design Guidelines), PPS (Project for Public Spaces), APA (American Planning Association), ASLA (American Society of Landscape Architects), Gehl Architects, UNEP (United Nations Environment Programme), Waterfront Toronto, PK3 Landskabsarkitekter & Tredje Natur, and Bruttomesso provide detailed criteria specific to coastal and waterfront designs, developing comprehensive principles and standards regarding sustainability, user orientation, and social interaction. Accordingly, coastal park design criteria are grouped under four main headings: accessibility, comfort and facilities, landscaping and planting, and social interaction.

The table analysis, the basic criteria that directly respond to the spatial and social needs of urban life, which are adopted as a priority in coastal park design by internationally respected institutions and authors, and also take into account environmental and cultural sensitivities, have been determined. Among these criteria, the most common and widely emphasized are: accessibility, comfort and variety of equipment, landscaping and planting applications, and designing areas that support social interaction. These criteria will be analyzed through the selected case studies to assess the success of floating parks in implementing the design principles and to evaluate their ecological, social, and spatial contributions within urban coastal environments.

12.2 Method

12.2.1 Site Characteristics

In this study, seven case study parks were selected to analyze the integration of ecological, structural, and social design strategies in contemporary urban waterfronts. This selection was made to ensure a representative coverage of diverse typologies, structural systems, and ecological contexts, while maintaining comparability in terms of public accessibility, multifunctionality, and urban integration—thereby addressing potential concerns regarding case selection and comparability across sites. Each site provides unique insights into floating, semi-floating, or adaptive reuse waterfront parks that combine ecological restoration with social and recreational functions. The inclusion of seven cases allows for a comprehensive comparative analysis while maintaining methodological rigor, balancing both breadth and depth of investigation.

Little Island, New York, USA, designed by Thomas Heatherwick and Heatherwick Studio, is an iconic waterfront park opened in 2021 along the Hudson River. Its primary structural system consists of precast concrete elements of varying shapes, transported by barges and positioned on 132 concrete piles installed in the river (Heatherwick Studio, 2021). The pier's elevation varies to allow sunlight to reach the water surface and aquatic life, while high-sloped forms on other edges act as windbreaks to enhance urban comfort (Davidson, 2021). The park functions not only as a recreational green space but also as a cultural landmark integrating art and nature. The undulating terrain incorporates three performance spaces, the largest being a 700-seat amphitheater with natural stone seating oriented toward the Hudson River sunset. The central flexible space accommodates meetings and events of varying scales (Heatherwick Studio, 2021; Ouroussoff, 2021). Landscape design was executed by MNLA (Mathews Nielsen Landscape Architects). A total of 400 plant species and 100 tree species were selected, emphasizing evergreen varieties to provide year-round wind control, shade, and seasonal diversity (MNLA, 2021). Plantings and terrain access to water create habitats for birds, insects, and other small aquatic species, enhancing the park's ecological function. The park is connected to the mainland via two pedestrian bridges. Within the site, paths, wooden seating, playgrounds, and dining areas are integrated. Back-of-house functions are placed beneath the pier, preserving spatial cohesion, avoiding additional structures at upper levels, and maintaining unobstructed views (Davidson, 2021). The park was selected as a model of innovative waterfront design, integrating art, nature, and ecological landscape strategies, exemplifying adaptive reuse of historic industrial sites into contemporary cultural public spaces.

Elizabeth Quay, Perth, Australia, was designed by ARM Architecture and opened as part of a major waterfront redevelopment integrating the city with the Swan River (Government of Western Australia, 2016; Arm Architecture, 2025). The project's central feature is an artificial island connected to the shore, surrounded by a 730-meter terraced promenade. The island's sloped and undulating walls enhance its dynamic character and allow direct interaction with water. Landscape elements and urban furniture reinforce spatial organization and visual continuity. Play areas, seating, water features, and lighting create a multifunctional space for diverse users. Public art installations, including Stuart Green's light work, contribute to the cultural identity of the site. Native plant species support biodiversity and wildlife, with stone and sand ramps accommodating swans. Soft landscaping provides a natural oasis within the urban fabric, and an integrated water treatment system maintains surface water quality. The site hosts festivals, outdoor exhibitions, and seasonal events. Elizabeth Quay was selected to demonstrate how artificial islands and waterfront redevelopment enhance urban connectivity, support biodiversity, and provide multifunctional public spaces while integrating ecological and social sustainability.

Race Street Pier, Philadelphia, USA, was originally constructed in 1896 as "Municipal Pier 11," serving both freight and passenger transportation as a two-level pier and representing a key component of the industrial-era logistical infrastructure. The pier was redeveloped between 2009 and 2011 by the Delaware River Waterfront Corporation (DRWC) and James Corner Field Operations, transforming it into a publicly accessible waterfront park that strengthened Philadelphia's connection to the riverfront (DRWC, 2011; Field Operations, 2012). The pier preserves its historical two-level configuration while being adapted to contemporary public space standards. The lower level includes green areas, recreational zones, and social gathering spaces, enhanced with shaded seating and landscaping to improve user comfort. The upper level functions as a panoramic viewing platform, with accessible ramps ensuring continuous circulation in compliance with universal design principles (Gehl, 2010). Recycled synthetic wood seating on the north side rises 3.5 m, linking levels and providing amphitheater-style seating and observation areas. Along the promenade, 200 solar-powered LED light blocks extend park usability into night-time hours while promoting sustainable energy use. Landscape design incorporates native plant species to support the waterfront ecosystem while providing shade, wind protection, and spatial organization. This comprehensive approach also allows the park to host a variety of events, including concerts, exhibitions, and festivals, throughout the year. Race Street Pier was selected for analysis as an example of adaptive reuse of historic industrial infrastructure, demonstrating how multifunctional, accessible, and ecologically sustainable public waterfront spaces can be successfully created in an urban context.

Hudson River Park Pier 25, located in the Tribeca neighborhood of Manhattan, New York, USA, is an example of transforming a former industrial and commercial waterfront into a public recreational area. Designed by the landscape architecture firm Michael Van Valkenburgh Associates, renowned for innovative and environmentally sensitive designs, the project aims to reintegrate a long-industrial waterfront into the city's public life. The park includes children's play areas, mini-golf, sand volleyball courts, skate zones, viewing terraces, and a pier for sailing, providing diverse activities for all age groups (Hudson River Park Trust, 2021). Walkways and viewing terraces offer panoramic views of the Hudson River (Michael Van Valkenburgh Associates, 2010). Sustainability principles guided the pier's design. The adaptive reuse of existing infrastructure reduced new construction and carbon footprint, while the park operates with a low-energy, low-maintenance model. Planting schemes utilize native species with consideration for wind direction and sunlight, enhancing both user comfort and ecological resilience (Hudson River Park Trust, 2021). Revenue-generating uses, such as mini-golf and food and beverage areas, provide financial support for park maintenance. Ecologically, the landscape and waterfront design at Pier 25 provide habitat for birds, insects, and small mammals. The transition zones (ecotones) between aquatic and terrestrial areas support biodiversity and food web formation, reinforcing the integration of both terrestrial and aquatic ecosystems (Hudson River Park Trust, 2021). Hudson River Park Pier 25 was selected as a study site for its demonstration of how a post-industrial waterfront can be reintegrated into public life while combining social, economic, and ecological sustainability principles, achieving a balanced relationship between user experience and ecosystem restoration.

Merchant Square Floating Pocket Park, London, UK, located within the Paddington Basin of Merchant Square, represents the city's first floating public park and serves as an innovative example of integrating blue-green infrastructure into dense urban environments. Designed by Tony Woods, Royal Horticultural Society Young Garden Designer of the Year and Gold Medalist, the park aims to reconnect urban communities with nature, supporting mental and physical well-being in a highly urbanized context (Woods, 2020; Landezine, 2017). It offers diverse experiences for visitors through seating areas, activity plazas, and water access points for canal tours, promoting social interaction and recreational opportunities. The floating platform covers 730 m^2 and comprises 21 custom steel sections and 62 custom wooden modules, with a total weight of approximately 110 tons. It is anchored to the canal bed with blocks to prevent lateral or longitudinal movement (Woods, 2020). The modular design ensures structural stability, flexibility, and user safety. Ecologically, the park contributes to urban biodiversity. Native and aquatic plantings integrated into the floating modules provide food, shelter, and breeding sites for birds, insects, and small aquatic organisms. Transition zones (ecotones) between the floating structures and the water support local food webs and enhance connectivity between terrestrial and aquatic habitats. Vegetation and water surfaces also regulate microclimate conditions, including temperature moderation and wind reduction, promoting ecosystem resilience in a dense urban environment. The park's design prioritizes both user experience and aesthetic continuity. Wooden seating and open areas accommodate various age groups for social and recreational activities, while the platform's integration with the canal allows visitors to engage closely with nature. Merchant Square Floating Pocket Park was selected as a study site for its innovative floating design, urban ecological integration, and multifunctional programming that collectively support social, recreational, and ecological objectives.

Wild Mile Eco Park, Chicago, USA, located along the North Branch of the Chicago River, is a 17-acre floating ecological park designed to transform a previously neglected urban waterway into a public recreational and ecological corridor. Developed as part of the Chicago Riverwalk Revitalization Initiative, the project aims to enhance urban biodiversity, restore aquatic habitats, and provide public access to nature within a dense urban context (Urban Rivers, 2025). The park supports over 10,000 native plant species and a variety of fish, birds, and aquatic invertebrates, reinforcing habitat connectivity along the river (City of Chicago, 2024). The modular design consists of rectangular and triangular floating platforms constructed with eco-friendly geotextile fabrics and planting substrates. These modules are anchored to the riverbed and seawalls, allowing them to rise and fall with changing water levels while maintaining structural and functional stability (Urban Rivers, 2025). The ecological design integrates emergent vegetation, floating wetlands, and riparian buffers that enhance water filtration, increase dissolved oxygen, and provide refuge for aquatic species (Chicago Department of Planning and Development, 2024). Public accessibility and inclusivity are central to the design. The park includes pedestrian walkways, viewing decks, and educational nodes designed for universal access, accommodating people of all ages and physical abilities. Seasonal programming such as community gardening, kayaking, and environmental education promotes continuous public engagement and awareness of urban ecology (City of Chicago, 2024). From an ecological urbanism perspective, Wild Mile Eco Park exemplifies how adaptive floating systems can merge ecological restoration with public space design. Its integration of habitat creation, educational value, and sustainable infrastructure demonstrates a replicable model for resilient waterfront redevelopment. The site was selected for this research because it represents a pioneering example of floating ecological infrastructure that successfully reconciles urban recreation with habitat regeneration and climate adaptability. It provides valuable insight into how floating landscapes can function as both ecological and social infrastructures within contemporary urban riverfronts.

Pier C Park, located on the Hudson River in Hoboken, New Jersey, USA, was designed by Michael Van Valkenburgh Associates (MVVA) (MVVA, 2010). The park opened in 2010 following the adaptive reuse of a nineteenth-century industrial pier, transforming an underutilized waterfront structure into a publicly accessible urban space (New Jersey Department of Environmental Protection [NJDEP], 2011). Distinct from typical waterfront parks, Pier C features a floating platform anchored on steel piles, designed to accommodate tidal fluctuations and river currents. The reinforced concrete platform provides a safe public space while integrating seamlessly with the existing shoreline (MVVA, 2010). Covering approximately 0.9 hectares, the park offers diverse functions for different age groups, including children's playgrounds, picnic areas, seating terraces, open lawns, and walking and jogging paths (Hoboken Parks Department, 2012). At the southern end, a small beach and viewing terraces provide both visual and physical engagement with the water. Pier C Park is accessible from the land and via pedestrian bridges. Its floating character enables the creation of additional public open space in dense urban fabrics, such as Hoboken, where land is limited (Carr, Francis, Rivlin and Stone, 1992). MVVA's landscape design integrates recreational and ecological strategies: habitats for seabirds were established along the shoreline, and native planting enhances both microclimate and visual diversity (MVVA, 2010). Low-lying areas allow visitors to experience tidal effects directly, creating a dynamic waterfront experience. Pier C Park exemplifies a contemporary waterfront park that responds to urban public space needs, strengthens the city's cultural memory, and brings together diverse social groups. Its integration with water and floating design make it a strong representative of the floating park concept in an industrially defined urban context (Lynch, 1960; Carr et al., 1992).

12.2.2 Research Design

This study was conducted as a qualitative research aiming to identify the design principles of floating parks within the urban green space system and to relate them to practical examples. The research was planned and implemented in three stages:

12.2.2.1 First Stage: Literature Review and Sample Area Selection

At the beginning of the study, the current literature on coastal and floating parks was comprehensively reviewed. National and international sources on urban design, landscape architecture, coastal management, sustainable infrastructure, and recreational uses were evaluated (McHarg, 1969; Gehl, 2010; Field and Barros, 2014; WEDG, 2020).

The sample areas examined at this stage are internationally recognized floating parks representing different scales, uses, and design approaches (Table 12.2):

- NYC Little Island Park, New York, USA (Fig. 12.1).
- Elizabeth Quay, Perth, Australia (Fig. 12.2).
- Race Street Pier, Philadelphia, USA (Fig. 12.3).
- Hudson River Park Pier 25, New York, USA (Fig. 12.4).
- Merchant Square Floating Pocket Park, London, UK (Fig. 12.5).
- Wild Mile Eco Park, Chicago, USA (Fig. 12.6).
- Pier C Park, Hoboken, New Jersey, USA (Fig. 12.7).

Fig. 12.1 Aerial view of Little Island Park, New York City, USA, designed by Heatherwick Studio, showcasing the floating "tulip-shaped" piers and landscaped public spaces (Source: Little Island NYC (2025), https://www.azuremagazine.com/article/manhattans-little-island-opens-to-controversy-and-awe/ Accessed 10 October 2025).

Fig. 12.2 Aerial view of Elizabeth Quay Park, Perth, Australia, illustrating waterfront promenades, landscaped public spaces, and pedestrian connections (Source: Elizabeth Quay (2025), https://architectureau.com/directory/tcl/elizabeth-quay/, Accessed 12 October 2025).

Fig. 12.3 View of Race Street Pier, Philadelphia, USA, designed by James Corner Field Operations, integrating tiered seating and waterfront access along the Delaware River (Source: Race Street Pier (2025), https://www.fieldoperations.net/project-details/project/race-street-pier.html/, Accessed 10 October 2025).

Fig. 12.4 Aerial view of Hudson River Park Piers 25 and 26, New York, USA, designed by Michael Van Valkenburgh Associates (MVVA), illustrating adjacent recreational and ecological waterfront structures (Source: Hudson River Park Trust (2025), https://www.cityclubny.org/post/completing-hudson-river-park-1 Accessed 10 October 2025).

Fig. 12.5 View of the Merchant Square Floating Pocket Park, Paddington Basin, London, UK, designed by Floating Pocket Park Ltd. and Atelier One (Source: Merchant Square London (2025), https://merchantsquare.co.uk/see-and-do/floating-pocket-park, Accessed 10 October 2025).

Fig. 12.6 Aerial view of the Wild Mile Park, Chicago, USA, illustrating floating ecological corridors and public access platforms (Source: Urban Rivers (2025), https://www.archdaily.com/911890/worlds-first-floating-eco-park-planned-for-chicago-river, Accessed 10 October 2025).

The selected case studies were chosen to enable a systematic comparison of floating parks with different scales, functions, and design approaches within the framework of waterfront park design criteria. The current conditions of each park were evaluated through field observations, photographs, and online resources. This approach allows for meaningful comparisons between the cases and facilitates an analytical understanding of the design principles of floating parks.

Fig. 12.7 Aerial view of Pier C Park, New Jersey, USA, designed by Michael Van Valkenburgh Associates (MVVA) (Source: NJ.com (2025), https://www.nj.com/, Accessed 10 October 2025).

Table 12.2 Workspace Specifications (Heatherwick Studio, 2021; Davidson, 2021; MNLA, 2021; Government of Western Australia, 2016; Arm Architecture, 2025; DRWC, 2011; Field Operations, 2012; Gehl, 2010; Hudson River Park Trust, 2021; Michael Van Valkenburgh Associates, 2010; Manhattan Youth, 2022; Landezine, 2017; City of Chicago, 2024; Chicago Department of Planning and Development, 2024; MVVA, 2010; NJDEP, 2011).

Structure Name	Year of Construction	Country	City	Area (m^2)	Coordinates
NYC Little Island Park	2021	United States of America	New York	11,000	40.742108, –74.010278
Elizabeth Quay Park	2015	Australia	Perth	10,000	31.959520, 115.856361
Race Street Pier	1896 (redeveloped 2010)	United States of America	Philadelphia	50,000	39.953182, –75.138770
Hudson River Park Pier 25	2010	United States of America	New York	10,000	40.720436, –74.014081
Merchant Square Floating Pocket Park	2016	United Kingdom	London	730	51.518848, 0.171257
Wild Mile Floating Eco Park	2022 (completed 2024)	United States of America	Chicago	17,000	41.906666, –87.651667
Pier C Park	2010	United States of America	Hoboken, New Jersey	1,250	40.740289, –74.025876

12.2.2.2 Second stage: Determination of Design Criteria

Based on the literature review and field studies, design criteria for floating parks were determined. Criteria defined by international standards and institutions were considered: Bruttomesso (2001), American Planning Association (APA, 2006), PK3 Landskabsarkitekter & Tredje Natur (2011), American Society of Landscape Architects (ASLA, 2018), United Nations Environment Programme (2009), Public Space Project (PPS, 2020), and US National Park Service (2020).

Additionally, theses, periodicals, journals, and other academic resources from Turkey were analyzed in comparison with the Coastal Design Guidelines of the

New South Wales Government. As a result, the fundamental design criteria of floating parks were systematized under the categories of planning, design, ecology, activity, accessibility, and management.

12.2.2.3 Third Stage: Case Study and Applicability of Criteria

The selected floating parks were examined in detail according to the determined criteria. The design and functional features of each park were analyzed through coastal park design principles to reveal both prominent design principles in practice and challenges encountered during implementation. This method systematically demonstrates why the criteria were chosen and with which standards they can be associated.

12.2.3 Data Analysis

In the analysis process, both literature and field data were handled comparatively. The applicability of the previously defined design criteria and the performance of the selected floating parks were evaluated in terms of ecological sustainability, social use, accessibility, and multifunctionality. Each park's current conditions were systematically compared with coastal park design standards using observations, photographs, and online sources (McHarg, 1969; Gehl, 2010; Field and Barros, 2014; WEDG, 2020).

This method reveals the rationale behind the selection of criteria and identifies the measurable indicators linked to each design principle. Elements such as circulation paths, vegetation layout, water contact areas, seating arrangements, and activity zones were analyzed to comprehensively assess both functional and ecological performance.

12.2.4 Findings/Main Findings and Arguments

The comparative analysis of the seven selected floating parks revealed the following key findings:

1. Ecological Integration: All parks implemented strategies that integrate terrestrial and aquatic ecosystems. Native vegetation, floating wetlands, and transition zones (ecotones) support local biodiversity, maintain habitat connectivity, and contribute to microclimate regulation. Wild Mile Eco Park and Merchant Square Floating Pocket Park are particularly successful examples, demonstrating how floating infrastructure can support urban ecological networks.

2. Social and Recreational Use: Design strategies effectively combine recreation, cultural activities, and leisure. NYC Little Island Park and Hudson River Park Pier 25, for instance, provide performance venues, children's play areas, and interactive spaces, supporting diverse user groups while maintaining ecological sensitivity.
3. Accessibility and Inclusivity: Universal design principles were consistently applied across all examples. Ramps, wide circulation paths, and multi-level platforms ensure accessibility for users of different ages and physical abilities. This is especially evident in Race Street Pier and Pier C Park.
4. Multifunctionality and Flexibility: Parks are designed to accommodate both seasonal and daily use. Temporary events, modular design elements, and mixed-use spaces respond to ecological and social needs simultaneously.
5. Argument and Contribution: This study demonstrates that floating parks can be systematically analyzed through coastal park design criteria, providing a framework to evaluate ecological, social, and functional performance. Comparative analysis shows that integrating environmental, recreational, and accessibility standards enhances user experience and establishes a resilient model for contemporary urban waterfront design. This framework can serve as an international reference for future floating park designs.

12.3 Results

According to the evaluation presented in Table 12.3, it is observed that all examined floating parks are easily accessible and include urban reinforcement elements, walking paths, barriers and delimiters for users, planting and flower beds, as well as squares and piers for events. However, the areas allocated for bicycle paths and ramps were found to be more limited.

12.4 Discussion

The analysis of floating and coastal parks, including international and local examples (NYC Little Island Park, Elizabeth Quay Park, Race Street Pier, Hudson River Park Pier 25, Floating Pocket Park, Wild Mile Eco Park, and Pier C Park), demonstrates the implementation of various design criteria and highlights variations across sites (Table 12.3).

Accessibility and Pathways: All parks provide walkways that are unobstructed, continuous, and constructed with visible natural materials, such as concrete, granite, stone, or wood. Hard and soft surfaces are distinctly separated, and reflective materials are avoided. Walkways generally meet the minimum width requirement of 120 cm. Ramps are present in most parks, except Hudson River Park Pier 25 and Pier C Park, with flooring materials designed to be

Table 12.3 Evaluation of floating parks according to Coastal Park design principles: Comparative analysis (Criticisms developed and analyzed by the authors).

Coastal Park Design Criteria	NYC Little Island Park	Elizabeth Quay Park	Race Street Pier	Hudson River Park Pier 25	Floating Pocket Park	Wild Mile Eco Park	Pier C Park
Accessibility							
Presence of walkways	+	+	+	+	+	+	+
Unobstructed and uninterrupted continuation	+	+	+	+	+	+	+
Made with visible color and natural materials (concrete, granite, stone, wood)	+	+	+	+	+	+	+
Separation of hard and soft floors from each other, their distinctive feature	+	+	+	+	+	+	+
No use of reflected substances	+	+	+	+	+	+	+
Be at least 120 cm wide	+	+	+	+	+	+	+
Presence of a bicycle path	–	–	–	–	–	–	–
Be at least 170 cm wide	–	–	–	–	–	–	–
Using non-reflective and non-slippery materials	–	–	–	–	–	–	–
Difference in color and textural from pedestrian paths (blue acrylic flooring, which is the standard criterion on asphalt flooring)	–	–	–	–	–	–	–
The presence of ramps	+	+	+	+	–	+	–
Its width should be at least 120 cm	+	+	+	+	–	+	–
The use of flooring material that does not reflect light and provides good adhesion to the ground with its rough surface	+	+	+	+	–	+	–
The presence of a ladder	+	+	+	+	+	–	+
Floor covering material is non-slippery and non-reflective	+	+	+	+	+	–	+
Dock and step dimensions (15 cm/28–30 cm)	+	+	+	+	+	–	+
Accentuation of step edges with contrasting colors for unobstructed use	+	+	+	+	+	–	+
Comfort and Facilities							
The presence of lighting elements	+	+	+	+	+	+	+
Its height is between 3–6 m	+	+	+	+	+	+	+

Table 12.3 (continued)

Coastal Park Design Criteria	NYC Little Island Park	Elizabeth Quay Park	Race Street Pier	Hudson River Park Pier 25	Floating Pocket Park	Wild Mile Eco Park	Pier C Park
Planning park entrances, exits and dark areas in heavily used areas	+	+	+	+	+	+	+
Solar energy operation of lighting elements	+	+	+	+	+	+	+
Preference for salt water and moisture-resistant materials (stainless steel, aluminum or durable plastic)	+	+	+	+	+	+	+
The presence of seating elements	+	+	+	+	+	+	+
Placed at regular intervals between 100 and 200 cm	+	+	+	+	+	–	+
Positioned at least 60 cm from the road so as not to interfere with users	+	+	+	+	+	–	+
Be at least 60 cm wide for one person	+	+	+	+	+	+	+
The presence of armrests on both sides of the seat element	+	+	+	+	–	–	–
The height of the picnic tables is between 75–90 cm	+	+	+	+	–	–	–
Use of water-resistant stainless materials	+	+	+	+	–	–	–
The presence of canopies and awnings	+	+	+	+	+	–	+
Made of wind-resistant (steel, wood, etc.) material	+	+	+	+	+	–	+
Height between 250–300 cm	+	+	+	+	+	–	+
In the case where it is placed on the flooring, the use of non-slip material	+	+	+	+	+	–	+
The colors of the canopies and awnings, light colors are preferred in terms of temperature management	+	+	+	+	+	–	+
Open sides so as not to obstruct the air flow	+	+	+	+	+	–	+
The presence of trash cans	+	+	+	+	+	–	+
Positioning at access points or next to activity areas	+	+	+	+	+	+	+
Positioning every 50–100 m along the main road	+	+	+	+	+	+	+
Striking colors	+	+	+	+	+	+	+

(continued)

Table 12.3 (continued)

Coastal Park Design Criteria	NYC Little Island Park	Elizabeth Quay Park	Race Street Pier	Hudson River Park Pier 25	Floating Pocket Park	Wild Mile Eco Park	Pier C Park
Fixed to the floor	+	+	+	+	+	+	+
The height of the boxes should be 90 cm for unobstructed use	+	+	+	+	+	+	+
Use of attention-grabbing colors	+	+	–	+	–	–	–
Availability of bins for recycling	+	+	–	+	–	–	–
Presence of WC cabins	+	+	–	+	–	–	+
Toilets are accessible to visitors with disabilities	+	+	–	+	–	–	+
Easy to locate	+	+	–	+	–	–	+
Presence of signs and information signs	+	+	–	+	–	–	–
Suitable for visual access at a height of 120–150 cm	+	+	–	+	–	–	–
Presence at decision points (entrances, activity areas, road junctions)	+	+	–	+	–	–	–
Use of attention-grabbing colors	+	+	–	+	–	–	–
Suitable for anthropometric measurements, at the level of view	+	+	–	+	–	–	–
Presence of bicycle parking spaces	–	–	–	–	–	–	–
Being easily accessible and integrated into the circulation roads within the park	–	–	–	–	–	–	–
Providing adequate lighting for night-time use	–	–	–	–	–	–	–
Positioning in areas with high user density	–	–	–	–	–	–	–
Presence of sales units	+	+	–	+	–	–	–
Sizing according to the density and size of the park	+	+	–	+	–	–	–
Appropriate accessibility to all user groups, especially people with disabilities	+	+	–	+	–	–	–
Providing aesthetic integrity of the buildings with the landscape and natural environment	+	+	–	+	–	–	–
The presence of barriers and limiters	+	+	+	+	–	–	+

Table 12.3 (continued)

Coastal Park Design Criteria	NYC Little Island Park	Elizabeth Quay Park	Race Street Pier	Hudson River Park Pier 25	Floating Pocket Park	Wild Mile Eco Park	Pier C Park
Preference for materials that are resistant to water and moisture (stainless steel, wood, concrete or plastic)	+	+	+	+	–	–	+
Leaving gaps of 10–15 cm between the barriers and designing them in such a way that children cannot pass	+	+	+	+	–	–	+
For decorative limiters, it should be between 50–70 cm	+	+	+	+	–	–	+
At least 110 cm of waterside barriers	+	+	+	+	–	–	+
Presence of water elements	+	+	+	+	–-	–	+
It is in line of sight, but in places that do not obstruct pedestrian circulation	+	+	+	+	–	–	+
Water elements, designing to serve the principles of aesthetic, cultural and sociality	+	+	+	+	–	–	+
Presence of sculptures and installations	+	+	–	+	–	–	+
Comply with accessible design principles that will not threaten user safety	+	+	–	+	–	–	+
In a way that supports the cultural, aesthetic and functional identity of the park	+	+	–	+	–	–	+
Using long-lasting materials suitable for outdoor and coastal conditions	+	+	–	+	–	–	+
Landscaping and Planting							
The presence of afforestation	+	+	+	+	+	+	+
Tree and plant heights should be in accordance with the viewing angle of the users	+	+	+	+	+	+	+
The use of plants with seasonal discoloration that offer variety for the aesthetic appeal of the park	+	+	+	+	+	+	+
Using broad-leaved plants and trees to create shade	+	+	+	+	+	+	+
Planted at intervals of 3–5 m for healthy development and user circulation	+	+	+	+	+	+	+

(continued)

Table 12.3 (continued)

Coastal Park Design Criteria	NYC Little Island Park	Elizabeth Quay Park	Race Street Pier	Hudson River Park Pier 25	Floating Pocket Park	Wild Mile Eco Park	Pier C Park
The presence of planting and flower beds	+	+	+	+	+	+	+
Leaving a transition distance between plants (at least 120 cm) so that access is not obstructed	+	+	+	+	+	+	+
Suitable for coastal climate, salt resistant, resistant to wind and sun	+	+	+	+	+	+	+
Selection of local and natural species	+	+	+	+	+	+	+
Creating boundaries and ensuring functional separation of areas	+	+	+	+	+	+	+
Social Interaction							
The presence of game elements	+	–	–	+	–	+	+
Selection of suitable colored material for game elements	+	–	–	+	–	+	+
Using soft upholstery material such as sand and rubber	+	–	–	+	–	+	+
Having a playground for 0–5 year olds	+	–	–	+	–	–	+
Having a playground for 6–14 year olds	+	–	–	+	–	+	+
Presence of seating elements for resting around	+	–	–	+	–	+	+
The presence of sports elements	–	–	–	+	–	–	–
Stimulating warm colors are preferred in fitness equipment	–	–	–	+	–	–	–
Suitable for barrier-free use	–	–	–	+	–	–	–
The use of grass, rubber or tartan as a floor material	–	–	–	+	–	–	–
The presence of squares and an observation deck	+	+	+	+	–	–	+
Planning in easily visible and accessible areas	+	+	+	+	–	–	+
Adapting to events such as concerts, exhibitions, festivals	+	+	+	+	–	–	+
Directed to the landscape	+	+	+	+	–	–	+
Anthropometric value of seating steps	+	+	+	+	–	–	+
Presence of water access points	+	–	–	+	–	+	+

Table 12.3 (continued)

Coastal Park Design Criteria	NYC Little Island Park	Elizabeth Quay Park	Race Street Pier	Hudson River Park Pier 25	Floating Pocket Park	Wild Mile Eco Park	Pier C Park
Suitable for barrier-free use	+	–	–	+	–	+	+
Suitable for connecting transportation vehicles such as boats, etc.	+	–	–	+	–	+	+
Water-resistant, non-slip material	+	–	–	+	–	+	+
Suitable for fishing	+	–	–	+	–	+	+

non-reflective and provide adequate traction. Ladders are available in all parks except Wild Mile Eco Park, with non-slip, non-reflective surfaces, appropriately dimensioned docks and steps (15 cm height/28–30 cm depth), and accentuated edges to ensure safety and usability.

Bicycle Infrastructure: Dedicated bicycle paths and bicycle parking facilities are absent in all parks. Consequently, minimum width, material selection, and visual differentiation criteria are not implemented, indicating a gap in multimodal transport provision.

Lighting: Lighting elements are present in all parks, with heights ranging from 3 to 6 m. Park entrances, exits, and darker areas in heavily used zones are appropriately considered. Lighting operates via solar energy and utilizes materials resistant to saltwater and moisture, including stainless steel, aluminum, or durable plastics.

Seating and Comfort: Seating elements are widely provided, spaced at intervals of 100–200 cm (except in Wild Mile Eco Park), positioned at least 60 cm from circulation routes, and a minimum of 60 cm wide per user. Armrests are present on both sides in most parks, except Hudson River Park Pier 25 and Floating Pocket Park. Picnic tables, where provided, are 75–90 cm in height and use water-resistant materials, except at Hudson River Park Pier 25 and Floating Pocket Park.

Canopies and Awnings: Canopies and awnings are present in most parks, constructed from wind-resistant materials and installed at heights of 250–300 cm. Flooring, color, and side openness criteria are applied where these elements exist, with Hudson River Park Pier 25 lacking some provisions.

Trash Management: Trash cans are generally provided, positioned at access points and along main roads every 50–100 m, fixed to the ground, and with heights of 90 cm. Use of attention-grabbing colors and availability of recycling bins are inconsistent, with some parks, including Wild Mile Eco Park and Hudson River Park Pier 25, lacking these features.

Sanitation and Signage: WC cabins are present in most parks and are accessible to users with disabilities. Signage and information boards are inconsistently implemented; appropriate visual height (120–150 cm) and placement at decision points are observed primarily in NYC Little Island Park, Elizabeth Quay Park, Race Street Pier, and Pier C Park. Use of attention-grabbing colors and anthropometric considerations follows a similar pattern.

Sales Units: Sales units are available in NYC Little Island Park, Elizabeth Quay Park, Race Street Pier, and Pier C Park, considering size, accessibility, and aesthetic integration.

Barriers and Limiters: Most parks include barriers and limiters, constructed with water- and moisture-resistant materials. Design considers gaps of 10–15 cm to prevent child passage, decorative limiter heights of 50–70 cm, and waterside barrier heights of at least 110 cm. Hudson River Park Pier 25 and Floating Pocket Park do not fully meet these criteria.

Water Elements: Water features are present in all parks except Hudson River Park Pier 25 and Floating Pocket Park. Placement ensures visibility without obstructing pedestrian circulation and supports aesthetic, cultural, and social functions.

Sculptures and Installations: Sculptures and installations are present in NYC Little Island Park, Elizabeth Quay Park, Race Street Pier, and Pier C Park. Designs comply with accessibility standards, reinforce cultural and aesthetic identity, and utilize durable materials suitable for coastal environments. Merchant Square Floating Pocket Park and Wild Mile Eco Park lack such installations.

Afforestation and Planting: All parks exhibit afforestation and planting practices, with tree and plant heights aligned with user viewing angles, broad-leaved species for shade, seasonal color variations, planting intervals of 3–5 m, flower beds, minimum spacing of 120 cm for circulation, selection of coastal climate-tolerant and local species, and functional area separation.

Play and Sports Elements: Play elements are present in NYC Little Island Park, Hudson River Park Pier 25, Floating Pocket Park, and Pier C Park, including age-specific playgrounds (0–5 years and 6–14 years), soft materials (sand, rubber), seating around play areas, and color-coded equipment. Sports elements are only present at Hudson River Park Pier 25, including warm-colored fitness equipment, barrier-free access, and appropriate floor materials (grass, rubber, or tartan).

Squares, Observation Decks, and Water Access: Squares and observation decks are present in most parks, except Hudson River Park Pier 25 and Floating Pocket Park. Design prioritizes visibility, accessibility, event adaptability, landscape orientation, and anthropometrically suitable seating. Water access points exist in NYC Little Island Park, Hudson River Park Pier 25, Floating Pocket Park, and Pier C Park, supporting barrier-free use, connectivity with transport, non-slip surfaces, and suitability for fishing.

Overall, the analysis demonstrates that fundamental accessibility, landscaping, and seating criteria are largely implemented across all parks, whereas bicycle infrastructure, certain sanitation and signage features, play and sports elements, and some aesthetic and recreational installations exhibit variability. These findings underscore the importance of integrated planning to balance user accessibility, ecological performance, and social and recreational functions in floating park design.

12.4.1 Transformative Growth in Floating Park Design

Based on the analysis of these case studies, Transformative Growth emerges as a conceptual lens to guide the evolution of floating park design beyond conventional frameworks. Traditional approaches often remain limited to incremental eco-efficiency or isolated social interventions, sometimes resulting in constrained practical and political outcomes. In contrast, Transformative Growth emphasizes simultaneous enhancement of ecological, social, and technological productivity, offering a holistic perspective for evaluating park design (Table 12.1). The parks studied (NYC Little Island Park, Elizabeth Quay Park, Race Street Pier, Hudson River Park Pier 25, Floating Pocket Park, Wild Mile Eco Park, Pier C Park) illustrate this potential. While North American parks demonstrate strong attention to social interaction and aesthetics, the Turkish example, Pier C Park, shows notable ecological and structural adaptability. Gaps in bicycle infrastructure, signage, play/sports elements, and certain aesthetic installations highlight opportunities for adaptive, multifunctional interventions that enhance user-centered design. Ecological productivity is reinforced through diverse planting schemes, shade provision, and support of local biodiversity. Social productivity is achieved via interactive spaces, such as observation decks, squares, and age-appropriate playgrounds, fostering inclusive engagement. Technological and design innovation is reflected in modular infrastructure, solar-powered lighting, durable materials, and flexible canopies/awnings, increasing functional resilience. By framing these design interventions within a Transformative Growth paradigm, floating parks are recognized not only as recreational or aesthetic interventions but also as instruments for urban resilience and sustainability. Integrating participatory planning, modular infrastructure, and multifunctional programming ensures that future floating parks can balance ecological boundaries, social inclusivity, and technological innovation, offering policy-relevant and transformative urban solutions.

12.5 Conclusion

This study examined floating and coastal parks through a comparative analysis of international and local case studies, including NYC Little Island Park, Elizabeth Quay Park, Race Street Pier, Hudson River Park Pier 25, Floating Pocket Park, Wild Mile Eco Park, and Pier C Park (Table 12.1). The findings demonstrate that while basic design criteria such as accessibility, landscaping, seating, lighting, and water management are largely implemented, significant variations exist in bicycle infrastructure, playgrounds and sports facilities, information signage, and certain aesthetic installations.

The novelty of this study lies in its multidimensional and comparative approach, integrating ecological, social, and technological perspectives to evaluate the functionality and adaptability of floating park design. Applying the concept of Transformative Growth, the study illustrates how urban park interventions can transcend conventional dichotomies of ecological modernization and degrowth. Floating parks can enhance not only economic productivity but also ecological productivity, social inclusion, and technological innovation, demonstrating the potential to reconcile environmental, social, and design objectives within urban coastal spaces.

The contributions of this study can be summarized in three main points. First, it provides empirical evidence on the current status of floating park design, highlighting both strengths and deficiencies. Second, it proposes a conceptual framework—Transformative Growth—to guide future park projects toward integrated ecological, social, and technological outcomes. Third, it offers practical implications for urban planners, landscape architects, and policymakers by emphasizing the importance of modular infrastructure, participatory planning, and multifunctional programming to foster resilient and sustainable waterfront environments.

In conclusion, floating parks are more than recreational or aesthetic interventions; they represent instruments of urban transformation. By combining user-centered design with ecological sustainability and social functionality, these parks demonstrate the potential to support inclusive, adaptive, and innovative urban coastal spaces. Future projects can build upon these findings to create resilient waterfronts that enhance the ecological, social, and aesthetic value of cities, contributing to the long-term sustainability and livability of urban coastal environments.

References

Americans with Disabilities Act (2010) Americans with Disabilities Act accessibility guidelines. U.S. Department of Justice

American Planning Association (2006) Planning and urban design standards. American Planning Association

American Society of Landscape Architects (2013) Sustainable sites initiative: Guidelines and performance benchmarks. ASLA, Washington, DC

American Society of Landscape Architects (2018) Smart policies for a changing climate. ASLA, Washington, DC

Arm Architecture (2025) Elizabeth Quay redevelopment project ARM Architecture

Beatley T (2011) Biophilic cities: Integrating nature into urban design and planning. Island Press, DC

Bolonkin A (2010) Land reclamation and urban challenges in coastal cities. J of Coast Eng 57(2):101–114

Bruttomesso R (2011) Urban waterfront regeneration and planning strategies. Routledge, London

Carr S, Francis M, Rivlin LG, Stone AM (1992) Public space. Cambridge University Press, Cambridge

Chicago Department of Planning and Development (2024) Wild Mile Eco Park: Planning report. City of Chicago

City of Chicago (2024) Wild Mile Eco Park overview. City of Chicago

CROW (2016) Design manual for bicycle infrastructure. CROW Publishing, Ede

Davidson, A. (2021). Little Island Park case study. Landsc J 40(3):45–62

Delaware River Waterfront Corporation (2011) Race Street Pier redevelopment report. Delaware River Waterfront Corporation

Deutsches Institut für Normung 18040 (2010) Accessibility standards for buildings. Deutsches Institut für Normung

Elizabeth Quay (2025) Elizabeth Quay Park, Perth, Australia. https://architectureau.com/directory/tcl/elizabeth-quay/. Accessed 12 Oct 2025

Fédération Internationale de Basketball. (n.d.) FIBA sports court standards

Fédération Internationale de Football Association. (n.d.) Football pitch dimensions and standards

Field Operations (2012) Race Street Pier landscape design documentation. Field Operations
Field S, Barros M (2014) Waterfront urbanism and public space integration. J of Urban Des 19(4): 521–539
Gehl Architects (2010) Urban public space guidelines. Gehl Architects
Gehl J (2010) Cities for people. Island Press, Washington, DC
Heatherwick Studio (2021) Little Island Park project documentation. Heatherwick Studio
Hoboken Parks Department (2012) Pier C Park public access and planning report. City of Hoboken
Hudson River Park Trust (2021) Pier 25 park management report. Hudson River Park Trust
Hudson River Park Trust (2025) Hudson River Park Piers 25 and 26, New York, USA. https://www.tribecatrib.com/content/eye-tribecas-waterfront-one-photographers-40-year-view. Accessed 10 Oct 2025
Intergovernmental Panel on Climate Change (2021) Climate change 2021: The physical science basis
International Dark-Sky Association (2019) Lighting design guidelines for outdoor public spaces. IDA
International Organization for Standardization (1998) ISO 9241-5: Ergonomic requirements for office work with visual display terminals
Landezine Landscape Architecture Platform (2017) Merchant Square Floating Pocket Park
Little Island NYC (2025) Little Island Park, New York City, USA. https://www.azuremagazine.com/article/manhattans-little-island-opens-to-controversy-and-awe/. Accessed 10 Oct 2025
Loukaitou-Sideris A (2013) Urban design for the public realm. Routledge, London
Mathews Nielsen Landscape Architects (2021) Little Island Park landscape design report
McHarg I (1969) Design with nature. Doubleday, Garden City, NY
Merchant Square London (2025) Merchant Square Floating Pocket Park, Paddington Basin, London, UK. https://merchantsquare.co.uk/see-and-do/floating-pocket-park. Accessed 10 Oct 2025
Michael Van Valkenburgh Associates (2010) Pier C Park design documentation
NEVER. (2017). Coastal vegetation and urban sustainability. Coast Ecol Rev 22(1), 11
New Jersey Department of Environmental Protection (2011) Pier C Park environmental assessment
NJ.com (2025) Pier C Park, New Jersey, USA. https://www.nj.com/. Accessed 10 Oct 2025
Ouroussoff, N. (2021). A new island rises in the Hudson. The New York Times. Available at: https://www.nytimes.com/2021/05/20/arts/design/little-island-pier-55-heatherwick-barry-diller.html
Perkins H, Leal W (2010) Urban waterfronts: Ecology and design. J of Landsc Archit 5(2):87–104
PK3 Landskabsarkitekter & Tredje Natur (2011) Waterfront planning guidelines. PK3 Landskabsarkitekter & Tredje Natur
Project for Public Spaces (2012) The social life of small urban spaces. Project for Public Spaces
Project for Public Spaces (2015) Design guidelines for public waterfronts. Project for Public Spaces
Project for Public Spaces (2016) Coastal parks and urban sustainability. Project for Public Spaces
Project for Public Spaces (2020) What makes a successful public space? Project for Public Spaces, New York
Race Street Pier (2025) Race Street Pier, Philadelphia, USA. https://www.fieldoperations.net/project-details/project/race-street-pier.html/. Accessed 10 Oct 2025
Selman P (2009) Planning for landscape multifunctionality. Sustain: Sci, Pract and Polic 5(2):45–52. https://doi.org/10.1080/15487733.2009.11908035
Selena, J. (2020). Ecological integration in urban waterfronts. Urban Ecol J 33(1)
Sha X (2010) Urban waterfront development in Australia and Asia. J of Urban Plan 36(3):211–230
Sungsine K (2011) Floating and adaptive urban infrastructures. Int J of Urban Sustain Deve 3(2):101–121
Tredje Natur & PK3 (2015) Climate adaptation strategies for urban waterfronts. Tredje Natur & PK3, Copenhagen
United Nations Environment Programme (2008) Sustainable coastal zone management
United Nations Environment Programme (2009) Sustainable coastal tourism: An integrated planning and management approach. UNEP, Nairobi
United Nations Environment Programme (2021) Guidelines for urban green and waterfront spaces

Urban Rivers (2025, October 10) Wild Mile Park, Chicago, USA. ArchDaily. https://www.archdaily.com/911890/worlds-first-floating-eco-park-planned-for-chicago-river.
US National Park Service (2020) Outdoor performance spaces design Guidelines.https://www.nps.gov/subjects/policy/mp-8-use-of-parks.htm
Waterfront Edge Design Guidelines (2018) Accessibility and infrastructure standards for waterfronts. WEDG Publications
Waterfront Edge Design Guidelines (2020) Waterfront ecological design guide. WEDG Publications
Waterfront Toronto (2021) Waterfront planning and design standards. Waterfront Toronto
Woods T (2020) Merchant Square Floating Pocket Park: Project overview. R Horticultural Soc

Chapter 13
Transforming Cities: From Managed Decline to Transformative Growth

Ali Cheshmehzangi, May Tan-Mullins, Martin de Jong, and Rob Raven

Abstract This book argues that the future of urban development hinges on a decisive shift from incremental, transactional growth toward transformative growth. This approach aligns spatial design, ecological systems, governance reform, and community involvement. Drawing on diverse case studies from Japan, Iran, China, France, Indonesia, Pakistan, and the U.S., the book demonstrates how cities under pressure are already finding new pathways for regeneration. From seismic-resilient morphology and brownfield reclamation to thermal adaptation, nature-based cooling networks, heritage-sensitive stormwater strategies, and the conflicts between ecological balance and tourism, the chapters show a common pattern: ecological integration, spatial intelligence, and governance that can coordinate rather than control are necessary for effective transformation.

In the face of climate volatility, growing urbanization, and institutional fragmentation, the book presents transformative growth as the only practical alternative to managed decline. It demonstrates how decisions about land use serve as climate choices, how community-centered adaptation can be supported by digital

---

A. Cheshmehzangi (✉)
School of Architecture, Design and Planning, The University of Queensland, Brisbane, QLD, Australia
School of Architecture and Urban Planning, Hunan University, Changsha, Hunan, China
e-mail: a.chesh@uq.edu.au

M. Tan-Mullins
University of Reading Malaysia, Johor Bahru, Malaysia

M. de Jong
Rotterdam School of Management, Erasmus University Rotterdam, Rotterdam, The Netherlands Rotterdam, The Netherlands
Smart City Institute, HEC Liège, University of Liège, Liège, Belgium
Institute for Global Public Policy, Fudan University, Shanghai, China

R. Raven
Monash Sustainable Development Institute, Monash University, Victoria, Australia

A. Cheshmehzangi et al. (eds.), *Transforming Cities*, Urban Sustainability,
https://doi.org/10.1007/978-981-95-7378-3_13

tools, and how continuity and long-term value must be given priority in rebuilding. In the end, *Transforming Cities* provides a forward-thinking agenda for legislators, planners, designers, and communities dedicated to creating successful cities that are also regenerative, climate-ready, and based on the welfare of people and the environment.

Keywords Transformative Growth • Urban Transformation • Transforming Cities • Technological Innovation • Transformation • Resilience • Sustainable Cities

13.1 Introduction: Entering the Era of Transformative Growth

In this challenging era, cities are at a pivotal juncture. The pressures driving urban life, such as climate volatility, demographic acceleration, technological upheaval, fragmented governance, expanding inequities, have combined into an era where the margin for error is narrowing. Urbanization is now the main factor influencing social, ecological, and economic realities worldwide rather than merely a background phenomenon. Climate hazards are no longer distant warnings; they are the daily experience of coastal inundation, fatal heat, catastrophic floods, and shattered infrastructures. Failures in governance are now active barriers to resilience and long-term prosperity rather than administrative annoyances. Cities must function within this framework, which does not allow for either business as usual or complacency. Although the urban century has begun, its course is still deeply contested.

In this context, the concept of transformative growth becomes more than just a theoretical theory. It is the main contention of this book, the thread that unites several case studies and academic debates into a cohesive plan for practical transformation. The traditional belief that environmental integrity and development must be compromised is rejected by transformative growth. It rejects the antiquated notion that growing cities must unavoidably devour land, destroy ecosystems, and uproot communities. Rather, it reframes growth as a process that, given the correct priorities, structures, and governance mechanisms, may enhance social cohesion, create long-term value, and repair ecological systems. When ecological innovation, climatic resilience, and socially conscious design become the cornerstones of development rather than its accessories, cities can evolve differently, as demonstrated by the accumulating evidence presented throughout the chapters.

The purpose of this final chapter is not merely to summarize what has come before but to crystallize the implications of the book's core message. As we have seen in the previous chapters, cities dealing with intense environmental pressures, i.e., from heat stress in subtropical regions to coastal erosion in small islands, are experimenting with both established methods and new technologies. Innovations such as integrated coastal landscapes, community-centered solar design, nature-based cooling structures, seismic-resilient form, and brownfield regeneration are not isolated. They embody a new urban language based on ecological intelligence, flexibility, and coordination. This conclusion steps back from the details and looks at the larger pattern that has been taking shape across the research, revealing how transformative growth is becoming not only possible but necessary.

Today's leaders must be able to see past short-term cycles and see the fundamental dangers of incrementalism. Cities will become vulnerable if they stick to transactional modes of urbanization, such as fragmented land management, parcel-by-parcel monetization, and symbolic sustainability tactics. When cities view growth as a financial calculation instead of a socio-ecological investment, managed decline becomes the default result. The data indicates that growth becomes transformative when it is linked to social inclusion, ecological performance, and cooperation at the urban level. It changes from adding additional structures to revamping the mechanisms that sustain urban life. Anticipatory design replaces reactive adaptability. It shifts from a model of compliance-driven governance to one in which communities, governments, and the business sector work together to co-regulate the city's destiny.

This book has consistently emphasized that transformative growth necessitates courage, not in a rhetorical sense, but in the readiness to question long-standing conventions, reject antiquated planning cultures, and make long-term resilience investments. When cities must decide between densification and sprawl, adaptation and recovery, ecological corridor preservation and immediate land profit, or community involvement and top-down control, courage is needed. When cities adapt their financial, regulatory, and geographical systems to the reality of a climate-constrained world, transformative growth is successful. That alignment is ultimately a question of leadership, institutional maturity, and societal imagination.

As this concluding chapter unfolds, it synthesizes the major lessons drawn from the book's cross-disciplinary investigations and case studies. From neighborhood-level spatial rearrangement to regional ecological planning, it provides a paradigm for comprehending how transformative growth might be used at various scales and situations. It also describes the opportunities and responsibilities that communities, planners, and city leaders must face to guarantee that cities not only survive but flourish in the twenty-first century.

Entering the era of transformative growth means accepting that transformation is no longer optional. It is the new standard for cities that want to stay sustainable, equitable, and rooted in the environment. This chapter's subsequent sections emphasize that message, make clear the future directions, and re-affirm our shared obligation to reshape urban futures while time is still on our side.

13.2 Lessons from the Case Studies: Transformative Growth in Practice

Examining the tangible, situated experiences of cities negotiating their particular challenges, weaknesses, and development realities makes the value of transformative growth completely apparent. A common theme emerges from the case studies in this book: cities are not changing because of isolated efforts or token acts of sustainability. They are changing through complex, frequently challenging recalibrations of ecological logic, governance, and design. Although each chapter focused on a distinct region and issue, taken as a whole, they show how transformative growth functions as a workable framework that is flexible in varied situations but rooted in quantifiable gains in resilience, ecological performance, and urban growth and equity.

The exploration of brownfield redevelopment in Musashi Kosugi, Japan, provides the first clear demonstration. The region, which was once plagued by post-industrial decline and dispersed land ownership, serves as an example of how targeted legislative tools and hierarchical planning frameworks can revive contaminated property without making the same mistakes twice. The Japanese example demonstrates a unique type of transformative growth that is based on close coordination between national policy and local initiatives. It demonstrates that redevelopment is successful not just through building but also through governance changes, such as defining roles, organizing land reclamation, and offering appropriate financial incentives. However, it also shows that even sophisticated planning cultures must deal with enduring gaps in policy integration and environmental rehabilitation. Here, transformative growth takes the shape of institutional learning, modifying traditional redevelopment to meet modern ecological and legal requirements.

In contrast, the discussion moves from regulatory structures to spatial layout in the chapter on earthquake-resilient urban form. It illustrates how morphological accuracy affects the city's capacity to react to seismic disasters by a thorough examination of block layout, open-space distribution, and street connection. What results is a sort of transformative growth based on spatial intelligence: anticipatory design, rather than post-disaster restoration, makes cities safer, more flexible, and more recoverable. Accessible evacuation lanes, dispersed public areas, and adaptable block layouts demonstrate how resilience is integrated into the city's geometry rather than being an add-on. This illustrates a recurrent theme in the book: transformative growth necessitates addressing cities' physical structure as well as their policies.

This morphological focus is extended by the study on livable streets, which demonstrates how the potential for urban growth is limited by small-scale physical flaws including narrow sidewalks, poor lighting, and poor integration between buildings and streets. The case serves as a reminder that expansion is dependent on everyday areas' usability, safety, and quality rather than just being a macro-scale occurrence. In this situation, attention to design elements that facilitate densification, social interaction, and mobility leads to transformative growth. According to the research, poorly designed streets impede the possibility of redevelopment, lower economic vibrancy, and erode social cohesiveness. Thus, enhancing the streets is essential to the city's ability to develop; but more importantly, it is not merely aesthetic.

An alternative viewpoint is provided by the Tehran cooling structures chapter, which integrates spatial arrangement with ecological function. The study reveals how misaligned green space networks impair cooling performance in a city experiencing high heat using space syntax and landscape measures. By combining quantitative spatial analysis with nature-based solutions, the suggested ecological network, which connects parks, corridors, and spatial flows, exemplifies transformative growth. The lesson is straightforward: when ecological interventions are informed by spatial evidence and in line with the structural logic of the city, they become transformative.

The Dunkirk chapter provides a post-industrial perspective, demonstrating how integrated planning has helped a city long characterized by pollution, decline, and maritime separation reconstruct its identity. The emphasis on urban narratives, aquatic landscapes, and mobility centers demonstrates a kind of change that extends beyond infrastructure. Dunkirk serves as an example of how cultural recalibration is necessary for regenerative urbanism, changing how citizens view their city and the social and ecological connections between its various areas. The case illustrates transformative growth as a relational process with roots in identity formation, movement, and heritage.

The Bali case study illustrates the interplay between ecological stability and the demands of international tourism by looking at the conflict between rice fields and resort-driven (urban) expansion. In this case, transformation necessitates facing strong market forces and developing spatial methods that protect ecological systems while allowing for change, such as mixed-use zoning, natural corridors, and community land management. This illustration demonstrates how transformative growth functions in contentious environments where global capital, livelihoods, and government converge.

Additional depth is provided by the chapters on Suzhou's heritage resilience, subtropical thermal resilience, and solar-optimized community blocks. They demonstrate how place-specific transformation is supported by digital technologies, performance modelling, and participatory techniques. In these situations, community involvement, real-time data, collaborative design, and hybrid governance all contribute to progress. These examples, which range from maximizing solar access at the block level to coordinating public involvement with stormwater management, demonstrate that social acceptance and technical intelligence must be synchronized for transformation to occur.

Taken together, the case studies confirm that transformative growth is not a universal formula. Japan's controlled redevelopment, Tehran's ecological realignment, Bali's cultural-ecological negotiation, and Dunkirk's identity reconstruction are just a few examples of how it manifests itself. Nonetheless, recurring characteristics appear in every setting: digital augmentation, ecological integration, design accuracy, democratic governance, and ongoing adaptation. The book's main thesis, that transformative growth is already taking place, but only in cities prepared to reconsider their underlying presumptions and collaborate across scales, sectors, and systems, is anchored by these trends.

13.2.1 Key Takeaways on Transformative Growth

The case studies in this book demonstrate the significant change in urban practice that is taking place, i.e., a change from cities unwillingly adjusting to demands to cities reinventing themselves with clarity, intention, and ecological purpose. Transformative growth is a pattern rooted in the lived reality of Musashi Kosugi,

Dunkirk, Tehran, Suzhou, Bali, and the subtropical villages struggling with escalating heat, not a catchphrase arising from theory. These many settings highlight a common set of ideas that ought to direct urban change in the coming decades. These ideas are summarized as follows: they are fundamental, practical, and unambiguous for cities that want to prosper rather than just survive.

The first and most decisive lesson is that "growth must restore ecological function rather than consume it." Every case study reinforces this point. The message is the same whether it comes from Suzhou's integration of stormwater management with heritage resilience, Bali's attempt to maintain agricultural landscapes in the face of tourism pressure, or Tehran's ecological network of cooling structures: ecological systems are urban infrastructure rather than amenities. When green networks, water systems, coastal ecologies, and microclimates are considered key assets, cities flourish more efficiently, sustainably, and fairly. Moreover, ecological performance must be measured, tracked, and included into decision-making in order to achieve transformative growth; it cannot be considered an afterthought once development outcomes are established. Regenerating the biological systems that make cities liveable is more important for urban growth in the future than expanding asphalt.

The second principle is that "resilience must be embedded through spatial design, not retrofitted after disruption." This is clearly illustrated in the chapters on subtropical thermal resilience and earthquake-resilient form. Long before calamities strike, block layout, open-space distribution, ventilation routes, evacuation networks, and sunlight access patterns affect emergency results. Resilience is a spatial condition, and cities that approach it as a technical checklist fail to recognize this. Designing neighborhoods, blocks, and structures that can withstand shocks, continue to function, and aid in recovery is necessary for transformative growth. This necessitates anticipatory design cultures and planning processes that can incorporate resilience into shape in advance rather than rushing to include it afterwards.

A third key takeaway is the need for "governance that shifts from control to coordination and co-regulation." The Musashi Kosugi case explains how hierarchical but coordinated governance enables brownfield regeneration, while Dunkirk shows how integrated planning knits mobility, water systems, history, and urban identity into a coherent framework. In contrast, Bali's swift changes in land use highlight the effects of disjointed administration in the face of demand from international investment. Adaptive governance, which unites public agencies, communities, and commercial actors through frameworks that manage shared duties rather than strict command structures, is essential for transformative progress. Lastly, urban transformation is a governance challenge: the ability to coordinate actors across scales is just as important as construction or landscape design.

Fourth, the case material highlights "redevelopment that prioritizes adaptive reuse, continuity, and identity." Replacement is not the same as transformative growth. Suzhou's adaptive repositioning of canal legacy, Dunkirk's waterfront transformation, and Musashi Kosugi's growth all demonstrate the strength of moving forward without losing the past. Redevelopment that respects identity and makes use of what is already effective helps cities become more resilient. Reusing land, buildings, or cultural narratives in an adaptive way lessens the impact on the environment and promotes community legitimacy. In times of rapid change, the continuity of place acts as a stabilizing influence, connecting cultural grounding with ecological innovation.

The fifth principle concerns technology: "digital tools should enable community-centred resilience, not replace community agency." The chapters on thermal performance modeling and solar-optimized community blocks demonstrate how data, simulations, and participatory design may improve local decision-making. Only when combined with collaborative governance and community understanding can digital twins, spatial measurements, and modelling platforms become really transformational. In this way, technology enhances rather than diminishes local intelligence. Urban futures built by co-production, made possible by digital tools, become more durable and adaptable, while those shaped only by digital optimization run the risk of becoming fragile and exclusive.

The sixth takeaway is unmistakable: "land-use choices are climate choices." The Bali scenario illustrates how food security, water systems, and ecological stability are all directly impacted by land conversion. Tehran demonstrates how cooling performance is determined by spatial alignment. Suzhou shows that heritage-sensitive zoning and land distribution are key factors in stormwater resilience. It takes guts to acknowledge that redevelopment objectives, density choices, and zoning regulations are essentially climate policies to achieve transformative growth. The long-term resiliency of the city is either strengthened or weakened by each parcel selection.

The final principle is that "transformative growth must be performance-based, not symbolic." While avoiding real reform, several cities have nominally embraced sustainability. However, the case studies in this book show that true adaptation is driven by performance-led techniques that make use of ecological indices, social perception data, heat measures, seismic reaction modelling, and landscape connectivity analysis. While performance changes cities, symbolic gestures may make headlines.

Taken together, these principles form a logical framework for cities that are committed to overcoming managed decline. They demonstrate how the convergence of ecological logic, spatial intelligence, innovative governance, and community involvement may lead to transformative progress. This is the essence readers should carry forward: the future of cities depends on the choices they make now, choices rooted in design, data, identity, and courage.

13.3 Applying Transformative Growth Across Scales and Contexts

As mentioned earlier, there is no set formula for transformative growth. It is a style of thinking about how to reimagine urban development such that social purpose, spatial intelligence, and ecological logic influence how cities develop in the future. This book's case studies show its effectiveness in some contexts, but its mobility is what really makes it valuable. It is possible to scale transformative growth up, down, or sideways. It can negotiate climate boundaries, market forces, cultural standards, and political realities. To reflect on these, this section examines how cities of various sizes, locations, and institutional cultures might apply the concepts of transformative growth.

At the building and block scale, transformation begins with the fundamentals of form and performance. The smallest urban units contain the DNA of resilience, as demonstrated by solar-optimized community blocks, thermally resilient building clusters, and seismically-attuned designs. Buildings become anchors of adaptive systems when they are designed to absorb seismic stress, improve ventilation, gather energy, or reduce heat. This effect is enhanced by blocks intended for social cohesion, shared ownership, and permeability. At this scale, transformative expansion entails making sure that each new structure and each rearranged block contribute to a broader ecological and spatial agenda. Buildings become contributors to communal resilience rather than distinct things.

Transformation becomes more relational at the neighborhood and district levels. In this case, whether a district develops into a regenerative system or becomes stuck in dysfunctional patterns depends on how streets, public areas, open-space networks, transportation flows, and ecological corridors interact. The logic of the neighborhood affects health, social life, and vulnerability through liveable streets that promote walking, accessible open spaces that disperse cooling, and interconnected green networks that guide biodiversity and runoff. District-level interventions in Tehran and Dunkirk demonstrated how transportation hubs, green infrastructure, and redesigned public spaces can transform identity and performance. Stitching together the social and physical fabric is necessary for transformative growth at this scale so that locations become cohesive systems rather than dispersed groups.

When transformative growth reaches the metropolitan level/region, it becomes a question of ecological integrity, land-use allocation, and governance. The Bali case illustrates how regional pressures, driven by tourism investment, cultural landscapes, and global mobility, collide with local ecosystems in ways that necessitate tough decisions. If cities are to prevent irreversible decline, industrial zones, heritage corridors, agricultural belts, coastline margins, and ecological reserves must be arranged into an integrated regional strategy. Clarity is necessary for metropolitan-scale transition, including where development should accelerate, where natural functions should be preserved, where transportation infrastructures should coincide, and where cultural landscapes should continue. Land-use decisions most clearly become climatic decisions in regional urbanism.

In the meanwhile, in order to survive, communities along the coastal interface must embrace hybrid land-water urbanism, adaptive sovereignty, and new forms of government. While this book's focus is primarily urban, lessons from coastal cases elsewhere remind us that transformative growth must anticipate rising seas, shifting shorelines, and the necessity to protect both ecological and political identities. Amphibious infrastructures, adaptable waterfronts, and floating parks provide opportunities to break down the traditional distinction between land and water. The basic foundation of cities is being reimagined by this change.

Across different global contexts, the interpretation of transformative growth naturally shifts. Here are some examples:

- **Global South megacities**, where informal settlements grow faster than formal planning can respond, require strategies rooted in community co-production and incremental upgrading rather than wholesale redevelopment. Transformative growth must build from the ground up, engaging residents as designers and stewards.
- **Mid-sized industrial cities** like Dunkirk demonstrate how post-industrial decline can become a platform for ecological regeneration and identity renewal. Here, transformation is a cultural and economic recalibration.
- **Island communities** face existential risk and must connect climate adaptation with governance innovation, integrating ecological design with political continuity.
- **Subtropical regions** battling extreme heat must prioritize cooling networks, ventilation pathways, and building-scale thermal performance.
- **Tourist-driven economies** like Bali must reconcile global demand with ecological protection by adopting zoning innovation, ecological corridors, and community-based land management.
- **Heritage landscapes** require synthesis: protecting cultural identity while embedding resilience through nature-based and hydrological strategies.

The institutional context is also very important. Through long-term land-use discipline and regulatory coherence, cities with robust planning regimes (such as Japan as a classic example) can achieve transformative growth. To avoid walled decision-making, cities with fragmented governance must instead rely on cooperative agreements, shared data infrastructures, and coordination platforms. To deal with informal urbanization, cities must implement hybrid governance, establishing adaptable frameworks that both guide evolutionary development and respect lived realities. Instead of rejecting these institutional obstacles, transformative growth adapts to them.

Delivering transformative growth also requires clear implementation pathways. The enabling conditions are set by policy levers including climate-responsive zoning, ecological performance regulations, and financial incentives for adaptive reuse. Financing methods need to change to blended approaches that prioritize ecological results and resilience over immediate profits. The legitimacy and durability of transformation are guaranteed by community co-production. Using digital twins, geographical analysis, and interactive modelling, digital augmentation aids in the institutionalization of evidence-based planning. Additionally, green infrastructure needs to be integrated as a structural framework that can cool cities, direct water, improve biodiversity, and fortify identity rather than as an ornament.

Ultimately, transformative growth's scalability depends on its capacity to uphold local environments while pursuing global imperatives. The principles hold, but their expressions change. This flexibility is its strength: transformative growth is not a universal prescription but a regenerative pathway that any city can adapt to its own terrain, culture, governance realities, and climatic pressures.

13.4 Conclusion: The Imperative to Transform

Cities today operate in conditions defined by urgency, constraint, and accelerating risk. Once the standard approach to urban administration, incremental changes are no longer able to keep up with the unpredictability of the climate, the fragility of infrastructure, and the social divisions exacerbated by unequal development. This book's main thesis is unambiguous: visionary cities are no longer the only ones who can achieve transformative growth. It is the bare minimum needed to prevent managed deterioration. A city's future will be molded by compounding losses rather than increasing prospects if resilience, ecological systems, and social cohesion are still viewed as secondary concerns.

The chapters across this volume have shown that prosperity is a myth in the absence of ecological stability. The future is simply mortgaged by economic progress that compromises natural systems, erodes public confidence, or increases risk exposure. Cities can develop in diverse ways, as demonstrated by the case studies, which range from Tehran's cooling networks to Bali's disputed ecological corridors, and from Musashi Kosugi's recalibrated brownfield redevelopment to Dunkirk's integrated coastal identity. They can utilize ecological logic to restore degraded landscapes, governance reform to align institutions, and spatial design to integrate resilience. Coordinated methods that can improve metabolic, social, and climatic performance can take the role of fragmented, transaction-driven urbanism. That is the essence of transformative growth.

If there is one lesson to be learned, it is that cities have a remarkable ability to rebuild and reimagine themselves when they decide to take decisive action. They are active engines of regeneration rather than passive recipients of risk; they are locations where communities can flourish despite climate constraints, living systems can be restored, and new forms of value can be produced. This book's case study examples demonstrate that when design, policy, environment, and society all move in unison, transformation is realistic, quantifiable, and attainable.

The decades ahead will not reward hesitation. Redesigning underperforming areas, investing in adaptation before a crisis arises, and elevating ecological intelligence as a fundamental planning concept are all tasks that urban leaders must be prepared to undertake. They must acknowledge that the means for change are already in place: digital twins that can predict risk and opportunity, spatial designs that incorporate resilience into form, long-term value-rewarding funding methods, and community collaborations that increase capacity and legitimacy. The difficulty lies not in creativity but in dedication.

This book ends with an appeal to communities, planners, designers, engineers, and policymakers who are responsible for the future of cities. Incremental improvements won't be enough for the next ten years. It necessitates a radical change toward cities that recognize and respond to their place in broader ecological systems. Cities that embrace transformative growth have the potential to become platforms for the well-being of people and the planet, not in spite of the challenges they encounter, but rather because they choose to address them head-on.

Although the road ahead is difficult, it is also full of opportunities. Cities that are prepared to rethink development, restructure governance, and reestablish their connection to the natural world will prosper. The need to change is not a cautionary tale, but rather an opportunity to create resilient, just, and promising urban futures.

GPSR Compliance

The European Union's (EU) General Product Safety Regulation (GPSR) is a set of rules that requires consumer products to be safe and our obligations to ensure this.

If you have any concerns about our products, you can contact us on ProductSafety@springernature.com

In case Publisher is established outside the EU, the EU authorized representative is:

Springer Nature Customer Service Center GmbH
Europaplatz 3
69115 Heidelberg, Germany

Batch number: 10837251

Printed by Printforce, the Netherlands